Springer Series in
Surface Sciences

23

Springer Series in Surface Sciences

Editors: G. Ertl, R. Gomer and D. L. Mills   Managing Editor: H. K. V. Lotsch

D. J. O'Connor   B. A. Sexton
R. St. C. Smart (Eds.)

Surface Analysis Methods in Materials Science

With 250 Figures

Springer-Verlag
Berlin Heidelberg New York
London Paris Tokyo
Hong Kong Barcelona
Budapest

Associate Professor D. John O'Connor

Department of Physics,
Newcastle University, Callaghan,
Newcastle, NSW 2308, Australia

Dr. Brett A. Sexton

CSIRO Division of Materials Science
and Technology, Locked Bag 33,
Clayton, Victoria 3168, Australia

Professor Roger St. C. Smart

South Australian Surface Technology Centre,
University of South Australia, The Levels,
Adelaide, South Australia 5095, Australia

Series Editors

Professor Dr. Gerhard Ertl

Fritz-Haber-Institut der Max-Planck-Gesellschaft, Faradayweg 4–6,
1000 Berlin 33, Fed. Rep. of Germany

Professor Robert Gomer, Ph.D.

The James Franck Institute, The University of Chicago, 5640 Ellis Avenue,
Chicago, IL 60637, USA

Professor D. Mills

Department of Physics, University of California,
Irvine, CA 92717, USA

Managing Editor: Dr. Helmut K. V. Lotsch

Springer-Verlag, Tiergartenstrasse 17,
W-6900 Heidelberg, Fed. Rep. of Germany

ISBN 3-540-53611-6 Springer-Verlag Berlin Heidelberg New York
ISBN 0-387-53611-6 Springer-Verlag New York Berlin Heidelberg

Library of Congress Cataloging-in-Publication Data. O'Connor, D. J. (D. John), 1952–. Surface analysis methods in materials science / D. J. O'Connor, B. A. Sexton, R. C. Smart. p. cm. – (Springer series in surface sciences ; 23). Includes bibliographical references (p.) and index. ISBN 3-540-53611-6 (Springer-Verlag Berlin Heidelberg New York : alk. paper). – ISBN 0-387-53611-6 (Springer-Verlag New York Berlin Heidelberg : alk. paper). 1. Surfaces (Technology)–Analysis. I. Sexton, B. A. (Brett A.), 1950–. II. Smart, R. C. (Roger C.), 1944–. III. Title. IV. Series.   TA418.7.028 1991   620 .44–dc20   91-18390

The use of general descriptive names, registered names, trademarks, etc. in this publication does not imply, even in the absence of a specific statement, that such names are exempt from the relevant protective laws and regulations and therefore free for general use.

Typesetting: Springer TEX in-house system

54/3140-5 4 3 2 1 0 – Printed on acid-free paper

Preface

The idea for this book stemmed from a remark by Philip Jennings of Murdoch University in a discussion session following a regular meeting of the Australian Surface Science group. He observed that a text on surface analysis and applications to materials suitable for final year undergraduate and postgraduate science students was not currently available. Furthermore, the members of the Australian Surface Science group had the research experience and range of coverage of surface analytical techniques and applications to provide a text for this purpose. A list of techniques and applications to be included was agreed at that meeting. The intended readership of the book has been broadened since the early discussions, particularly to encompass industrial users, but there has been no significant alteration in content. The editors, in consultation with the contributors, have agreed that the book should be prepared for four major groups of readers:

- senior undergraduate students in chemistry, physics, metallurgy, materials science and materials engineering;
- postgraduate students undertaking research that involves the use of analytical techniques;
- groups of scientists and engineers attending training courses and workshops on the application of surface analytical techniques in materials science;
- industrial scientists and engineers in research and development seeking a description of available surface analytical techniques and guidance on the most appropriate techniques for particular applications.

The contributors mostly come from Australia, with the notable exception of Ray Browning from Stanford University. Australia is a very large country with a relatively small population, so it is inevitable that the Australian surface science community is spread rather thinly across the country. One aim in producing this book has therefore been to bring together the breadth of expertise within this Australian scientific community. All of the authors have made significant contributions to the techniques and their applications, in many cases over a period of more than 20 years. A second aim is to emphasise by example the very wide spectrum of information that can now be obtained from the use of a variety of surface analytical techniques applied to the same material. Here, the intention has been to encourage people involved in research, development and process control to become aware of the increasing usefulness of surface analysis in their own fields of materials science.

Finally, we believe that the approach adopted here – namely descriptions of the basic techniques, their limitations and their applications – will be accessible and beneficial to people in any of the four groups of potential readers in any country of the world. The authors are all closely involved with the international scientific community in their own areas of research and this is reflected in their selection of examples. We hope that this book will fulfil a need by extending the range of techniques from those covered by more specialised texts confined to one or two techniques, and by providing examples of real applications to research, development and problem solving in materials science.

The strategy and structure of the book is as follows: the book is designed for use by all those interested in the surface characterisation of materials. This group is expected to include scientists (e.g. chemists, physicists, biochemists, biologists, geologists, geochemists), technologists, metallurgists, engineers and workers in the various biomedical and microelectronics applications areas. It is also intended for those requiring an introductory overview of surfaces and techniques for their analysis, in particular final year undergraduate and graduate students in any of the specialties listed above. It is important to state clearly that it is not intended for experienced practitioners in particular techniques of surface analysis. It does not attempt to critically review all techniques or all variations and restrictions on the use of a particular technique. For instance, we are well aware that there are many difficulties in procedures for quantification, chemical mapping and depth validation for particular types of sample that are not explained in the short presentations on each technique given in this book. They will need to be learnt when undertaking a specific investigation of a material using one of the techniques described here and will probably require further reading in more advanced, single-technique books.

The book has been deliberately structured in three parts.

Part I provides a descriptive overview of different materials, the properties of their surfaces, the range of techniques available to study their structure and composition and some guidance to the choice of particular techniques based on information they can provide and major limitations to their use. It also includes a description of the essential elements of vacuum technology required for many of these techniques.

Part II presents short descriptions of major techniques for use in materials science and technology. The selection has been made on the basis of the accessibility of the technique and the universality of its application to different types of materials, and is certainly not unequivocal. Several techniques mentioned in Part I are not included in Part II. It is likely that the relevance of the various techniques will change with time and that, in later editions, it may be necessary to delete or add techniques. The descriptions concentrate on the types of information provided by the technique, samples most readily analysed and limitations to their use. The chapters of Part II can be read separately and, in many cases, it will not be necessary for the reader to complete all chapters in the first instance.

Part III describes some major applications. Again, some judgement has been exercised in the choice of these applications, with emphasis placed on major areas in materials science and technology. Options for addition and deletion also exist in this section.

The book is best approached by initially reading the whole of Part I and selected sections from Parts II and III. The reader may wish to focus on particular materials of direct interest to his or her research, development or technological application. This itself will suggest particular techniques as being most appropriate initially. It may also require more detailed reading of single-technique descriptions in other books: the references in Parts I and II give specific guidance to books most suitable for this purpose.

Newcastle   *John O'Connor*
Clayton   *Brett Sexton*
Adelaide   *Roger Smart*
December 1991

Contents

Part IV   Appendix

Contributors

Baker, Bruce G.
School of Physical Sciences, Flinders University, Sturt Road,
Bedford Park, SA 5042, Australia

Browning, Ray
Integrated Systems, Stanford University, Stanford, CA 94305, USA

Cocking, Janis L.
DSTO Materials Research Laboratory,
PO Box 50, Ascot Vale Victoria 3032, Australia

Dell, John M.
Telecom Australia Research Laboratories,
770 Blackburn Road, Clayton, Victoria 3168, Australia

Jennings, Philip J.
School of Mathematical and Physical Sciences, Murdoch University,
Murdoch, Western Australia 6150, Australia

Johnston, Graham R.
DSTO Materials Research Laboratory,
PO Box 50, Ascot Vale, Victoria 3032, Australia

Kibel, Martyn H.
Telecom Australia Research Laboratories, 770 Blackburn Road,
Clayton, Victoria 3168, Australia

King, Bruce V.
Physics Department, University of Newcastle, Newcastle, NSW 2308, Australia

Klauber, Craig
CSIRO Division of Mineral Products, c/o Curtin University of Technology,
GPO Box U1987, Perth, WA 6001, Australia

Leckey, Robert
Department of Physics, LaTrobe University, Bundoora, Victoria 3083, Australia

MacDonald, Ron J.
Department of Physics, University of Newcastle, Callaghan,
Newcastle, NSW 2308, Australia

Morris, Graeme C.
Department of Chemistry, University of Queensland, St. Lucia,
Queensland 4072, Australia

O'Connor, D. John
Department of Physics, University of Newcastle, Callaghan,
Newcastle, NSW 2308, Australia

Payling, Richard
BHP Coated Products Division, Research and Technology Centre,
PO Box 77, Port Kembla, NSW 2505, Australia

Price, Garth L.
Telecom Australia, Research Laboratories, 770 Blackburn Road,
Clayton, Victoria 3168, Australia

Roberts, Noel K.
Department of Chemistry, University of Tasmania,
Hobart, Tasmania 7001, Australia

Robins, John L.
Department of Physics, University of Western Australia,
Nedlands, Western Australia 6009, Australia

Sexton, Brett A.
CSIRO, Division of Materials Science and Technology, Locked Bag 33,
Clayton, Victoria 3168, Australia

Sie, Soey H.
CSIRO, Division of Exploration Geoscience,
PO Box 136, North Ryde, NSW 2113, Australia

Smart, Roger St.C.

South Australian Surface Technology Centre, University of South Australia, The Levels, Adelaide, South Australia 5095, Australia

Turner, Peter S.

CSIRO Division of Wool Technology, Geelong Laboratory, PO Box 21, Dalmore, Victoria 3981, Australia

Part I

Introduction

1. Solid Surfaces, Their Structure and Composition

C. Klauber and R. St. C. Smart

With 25 Figures

1.1 Importance of the Surface

> "When a plate of gold shall be bonded with a plate of silver, or joined thereto, it is necessary to be beware of three things, of dust, of wind and of moisture: for if any come between the gold and silver they may not be joined together ..."
>
> *Translated extract from "De Proprietatibus Rerum" (The Properties of Things)*
> *1323 A.D. by Bartholomew [1.1]*

Awareness of the important role played by surfaces in technology has existed for some time, although it is only in the past three decades that we have been able to establish an improved understanding of their properties. In everyday life our perceptions of solid materials, and in particular their surfaces, are strongly distorted by the limitations of visible light. These wavelengths are a thousand times larger than dimensions of the surface region in which well understood bulk properties of materials break down, making way for the transitional interface with another phase, which may be gaseous, liquid or solid. Such are the alteration of bulk properties, structural and compositional, that it is not unreasonable to consider surfaces as an additional phase of matter [1.2]. Whilst this may serve as a useful general concept for surface scientists, in the various fields of technological endeavour what is thought of as a surface varies enormously, particularly in depth characterisation. Accepting the simplest definition of the surface, as the boundary defined by the outermost atomic layer separating the bulk solid from an adjacent phase, is thus inadequate in the area of practical surface technology. A more meaningful approach is to consider a selvedge layer of variable depth. In fact, the different depth regimes of the surface are defined by *that* depth which actually plays *the* definitive role in the technological application (see Table 1.1). Obviously the boundaries between these surface selvedge depths are not always clear and overlap will exist between adjacent categories. The point of particular importance is to view the "surface" with a degree of flexibility, depending not only upon the nature of the material and its environmental history, but the role that it plays. Wide variations in selvedge depth, from the Ångstrom to the millimetre, fortunately do not pose insurmountable problems in surface analysis. This is due to the complementary nature of the techniques that are employed. Indeed, from

3

Table 1.1. The importance of the various depth regimes

Selvedge* depth	Examples of the definitive role
Outer monolayer ~ 0.1 nm	Heterogeneous catalysis, surface tension (contact angle) control, selective adsorption, electrochemical systems, biological systems, sensors
Thin film ~ 0.1-100 nm	Emulsions, tribological (friction) control, anti-reflection coatings, lipid membranes, Langmuir-Blodgett films, interference filters, release agents
Near surface 0.1-10 μm	Semiconductor devices, surface hardening, polymer biodegradation, controlled release membranes, osmosis devices, photographic film, optical recording media, aerosols
Thick layers > 10 μm	Anti-corrosion coatings (anodizing, electroplates, paints), phosphors, adhesives, electrowinning, magnetic recording media, surface cladding

*We use the term "selvedge" to emphasize that the surface is often a region (with depth) rather than a two-dimensional layer.

the investigator's viewpoint, the "surface" in a practical sense is often dependent on the analytical technique being used.

In attempting to understand the surfaces of materials and methods by which they may be structurally and compositionally characterized, it is useful to remind ourselves of the basic categories of bulk solid materials. Structurally a material will either be crystalline or amorphous depending upon the extent of internal order, a crystalline solid having the same unit cell structure repeating over macroscopic dimensions, whereas the amorphous material possesses order only on the nearest neighbour or molecular dimension scale. The disorder in amorphous materials, such as glasses and some polymers, ensures a continuum in their structural nature, quite different from crystalline solids, where, due to thermodynamic constraints, the individual crystallites are generally small. Crystalline materials are usually encountered in a polycrystalline form, each crystalline grain abutting another of differing crystal orientation. This intergranular region is a classic solid-solid interface with the grain boundary surface properties having a direct bearing on the gross mechanical and thermal properties displayed by the material. The solid may, of course, be multi-phase, each in either amorphous or crystalline form as in ceramics. The polycrystalline material may be an element, compound or alloy, the alloy in turn a continuous solid solution or a mixture of precipitated phases. The "modern" engineering practice of using composite materials (laminates, fibre reinforcing etc.) to create particular properties is merely an extension of phase precipitation.

Technologically the importance of materials' surfaces cannot be understated since they influence so many facets of modern industry. Surfaces are found to:

- control material stability via corrosion and friction/wear characteristics in everything from automobile components to medical prostheses
- determine material adhesion characteristics

- be crucial to systematic process control in materials fabrication such as electronic devices and thin films
- play a vital role in heterogeneous catalytic processes for compound synthesis as in petrochemicals and fertilizers
- control mineral beneficiation via selective flotation and adsorption
- be germane to membrane processes important in numerous diverse fields from environmental control to medicine.

The application of surface analysis to such a variety of materials science and technology problems has several purposes. Sometimes the technology may be old and operationally understood only on a folklore basis. If we are to progress it is essential that the processes be understood in detail even if no direct economic benefit is obvious. As part of industrial process control, problems may arise that require immediate solutions. Without knowledge of the factors and mechanisms involved, solutions must still be based on empirical testing and guesswork. The area of system and materials design is increasingly important. We may know the surface properties that we want, but how can they be achieved and how do we know if we have achieved them? It is the purpose of this and following sections to explain and evaluate some of the more common methods and applications of surface analysis used to achieve these purposes.

The primary aim of materials science is to engineer materials with specific mechanical, electrical, magnetic, optical, thermal and chemical properties. These materials need to be stable under the environmental pressures that they normally encounter. Provided that the material is not intrinsically unstable or that extremes of mechanical force or radiation are not disrupting bonds within the bulk, all processes for breakdown of that material will be initiated at a surface. That surface will either be directly exposed or will lie within an accessible pore or within an intergranular region. Each class of material can exhibit a characteristic set of surface selvedge features, with particular physical, chemical and electronic properties. Examples of these are summarized in Table 1.2.

More general structural features, such as epitaxial layer growth, superstructures, nucleation, coatings (e.g. adhesives, passivation layers), and thin films apply to a variety of materials. Additionally, there is another kind of surface structure, which can be given the generic term *electronic* surface structure, that is of central importance to many applications in such fields as microelectronics, catalysis and hydrometallurgy.

These features all play a role in the overall engineering properties of those classes of materials. A common and obvious example of a surface-initiated phenomenon is the corrosion of iron and steels. This is particularly dramatic since the corrosion product typically has an iron content molar volume up to four times that of the original reduced metal. Such enormous expansion accentuates the material's failure. Whilst such massive corrosion is controlled by mass and charge transport in the solid state, its initiation is certainly a surface process. The corrosion need not be visibly dramatic in order to eventually affect the materials performance. Steels are typically protected from corrosion by painting, yet these coatings will also eventually oxidize, crack and fail to perform their intended task.

5

Table 1.2. Surface features of a variety of materials

Material	Surface Features
Metals, alloys	Oxide coating (thickness, type); faceting; relaxation; reconstruction, elemental segregation; defects; fracture faces; corrosion layers and regions; adsorbed layers; interdiffusion and reaction (joining technology)
Semiconductors (including organic semiconductors)	Space-charge region; reconstruction; relaxation; defects; low coordination sites; segregation (multi-element); adsorbed layers; reaction profiles
Ceramics (including high T superconductors)	Grain structure; phase separation; intergranular regions; fracture faces; pores; amorphous regions; triple points; reaction products; elemental distributions; leaching profiles
Minerals (including soils, oxides, salts)	Grain structure; phase separation; altered surface layers (weathered, deposited, oxidised, reduced); faceting; reconstruction; defect sites (low coordination); intergrowth structures; leaching profiles; adsorbed layers (minerals processing)
Glasses	Hydrolysed layer; crystalline regions; inhomogeneities; elemental segregation; reaction products (distribution); mould materials transfer
Polymers	Altered surface layers (oxidised, reacted); excluded layers (lubricants, catalysts, unreacted monomer); phase separation (elastomers, monomers); surface segregation of reactive groups
Composites and natural materials (including wood, paper, paints, cement, fibreglass, tissue, blood, bone)	Distribution of materials; modified surface layers (bonding compatibility); interphase regions; elemental distributions; reaction profiles; leaching profiles

Not only chemical stability, but mechanical stability is of engineering importance. Wear can be minimized not only be lubrication, but also by surface hardening. This might be achieved by methods of nitridization or carburization to create a refractory surface phase of extreme hardness. The surface modification of materials by ion beams is one area of active current interest [1.3]. As an example it has been recently found by *Rabalais* and *Kasi* [1.4] that when mass selected C^+ ion beams (20–200 eV) impinge upon atomically clean surfaces (Si, Ni, Ta, W and Au), they initially create a bonded carbide structure. However, of particular interest is the observation that, with continued deposition, the carbon layers build up into a diamond-like structure. This is significant because of the particular hardness of diamond combined with its low electrical and high thermal conductivity.

An interesting use of active surface chemistry to maintain a material's stability can be found in the simple quartz-halogen incandescent lamp, first marketed in 1959. In a conventional incandescent lamp the brightness is governed by the filament temperature, which is necessarily limited so as to maintain a useful lamp lifetime. The quartz-halogen variety has the tungsten filament encased in a clear fused quartz envelope containing halogen (F_2, Br_2 or I_2) gas. Any tungsten which evaporates from the filament deposits on the cooler envelope wall where it reacts with incident halogen molecules to form a volatile halide. Halide production and desorption is enabled due to the quartz envelope wall temperatures, typically in the range 250–600°C. The tungsten halide then diffuses back to the filament where it dissociates to reform tungsten metal and halogen gas. Filament integrity is thus maintained at higher temperatures enabling more efficient light production and longer life – all due to two recycling surface chemical reactions. The internal surface of the envelope can be further improved by the addition of etch barriers (aluminium fluoride, aluminosilicate) and infrared reflectants (titanium silicate, zirconia) [1.5].

1.2 Solid Surfaces of Different Materials

Having reminded ourselves of the variety of forms in which solid materials can exist, it is a useful exercise to conceptually "create" a surface from one of these materials. Taking the simplest case of a crystalline solid, we might imagine an ideal case of a sudden termination of the crystalline periodicity at the solid-vacuum interface as in Fig. 1.1a. In layer, chain or sheet structures, where relatively weak dispersion forces hold the constituents of the solid together, such an "ideal" surface case might be envisaged. In strongly bonded solids such as metallic, covalent or ionic systems, the termination of the periodicity means that valence electrons will spill out into a continuum with no positive cores, freed covalent bonds will be "dangling" into space and the Madelung constant, eval-

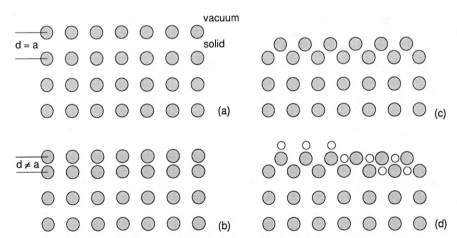

Fig. 1.1. Schematic of (**a**) a solid surface created by terminating the bulk of a crystalline solid. Bonding imbalances cause the outer layer to move (**b**) and even undergo reconstruction (**c**). When then exposed to a reactive medium, foreign atom adsorption can occur (**d**), possibly leading to surface compound formation

uated on the basis of 3-dimensional symmetry, will no longer apply. Invariably this breakdown of previously balanced bonding forces leads to a rearrangement of the outer layer or layers and to a periodicity, and possibly unit cell structure, different from that previously existing in the bulk. The surface in this sense becomes the ultimate defect that a solid material can have.

The simplest rearrangement is that of outer layer relaxation or contraction (Fig. 1.1b), the layer either moving away from or towards the inner layers. Dynamical low energy electron diffraction (LEED) calculations indicate that for metals this movement may be 0–25% of the normal interlayer spacing [1.6]. As no bond breaking is required this rearrangement is spontaneous. A more complex alteration occurs when the structure parallel to the surface undergoes what is known as reconstruction, shown schematically in Fig. 1.1c. This is particularly common in covalently bonded semiconductor materials. The surface atoms rearrange themselves so as to minimize the dangling bonds which arise from the original surface creation. Reconstruction can be spontaneous or may require heating to be induced. Surface reconstructions are dealt with further in Sect. 1.4 of this chapter and Part II of the book. A large amount of research work carried out in the discipline of surface science is concerned with the *in situ* creation of such virginal surfaces (atomically clean and ordered) and their subsequent controlled reaction with various gases. These can be created *in situ* by a number of methods:

- cleaving or fracturing in the vacuum system; useful for single crystal work, grain boundaries and fracture mechanics, but very limited for materials analysis
- evaporating of clean films (which can be substrate epitaxial); has application in nucleation, catalysis, adsorption and adhesion studies but is generally only useful for model systems
- high temperature flashing; generally limited to refractory materials and used to anneal out ion bombardment damage; heating usually causes surface segregation of bulk impurities
- field induced desorption; usefully limited to refractory metals in the form of a field tip
- ion beam cleaning (utilized in most surface analysis); has additional use of depth profiling, but surfaces and interfaces are damaged and smeared, often with preferential sputtering of a component
- gas-surface reaction to create an altered surface phase more easily removed by one of the above methods (surface titration).

The general techniques for the preparation of atomically clean surfaces have been reviewed by *Verhoeven* [1.7] and specific methods for selected elements by *Musket* et al. [1.8].

Our conceptual surface creation in Fig. 1.1 continues to a further stage if it is removed from vacuum and subjected to a reactive liquid or gas environment. The fluid molecules will interact with the surface to create an altered selvedge. This may begin with simple atomic adsorption, followed by incorporation into

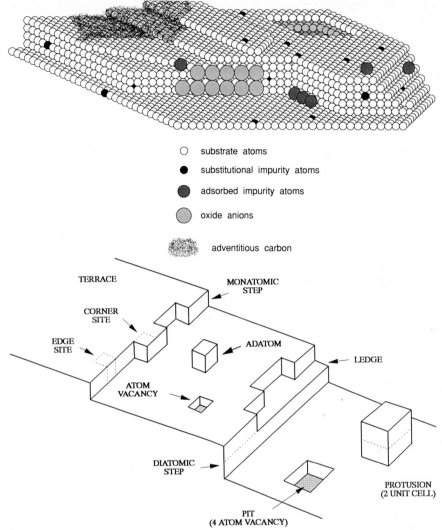

TERRACE

MONATOMIC
STEP

CORNER
SITE

EDGE
SITE

ADATOM

LEDGE

ATOM
VACANCY

DIATOMIC
STEP

PROTUSION
(2 UNIT CELL)

PIT
(4 ATOM VACANCY)

Fig. 1.2. Two representations of part of a solid surface depicting a variety of different surface sites. These sites are distinguishable by their number of nearest or coordinating neighbours. Low coordination sites (surface defects) are the preferred locations for adatom adsorption and can thus be the initiation point for a variety of surface processes. Also depicted are an oxide facet, a cubic etch pit and a region of extensive adventitious carbon contamination

the near surface region and surface compound formation (Fig. 1.1d). Due to the number of possible surface-fluid couples the variety of reactions is immense, as evidenced just by the number of adsorbed monolayer systems that have been studied [1.9]. Of course for most "real" surfaces the situation is considerably more complex than depicted in Fig. 1.1d. Polycrystallinity leads to microfacets and domains on grains riddled with structural defects following a roughened topography. A schematic "snapshot" of part of a surface is shown in Fig. 1.2.

9

The structure of a surface is inextricably linked to its reactivity and, in many cases, its composition. At one extreme, we know that defects, such as dislocations, protusions, edge and corner sites on facets, on metal surfaces provide sites of enhanced reactivity in chemical attack [1.10]. For an ionic oxide salt, like MgO, it has been shown [1.11–13] that simple molecules (e.g. H_2, CO, H_2O) will not adsorb on the fully coordinated, five-fold ions in a {100} face but will react readily with "kink" sites (i.e. edge and corner sites) of lower coordination as illustrated schematically in Fig. 1.2. At the other extreme, we find profound alterations in chemical composition between surface and bulk due to segregation (i.e. selective deposition of particular elements), surface rearrangement in the first few atomic layers, and consequent changes in the potential energy of surface sites e.g. [1.14, 15]. This occurs in metal alloys, salts, minerals, doped elemental semiconductors and glasses. In polymers, we often see surface layers of material excluded during polymerisation. In ceramics, segregation of particular elements into narrow amorphous films between grains of different phase is exposed at the surface after preferential fracture. The structure of corroded material reveals depth profiles in which the progress of the reaction can be followed, particularly at structural features like grain boundaries, intergranular films, pores, precipitates, impurity regions and bonded interfaces. All of these examples relate particular surface structures to particular forms of reactivity. They are a few grains of sand on the seashore of surface structural features.

1.2.1 A Material Under Attack: Aluminium

The illustration of our conceptual surface and interface behaviour can be enhanced by reference to aluminium. Its importance as the most widely utilized non-ferrous metal is due to its many useful physical, chemical and metallurgical properties, not the least of which is its resistance to corrosion. Consider for the moment pure aluminium, in particular oxygen reaction at the Al(111) surface. As aluminium has a face centred cubic structure the (111) surface is a close packed plane. A wide variety of adsorption studies, e.g. [1.16–19] and references therein, conclude that, upon initial interaction with oxygen, a (1 × 1) overlayer forms, i.e. for every surface aluminium atom a dissociated ionised oxygen atom adsorbs in registry. The registry site is not directly atop the aluminium atoms, but rather in the shallow 3-fold hollow so as to maximize coordination. As oxygen coverage increases, another state forms that is thought to correspond to an oxygen underlayer, i.e. a layer beneath the first aluminium layer. With still heavier oxygen exposures, the resultant three-dimensional oxide layer that forms resembles amorphous Al_2O_3. This parallels the result observed at room temperature in dry air in which polycrystalline aluminium rapidly forms an oxide layer of Al_2O_3. Since the oxide is stable, tightly adherent and highly impervious, further corrosion is naturally resisted and the reaction is self-limiting, ceasing after producing a selvedge of about 5–10 nm in thickness, i.e. the surface passivates. If the film is disrupted, it begins to reform immediately in most environments. Of course aluminium's many properties do vary significantly both with its purity and with alloying. It forms a large variety of commercially useful binary, tertiary and quaternary alloys containing intermetallic phases with Mg, Si, Mn, Fe, Ni, Cu, Zn, Sn and other metals [1.20].

Not surprisingly the corrosion which can and does occur with aluminium and its alloys can be of many different types, such as intergranular, exfoliation, stress-corrosion cracking or even uniform attack in which the whole oxide film is destroyed. However, by far the most common form of corrosive attack on aluminium alloys is that of pitting corrosion [1.21]. Not unexpectedly, these pits will form at localized discontinuities in the oxide film. Frequently they possess a characteristic habit, indicating preferential attack on particular crystallographic planes e.g. dry hydrochloric acid gas produces cubic pits because it etches faster in the $\langle 100 \rangle$ direction whereas, in the presence of moisture, corrosion is more rapid in the $\langle 111 \rangle$ direction leading to octahedral pits [1.22]. An etch pit is schematically illustrated in Fig. 1.2. The vast metallurgical research effort into corrosion control of aluminium and its alloys is concerned essentially with improved coherence of the surface selvedge oxide. This can range from chemically producing a variety of conversion coatings, principally aluminium oxyhydroxides with specific substitutions of other ions such as chromate and phosphate, to engineering solutions such as cladding. In cladding, a thin layer (typically 5–10% of the nett sheet thickness) of pure aluminium or alloy, which is anodic to the base alloy, is attached to the structural sheet. The cladding provides electrochemical protection by corroding preferentially [1.19]. Cladding finds use in products such as aircraft skins.

It is the very aluminium alloy composition, conferring the desirable engineering characteristics, which is also the key to the corrosion, being strongly affected

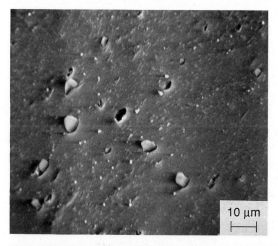

Fig. 1.3. Morphology of an electropolished aluminium alloy (6081) surface. The particles standing in relief are β-AlFeSi intermetallic phase particles with the Al and Fe components leached out [1.23]

by the type, amount and properties of the intermetallic phases present in the matrix. A particularly interesting experimental approach taken by *Nisancioglu* et al. [1.23] for alloys with Fe-rich surface inclusions is their selective dissolution and removal by a controlled electrochemical etching process (Fig. 1.3). As this can be done with minimal dissolution of the surrounding matrix, i.e. without exposing additional intermetallics beneath the surface inclusions, an essentially pure aluminium surface with enhanced corrosion resistance is produced.

1.3 Methods of Surface Analysis

1.3.1 Variety of Surface Analytical Techniques

To instrumentally probe a medium, in this case a solid surface, one of six basic probes may be applied to the surface: electrons, ions, neutrals, photons, heat or a field. The analysis consists of measuring the surface's response, also evident in one of these six ways. Combining the six probes and six responses gives 36 fundamental classes of experimental technique by which we may analyse a surface. By altering the energy and/or mass and/or character of these probes the variety of possible experimental methods increases even further. Obviously not all options will provide an experimentally viable pathway to gain useful surface information i.e. the surface structure (physical and electronic) and the surface composition (chemical). Whilst all possible variants have not yet been explored to the fullest extent, the number of surface analytical techniques currently in use is nonetheless quite impressive. Appendix A.1 lists over 130 technique acronyms; some of these are admittedly restricted to esoteric research endeavours, but many have gained widespread use in hundreds of laboratories throughout the world. As it is beyond the scope of this book to cover all of these, we will concentrate only on the more common and versatile, thirteen techniques in all. These are outlined in Table 1.3 within a subset of 12 of the 36 fundamental classes. There is also a variety of more classical surface-specific techniques such as basic measurements of contact angle, zeta potential, particle size distributions, surface areas and porosities and also electrochemical methods such as cyclic voltammetry which are not included. They are generally covered in a number of good texts on colloid and surface chemistry (e.g. refer to the bibliography at the end of the book).

Table 1.3. Summary of the various techniques considered in this book for analyzing the structure and composition of the surface selvedge layer. The 12 basic categories, each of which may encompass a range of techniques, represent a subset of a possible 36

Summary of surface techniques * surface < 10 nm	° near surface ~ 1 μm	
Emitted analyzed response		
Incident excitation probe	Electrons	Ions	Photons

	Electrons	Ions	Photons
Electrons	**AES*** Auger electron spectroscopy **SAM*** Scanning Auger microscopy **SEM°** Scanning electron microscopy **LEED*** Low energy electron diffraction **RHEED*** Reflection high energy electron diffraction		**EDAX°** Energy dispersive analysis of X-rays
Ions		**SIMS*** Secondary ion mass spectrometry **ISS*** Ion scattering spectroscopy **RBS°** Rutherford backscattering spectroscopy	**NRA°** Nuclear reaction analysis
Photons	**XPS/ESCA*** X-ray photoelectron spectroscopy Electron spectroscopy for chemical analysis **UPS*** Ultraviolet photoelectron spectroscopy	**FTIR°** Fourier transform infrared spectroscopy	
Electric field	**STM*** Scanning tunnelling microscopy		

1.4 Structural Imaging

In approaching a problem of surface characterization the first need is usually physical imaging.

1.4.1 Direct Physical Imaging

By far the most important information on surface structure, at least in the first stage of examination of a sample, comes from the techniques that provide *images* of structural differentiation in the surface layers.

13

The simplest and most accessible of these techniques is scanning electron microscopy (SEM). Chapter 3 covers this technique in more detail. Secondary electron images can be obtained on all materials identifying surface features, on most instruments, to a practical limit of ~ 100 nm. Despite the considerable depth of penetration of the incident primary electron beam (e.g. 0.5–$5\,\mu$m), the re-emitted electrons (as secondary and backscattered electrons) come from mean depths of 50 nm $- 0.5\,\mu$m depending on the density of the material. Hence, the technique is sensitive to the near-surface region. Scanning (or rastering) the beam over the surface minimises surface damage and surface charging. The surface features in this size range include extensive faceting, phase separation, morphology of crystals, the structure of fracture faces, precipitates, pores, distribution of materials in composites, bonding at interfaces and preferential reaction at different sites on the surface. Backscattered electron images can give contrast based on the average atomic number of the region or phase examined. Hence, it can be used to differentiate between grains in a multiphase ceramic, or mineral mixture, or between materials in a composite. Topographic images, obtained by combining different backscattered electron images, can reveal detail of pits and protusions, precipitates and altered regions on the surface. Figure 1.4 shows examples of the three types of images – secondary electron, backscattered electron and topographic – from the same area of a ceramic surface. The SEM is relatively easy to use requiring straightforward specimen preparation and conventional vacuum. Its disadvantages are damage by the electron beam to polymer surfaces, some minerals and insulating materials (a minor effect in the scanning mode) and limited resolution.

Transmission electron microscopy (TEM) and scanning transmission electron microscopy (STEM), also covered in Chap. 3, both give much higher resolution of surface features down to unit cell level or < 2 Å in most cases. Hence, atomic steps, ledges, corners, defect sites, fine grains, intergranular films and interphase regions can be imaged. Diffraction contrast, phase contrast and defocus contrast can be used to enhance the images, particularly at edges and surfaces. SEM can also provide selected area electron diffraction (SAED) from crystalline regions. This is particularly important in distinguishing crystal structures of phases with closely similar (or the same) chemical composition. The major difficulty with these techniques is that the sample must transmit electrons before they are focussed to form an image. The sample thickness has to be usually < 300 nm for this to be achieved and the transmission (and image) is highly sensitive to this thickness and the material itself. Specimen thinning using accelerated ion beams introduces severe surface damage. TEM and STEM are most effectively used on very small particles and on the thinner (wedge-shaped) edges of larger particles where structural detail of major importance can often be obtained. An example of this kind of detail can be seen in Fig. 1.5.

Scanning Auger microscopy (SAM), covered in Chap. 6, combines physical imaging of the surface, as in SEM, with chemical analyses of spots, individual areas and chemical imaging (i.e. mapping). SAM uses a focussed electron beam, with energies in the 3–50 keV range, to cause ionisation of core levels

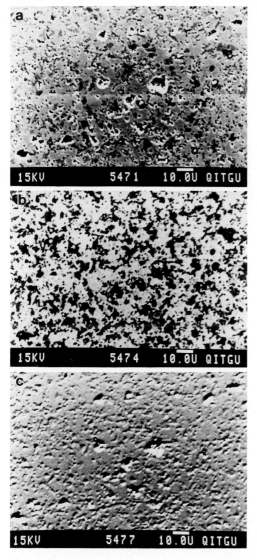

Fig. 1.4. Comparison of (a) secondary electron (b) backscattered electron (BSE) and (c) topographic images from scanning electron micrographs of the same area of a polished ceramic surface showing grain pull-out and porosity. The white bar indicates magnification of 10 μm. Note that the BSE image, in addition to porosity, indicates regions of low (dark) and high (white) average atomic number

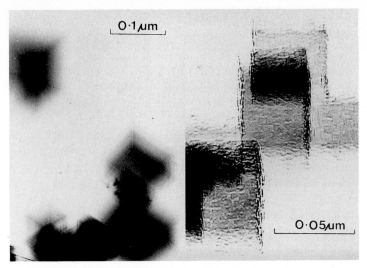

Fig. 1.5. Transmission electron micrographs of MgO "smoke" cubes formed by burning Mg in air. The left-hand image was recorded immediately after preparation; the right-hand image was recorded after < 5 sec immersion in pH 3 nitric acid. Phase contrast imaging reveals that the initially smooth cubes are roughened before dissolution commences. [Reprinted with permission from Fig. 13, C.F. Jones, R.L. Segall, R.St.C. Smart, and P.S. Turner, Proc. Roy. Soc. Lond. A 374, 141 (1981)]

in surface atoms. The ionised atoms relax via emission of Auger electrons in a two-electron process. Each element in these surface layers produces a unique set of Auger energies (from its set of energy levels) so that both chemical identification and composition (i.e. surface concentrations) can be determined. The electron mean free path in solids, discussed in detail in Sect. 1.5.1, reveals that Auger electrons with energies in the 20–1000 eV range escape from depths of only 5–20 Å. Additionally, of course, the primary electron beam produces secondary and backscattered electrons which can be used for imaging. Current instrumentation can achieve a beam size of 20 nm, rastered across the surface like a TV screen, to produce either a secondary (or backscattered) electron image or, using a selected Auger signal, an elemental distribution over a surface. Both types of image are illustrated in Fig. 1.6. SAM is an exceptionally useful tool also for spot (< 50 nm) analyses of the surfaces of individual particles, grains in ceramics, interiors of pores, precipitates, and other differentiated areas on a sample surface. The surface sensitivity of the technique, combined with structural images, gives us a powerful probe of the localised chemistry in surface layers. The limitations of the technique lie in three main areas:

- samples which are poor conductors or insulators cause major charging at high resolution (i.e. small beam size, high voltage and high current density) deflecting the beam onto adjacent areas
- chemical information on covalency, bonding or oxidation state of the element is limited or not available

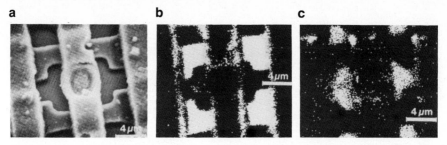

Fig. 1.6a–c. Scanning Auger microscopy/spectroscopy from an area of a 64 K random access memory MOS device. (a) Secondary electron image, (b) elemental Si (LMM transition, 89 eV) distribution, (c) oxidised Si (LMM, 72 eV) distribution. [Reprinted with permission from Figs. 3, 4 K. Bomben, W. Stickle, J.J. Hammond, Proc. 1st Elec. Conf., Vol. 1, SADME (CA, USA 1987)]

— high current densities and voltages can cause decomposition of some samples as in electron microscopy.

Generally, however, SAM can be applied to all materials, with greater or less resolution, provided they are stable in vacuum.

The application of Auger spectroscopy to determination of chemical composition is discussed in Sect. 1.5 and Chap. 6. It is sufficient to note here that the combination of structural and chemical identification allows SAM to give direct information on most of the surface features previously listed with the exceptions only of reconstruction, relaxation, and space-charge regions. It is indeed a powerful surface technique in materials science.

A new and equally exciting technique for imaging surfaces has become available since 1981. The scanning tunneling microscope (STM) uses the low field quantum mechanical tunneling effect, where the tunneling current is approximately related to the inverse exponential of the barrier thickness (gap width) and the square root of the work function, to produce images with resolution of individual atoms (see Chap. 10). The current dependence on gap width (d) is very sensitive (e.g. a ten-fold change for a change in d of 1 Å). An almost atomically sharp metal tip is used to scan areas as small as 10×10 Å to produce the most detailed images of atomic defects yet seen on material surfaces. The vertical resolution is even higher than the lateral resolution, in most cases being better than 0.1 Å. Figure 1.7 illustrates the level of detail seen in an STM image. The technique can also be used to study differences in local work function allowing, in principle, identification of chemically different species in the surface (e.g. impurities or defect sites) or adsorbed on the surface. The distribution of electronic states across surfaces can be defined in this way. The technical problems in achieving this resolution, like decoupling of the STM from external vibrations, controlling the tip scanning, measuring the current and graphically presenting the image, are non-trivial but now largely solved as explained in Chap. 10. The STM is rapidly becoming relatively accessible but, again, is limited in application to good semiconductors and metals. Contamination, normal on any real material surface, will obviously obscure information on the underlying material but may,

17

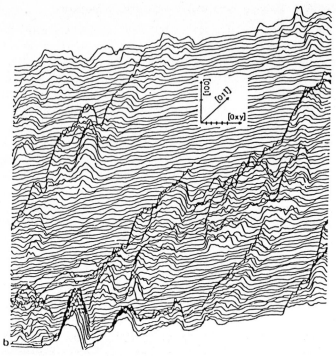

Fig. 1.7. Scanning tunneling microscope image of carbon islands on a reconstructed Au (100) surface (smooth parts). Divisions on the (100) axis are 0.5 nm apart. [Reprinted with permission from Fig. 2, G. Binning and H. Rohrer, Proc. 9th Int. Vac. Cong.; 5th Int. Conf. Solid Surf., Madrid, Spain (Sept. 1983) (Ed. L. de Segovia) pp. 77–79 (Spanish Vac. Soc., ASEVA Publisher, 1984)]

of course, be directly studied with the STM for identification and distribution as in Fig. 1.7. The recently developed combination of STM with SEM, to allow an extreme range of imaging, promises to be of great benefit to the surface scientist.

There are other techniques for surface imaging deserving of mention but not included in our compilation. Field emission (FEM) and field ion (FIM) microscopes [1.24, 25] both based on the tunneling of electrons under high field conditions, and capable of atomic resolution, are now older techniques (with a large literature) to some extent superceded by STM. They apply only to refractory metals and require special sample preparation not normally applicable to materials science. They have greatly increased our understanding of surface structure but must be regarded as essentially fundamental research tools.

Low energy electron reflection microscopy (LEERM) is a relatively new technique [1.26], used in combination with LEED, giving images from elastically reflected electrons in the 0–300 eV range. Its practical resolution is about 50 nm but this can allow a combination of image and diffraction patterns from differentiated surface features, a result of considerable importance for surface phase formation in catalysis, reactivity, leaching, corrosion and surface segregation. It is not yet sufficiently accessible for most materials scientists to warrant a full discussion but may well be a technique for the future.

1.4.2 Indirect Structural Imaging – Relaxation and Reconstruction

Provided that low energies are used, the scattering of electrons, atoms and ions can be restricted to the top few atomic layers giving surface sensitive probes for atomic positions (interatomic spacings), periodic structures (via diffraction of the scattered beams), faceting, defect sites and disorder. Low energy electron diffraction (LEED) uses electrons in the 10 to 500 eV range with escape depths of 5–10 Å (Chap. 13). Atomic beam scattering (ABS) uses atomic beams of low-mass, unreactive gases (e.g. helium) monoenergetic in the range 30 to 300 meV, which scatter and diffract from literally the top atomic layer. Ions with energies less than 5 keV i.e. low energy ion scattering (LEIS) also scatter predominantly from the top layer due to the very large cross sections for ion-atom elastic scattering (Chap. 11). They can be used for studies of shadowing (of one atom by another) and atomic displacement (i.e. interplanar spacing and defect) effects. Scattering of ions with higher energies, i.e. medium energy MEIS, (20–200 keV) and high energy HEIS, or Rutherford backscattering RBS (200 keV–2 MeV), can give information on atomic positions and defects (e.g. interstitials) in the bulk of the solid. Penetration is deeper and blocking effects due to displaced atoms are more evident (Chap. 9). Similarly, medium (MEED, 500 eV to 5 keV) and reflection high energy electron diffraction (RHEED, 5–200 keV), using the same principles as LEED, give information particularly on surface topography and faceting but with less surface sensitivity (up to 10 nm depth) (Chap. 12). All of these techniques give primary data which can only be transformed to the surface lattice (LEED), surface atomic positions (ABS, LEIS) or interlayer spacings (MEIS, HEIS) by calculation using a model for the dynamical processes occurring at the surface. Programs for routine analysis in LEIS, MEIS and HEIS are readily available and reliable to a high level of accuracy. In the other cases, the final structural information is only as good as the assumptions the model allows and particularly for ABS and LEED, relatively sophisticated models are required for reliable results.

The electrons elastically scattered from surface atoms (i.e. $\sim$ 1–10% of the primary beam) in LEED studies of crystal surfaces will diffract sharply only from well ordered domains of 100 nm^2 or greater as determined by the coherence length over which the electrons will remain in phase. Disorder in the surface can thus give diffuse spots and loss of intensity, providing data for analysis of this disorder. Faceting appears as extra spots in the diffraction pattern and steps on the surface can produce multiple splitting of the primary spots in the pattern. The height of the steps can be inferred from the "appearance voltage" of the splitting. The accuracy of interatomic spacings determined from LEED calculations is about 0.1 Å if the surface is well-ordered and a dynamical model is used.

One of the major applications of LEED to surface structural elucidation has been in the determination of the nature and magnitude of surface relaxation and surface reconstruction with particular reference to verification of some now very sophisticated and highly accurate computer models of these processes [1.27]. For instance, it is now known that the neutral (100) planes of simple ionic oxides with NaCl structure, like MgO and NiO, have surface spacings almost unaltered from the bulk after relaxation and "rumpling". By contrast, surface planes which are charged [e.g. the (111) plane in a fluorite structure] or have net dipole moments [e.g. the (111) rocksalt plane or (001) fluorite plane] can show relaxation up to half the bond length and even reconstruction of (111) faces to stepped (100) facets a few unit cells long. Dynamical processes resulting from reaction or heating involving changes in surface topography via reconstruction can be followed from time-dependent LEED patterns using video recorders [1.28]. Segregation of impurities into (or out of) the surface can often be detected in LEED patterns as ordered overlayers producing streaks or new spots.

The combination of RHEED with LEED is particularly powerful for studying topography and reaction profiles propagating from the surface into the bulk of the material as in leaching, oxidation, corrosion, interdiffusion, joining technology, thin film reactions, elemental segregation, interphase regions and recrystallised (e.g. weathered) surfaces.

The LEED and RHEED techniques, however, are unfortunately limited to reasonably large ($> 2\,\text{mm}$) single crystals with relatively well-defined faces presented at the surface. Hence, metals, semiconductors, thin films and minerals can be studied – some multiphase ceramics are also possible – but powdered samples, glasses, polymers, composites and natural materials are not normally possible. For this reason, they tend to be available more for fundamental research and materials development rather than for general materials examination. The LEED technique is not normally destructive but RHEED can degrade some samples.

More information on relaxation, reconstruction and the positions of atoms in ordered adsorbed layers has been obtained using ABS. Helium atoms with energies below $300\,\text{meV}$ have de Broglie wavelengths above $0.26\,\text{Å}$, they diffract from the top layer alone, and their angular intensities can be easily measured using a goniometer-mounted mass spectrometer. The measured intensities have to be matched with a calculated scattering contour of the surface. This contour is usually derived from a model of the surface structure and, if the model successfully predicts the diffraction pattern, this provides an indirect picture of the atomic arrangement in the surface. For instance, ABS has shown that substantial charge redistribution occurs in the NiO (100) (rocksalt structure) surface levelling the electron density contour with the oxygen $0.3\,\text{Å}$ above the nickel atoms, i.e. "rumpled" [1.29]. An ab initio surface density calculation is usually necessary to obtain precise values of atomic positions and bond lengths. The distribution and concentrations of atomic defect sites can also be estimated from ABS patterns [1.30]. The technique has the major advantages of being non-destructive to samples and being applicable to any sample, conducting or insulating, exhibiting an ordered surface layer, i.e. metals, semiconductors, ceramics, minerals and some crystalline polymers. ABS is not widely available mainly because the differentially-pumped high pressure nozzle sources to produce the monoenergetic atomic beam are not simple to make or operate.

Other techniques that can give indirect structural information include surface enhanced X-ray absorption fine structure (SEXAFS) [1.31], X-ray absorption near edge structure (XANES) [1.32] and surface extended energy loss fine structure (SEELFS) [1.33]. Both SEXAFS and XANES require synchrotron radiation (not accessible for most materials scientists) and very sophisticated calculations to obtain structural data. SEELFS can be used in a TEM, can be applied to almost all materials, and will certainly become more widely used as the instrumentation becomes available. At present, it is in a relatively early stage of development as described in Chap. 3.

1.5 Composition of the Surface Selvedge

Having established a picture of the simpler structural aspects of the outer atomic layers, the need for a more comprehensive understanding of the chemical composition of the surface becomes apparent. Defining the information that we require in order to determine a surface's composition is a simple exercise; actually obtaining that information is considerably more difficult. Indeed, very often only part of the picture will be fully understood. Beyond simply identifying the elements in the selvedge, their chemical states and atomic or molar proportions also need description. The lateral and depth distribution of each element is required i.e. a five dimensional problem per element. Up to atomic number 92 there are 90 elements which are naturally occurring. Even confining ourselves to the simple case of formal chemical states, those 90 elements represent about 250 possible variants [1.34]. Placing these variants in an atomically resolved matrix of macroscopic dimensions is not feasible. The pragmatic approach is thus to restrict our evaluation to the phases present and to the way they are distributed in the selvedge. This will be determined by the spatial resolution that the techniques applied can achieve. In order to establish phase identity a first approach might be to evaluate elemental stoichiometries. However, even the simple problem of ascertaining a mole percentage is subject to many errors. n the absence of due care these may be up to 20%. This would blur the distinction between, for example, a sesquioxide X_2O_3 and a dioxide XO_2. In such a case, the chemical state of X, i.e. whether it existed as X^{3+} or X^{4+}, would become a determining factor. Table 1.4 lists a comparison of the techniques examined in this book with respect to their ability to determine surface compositions [1.35].

The region represented by the selvedge is variable and is quite dependent upon the technique in question. For a low energy incident ion technique such as ion scattering spectroscopy (LEIS), projectiles entering below the outer surface are efficiently neutralised and are therefore, in effect, invisible to ion detection by electrostatic means. By contrast, a photon-excitation technique such as X-ray photoelectron spectroscopy (XPS), can reach thousands of nm's into the selvedge producing characteristic photoelectrons from a significant depth. However for a photoelectron from a specific element to be measured, characteristic electrons must reach the analyser without undergoing any energy loss. The inelastic mean free path (IMFP) of these electrons means that only electrons from the first few surface layers meet this requirement, ensuring the surface sensitivity of the technique.

Table 1.4. Comparison of the different surface analytical techniques (considered in further detail in Part II) with respect to the incident probe characteristics, their ability to obtain chemical information and experimental considerations [1.35]. For the latter categories the more dots the better. NA indicates not applicable

	Incident probe				
	Particle	Energy	Energy resolution	Current or flux	Beam diameter
AES	electron	0.5–10 keV	NA	0.1–500 μA	0.1–1 mm
SAM	electron	3–30 keV	NA	0.5 nA–2 μA	300–5000 Å
XPS	photon	1–15 keV	0.5–2 eV	10^{12}–10^{13} s^{-1}	0.2–6 mm
LEED/	electron	15–500 eV	0.5–1 eV	0.1–5 μA	0.2–1 mm
RHEED		2–30 keV		0.3–0.5 μA	50–100 μm
RBS	^{1}H, ^{4}He	1–3 MeV	keV		
NRA	^{1}H, ^{4}He, ^{19}F	0.3–6.4 MeV	keV		
FTIR	photon	0.05–0.5 eV	0.012–0.5 meV		$\sim$ mm
SIMS	ion	0.5–30 keV	NA	1 pA–100 μA	500 Å–2 mm
static				1 pA–10 nA	
dynamic				> 1 μA	
ISS	ion	1–3 keV	5–10 eV	10 pA–1 μA	200 μm–2mm
STM	electron	0.1 eV	NA	0.5–1 nA	2 Å
EM	electron	10–50 keV	eV	0.1 μA–10 pA	2 nm–1 μm
UPS	photon	10–50 eV	3–20 meV	10^{11}–10^{12} s^{-1}	1–3 mm

1.5.1 Electron Inelastic Mean Free Paths

The inelastic mean free path (IMFP or λ) of an electron travelling within a solid can be defined as the mean distance it traverses before undergoing an inelastic event, i.e. some interaction whereby it loses energy. It is evident that, in any of the surface analytical techniques utilizing electrons as the excitation probe and/or the analyzed response, it is this path length which will govern the technique's surface sensitivity. The expectation of a complex dependence of electron IMFP upon the nature of the solid material is not found empirically and the surface electron spectroscopist can resort to the very useful so called "universal" IMFP curve. *Seah* and *Dench* [1.36] have compiled the most comprehensive electron IMFP data base to date. Shown in Fig. 1.8 is the resultant universal curve for the elements. Note that, for electrons in the 10–1000 eV range, the IMFP varies from about 2 nm through a minimum of 0.45 nm and back up to about 1.6 nm. In terms of monolayer equivalents (λ_m) the semi-empirical relationships found for electrons of energy E (in eV) above the Fermi level (between 1 and 10,000 eV) are:

$$\lambda_m = \frac{538}{E^2} + 0.41 a^{1.5} E^{0.5} \quad \text{for elements} \tag{1.1}$$

$$\lambda_m = \frac{2170}{E^2} + 0.72 a^{1.5} E^{0.5} \quad \text{for inorganic compounds} \tag{1.2}$$

Chemical information			Experimental considerations					
Inner shells	Valence shells	Ease of interpretation	Ease of use	Ease of quantification	Lack of surface damage	Coping with charging	Speed and sensitivity of analysis	Elements of zero or low sensitivity
..	..	..	....	...	..	..	.....	H, He
..		..	...	..	..	.	..	H, He
....	..	...	...	...	....	...	...	H, He
Structural techniques			..					
–	–		....	...	...	.	..	Low z
–	–		.	..	...	.	.	High z
Vibrational states		..	....	..	....	....	...	Inorganics
Molecular			..		.	..	....	–
Elemental		..	...	..	NA	..	.....	–
Elemental		..	...	...	..	...	....	H, He
Structural technique		..	..				..	
Elemental			...	..	..	.	..	Low z
NA	....	...	...	..	......	...	...	

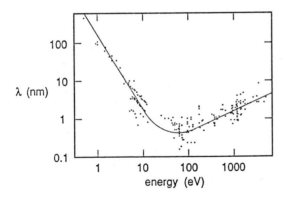

Fig. 1.8. Compilation of electron inelastic mean free path measurements (IMFP, λ) for elements from *Seah* and *Dench* [1.36]. Solid curve of best fit represents the "universal" IMFP applicable to all elements

where a, the solid material's particular monolayer thickness in nm, is given by:

$$a = \left(\frac{A}{\varrho n N_A}\right)^{1/3} \times 10^8 \, \text{nm} \quad , \tag{1.3}$$

where A is the molecular weight, ϱ the bulk density in $\text{kg}\,\text{m}^{-3}$, n the number of atoms in the molecule and N_A is Avogadro's number.

Note that the Seah and Dench approach is essentially *structureless* as single crystal, polycrystalline and amorphous examples of the same substance are only differentiated on the basis of their bulk densities. More recent comments on the Seah and Dench formulae have been made by *Ballard* [1.37] and *Tanuma* et al. [1.38].

The electron IMFP is particularly useful since, by assuming a process of *homogeneous* attenuation, a Beer-Lambert type expression results for the description of electron flux reduction:

$$I = I_0 \exp\left(\frac{-z}{\lambda \cos \theta}\right) \quad , \tag{1.4}$$

where I_0 and I are the incident and emergent intensities and $z/\cos\theta$ is the path length for electrons travelling θ off-normal through a material of depth z along the normal. Although, as *Ballard* [1.37] points out, such an expression strictly only applies to electromagnetic radiation, in practical terms I/I_0 plots are found to be sufficiently linear to be useful for approximate estimates. The total electron intensity reaching the surface at an angle θ from *above* the depth z, as a fraction of intensity reaching the surface from *all* depths, can then be established from a simple integration to be:

$$F_{z,\theta} = 1 - \exp\left(\frac{-z}{\lambda \cos \theta}\right) \quad . \tag{1.5}$$

Thus for $\theta = 0°$, 63% of all electron intensity reaching the surface comes from within one IMFP of the surface, 86% from within two IMFPs and 95% from within three. For the electron spectroscopies IMFP is not only relevant to surface sensitivity, but also to quantification of the analytical method.

Some of the implications of electron IMFPs in terms of the technique's analyzed response can be usefully illustrated by working through the case of a passivated aluminium surface. For simplicity let us consider a 3-phase planar film structure (Fig. 1.9) of bulk aluminium covered with a coherent but amorphous 25-monolayer film of Al_2O_3 capped in turn by a single monolayer of the monohydrate oxide AlO(OH). Only A, ϱ and E need to be first established. All other parameters arise from the above equations, including λ in nm from λ_m and

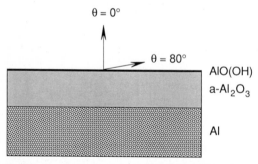

Fig.1.9. Model of a passivated aluminium surface consisting of a 25-monolayer film of a-Al_2O_3 on bulk metal, with an outer monolayer of AlO(OH). Calculated IMFP-dependent relative intensities of Al $2p$ and O $1s$ photoelectrons at the exit angles of $\theta = 0°$ and 80° are given in Table 1.6

Table 1.5. Data set for the three-phase model of a passivated aluminium surface

Phase	A (g mol^{-1})	ϱ (kg m^{-3})	n	a (nm)	Al $2p\,\lambda$ (nm)	O1s λ (nm)	d_{Al} (nm)	d_O	z (nm)
AlO(OH)	59.99	3.01×10^3	4	0.202	2.25	1.76	0.501	1.002	0.202
a-Al$_2$O$_3$	101.96	3.5×10^3	5	0.213	2.43	1.90	0.686	1.029	5.325
Al	26.98	2.70×10^3	1	0.255	1.81		1.000		∞

a. Approaching the analysis of this 3-phase selvedge with XPS, the characteristic photoelectrons of interest are the Al 2p and the O 1s. From literature values for their kinetic energy (1178.9 and 1180.9 eV for Al 2p in the oxides and the metal respectively and 722.0 eV for O 1s in the oxides [1.39]) and estimates of a, values of λ follow (Table 1.5).

As relative photoelectron intensities are a prime experimentally-measured quantity, it would be useful to estimate the effective relative photoelectron counts from each of the three phases. This consists of the initial photoelectron production modified by its transport to the detector. Since the X-ray attenuation with depth can be considered negligible compared with the effects of λ, photoelectron production will be proportional to the relative atom density d_i and the layer thickness z, with subsequent escape dependent upon λ and z. Equation (1.4) is a statement of the probability of an electron on trajectory θ reaching $z = 0$ from a depth z. Hence, by integrating I/I_0 from $z = 0$ to ∞ we derive a maximum electron escape factor of $\lambda \cos \theta$. The fraction for a film of $z < \infty$ is given by (1.5). Combining both with d_i yields an expression for relative photoelectron production and escape from a selvedge of thickness z:

$$I_z = d_i \lambda \cos \theta \left[1 - \exp \left(\frac{-z}{\lambda \cos \theta} \right) \right] \quad . \tag{1.6}$$

It follows that electrons from layer 2 passsing through layer 1 will emerge with a relative intensity of:

$$\exp \left(\frac{-z_1}{\lambda_1 \cos \theta} \right) d_{i,2} \lambda_2 \cos \theta \left[1 - \exp \left(\frac{-z_2}{\lambda_2 \cos \theta} \right) \right] \tag{1.7}$$

with an additional product term for every subsequent attenuating layer.

Combining the data from Table 1.5 for the three-phase model and considering electron escape trajectories along $\theta = 0°$ and $80°$ off-normal, the relative fractions of Al 2p and O 1s photoelectrons can be evaluated from (1.7) and its extension. These are given in Table 1.6, normalized for unit total Al 2p emission at $\theta = 0°$. As would be expected the principal Al 2p and O 1s signals come from the a-Al$_2$O$_3$ layer. The underlying metallic substrate contributes only 11% to the total Al emission at $0°$ and effectively disappears at $80°$. By contrast the relative contribution from the monohydrate outerlayer comparatively increases over six-fold for Al 2p and over seven-fold for O 1s (although the absolute signal strength decreases).

Table 1.6. Relative Al 2p and O 1s photoelectron intensities for two escape angles from the passivated surface. No correction has been made for relative cross-sections and values are normalized for unit Al 2p emission at $\theta = 0°$

	Phase	$\theta = 0°$	$\theta = 80°$
Al 2p photoelectrons from:	AlO(OH)	0.059	0.048
	a-Al$_2$O$_3$	0.827	0.105
	Al	0.113	$< 10^{-6}$
O 1s photoelectrons from:	AlO(OH)	0.117	0.090
	a-Al$_2$O$_3$	1.002	0.107

This simple quantitative model does not make allowance for complicating factors such as relative photoionization cross-sections or spectrometer characteristics. Also the simplicity overlooks aspects like film roughness, which can have a large bearing upon such a quantitative analysis [1.40]. However it does serve to illustrate the influence of λ upon surface analysis. Note that one consequence of λ is that isotropic electron production within a solid is transformed to an anisotropic $\cos \theta$ dependence with transport to the surface. Further general consideration to analyzing surface composition is given in the rest of Sect. 1.5, although formal treatment is left to the individual techniques in Part II.

Whether or not a component is detected thus depends initially upon its depth z from the outermost layer (assuming that it is within the lateral region being probed) and the probability of its responding with sufficient signal to be detected. If the concentration within the volume element falls below its sensitivity limit then that component will be invisible. The questions of varied elemental sensitivities, practical limits to detection, spatial resolution and the definition of chemical state information all affect how well the surface selvedge composition is determined. Before considering any examples of composition determination it is worth considering some of these basic constraints.

1.5.2 Variation of Elemental Sensitivities

For the unknown surface confronting the researcher, the first problem is that of detection of the elements present. Not all elements will be detected with equal ease and the variation of detectability changes between the various surface analytical techniques. Figures 1.10–12 illustrate the relative elemental sensitivities across most of the periodic table for three of the most common techniques i.e. AES, XPS and SIMS. The two electron spectroscopies, AES and XPS, both relying on ionization from particular energy levels and electron detection, have comparable relative sensitivity with variations of less than two orders of magnitude. However, things are quite different for ion spectrometry. In secondary ion mass spectrometry (SIMS), described in Chap. 5, an accelerated ion beam (e.g. Ar$^+$, Cs$^+$, O$_2^+$, 500 eV-5 keV) is focussed (down to < 40 nm) on the surface and the secondary ions, both positive and negative, sputtered from the surface are

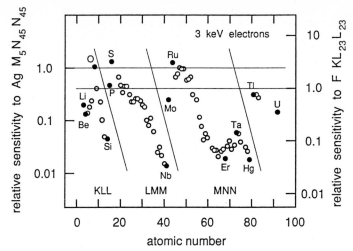

Fig. 1.10. Experimental elemental sensitivities relative to Ag $M_5N_{45}N_{45}$ and F $K L_{23}L_{23}$ across most of the periodic table for Auger electron spectroscopy (AES). Values are for an incident beam energy of 3 keV and include cylindrical mirror analyser transmission function. Not all transitions have been included. Those selected are the most intense, conveniently located transitions, e.g. intense, low-energy valence transitions are difficult to utilize quantitatively. [Compiled from L.E. Davis, N.C. McDonald, P.W. Palmberg, G.E. Riath, and R.E. Weber, Handbook of Auger Electron Spectroscopy, 2nd ed. (Perkin Elmer Corporation, Eden Prarie USA, 1976)]

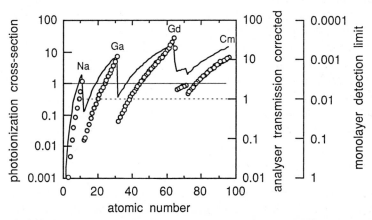

Fig. 1.11. Elemental sensitivities relative to F $1s$ across most of the periodic table for X-ray photo-electron spectroscopy (XPS). Values are based upon theoretical photoionization cross-sections with Mg K_α radiation for the most intense levels (–). The second data set (o) illustrates the influence of electron spectrometer transmission on the relative sensitivities. The transmission is for a concentric hemispherical analyser run at constant analyser energy (constant absolute resolution). [Compiled from J.H. Scofield, J. Electron Spec., Rel. Phen. **8**, 129 (1976); and A.E. Hughes and C.C. Phillips, Surf. Interface Anal. **4**, 220 (1982)]

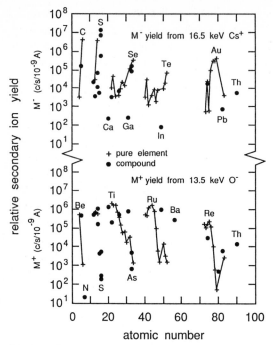

Fig. 1.12. Elemental sensitivities relative to F across most of the periodic table for secondary ion mass spectrometry (SIMS). Note that as these are for ion production they differ from the simple sputter yields shown in Fig. 1.13. [Compiled from H.A. Storms, K.F. Brown, and J.D. Stein, Anal. Chem. **49**, 2023 (1977)]

mass-analysed in a quadrupole or time-of-flight mass spectrometer. The beam can be rastered to produce chemical mapping, as in SAM, with very low (ppm) detection limits for most elements. So-called dynamic SIMS uses high current densities and sputters many monolayers per second whereas static SIMS uses low current densities and removes a single surface layer over periods up to an hour. The ease of ion detection and discrimination provides SIMS with an enormous dynamic range, easily encompassing five orders of magnitude in terms of relative sensitivity. Whilst that provides SIMS with its greatest strength, because atom *removal* is involved (rather than an intra-atomic excitation), these relative sensitivities are dramatically matrix dependent. For example a SIMS spectrum of GaAs based on positive ion detection would, on inspection, suggest a sample of pure Ga and a 0.1 atom % trace impurity of As. Conversely, the negative ion spectrum would suggest a sample of pure As [1.41]. Such widely varying matrix sensitivities have given SIMS a reputation of being nonquantitative. However, quantification can be achieved by the preparation of suitable standards, e.g. by moderate dose ion implantation into the correct matrix [1.42].

The problem of overlooking an elemental component remains even with a relatively easily-quantified technique such as XPS because elements with low photoionization cross-sections could easily be missed. For example, a 1 mole

% content of Al or Si in a metal oxide matrix is detectable. However, the O1s signal is > 4 times more sensitive than either the Si $2p$ or Al $2p$ signal. With a signal-to-noise ratio S/N (i.e. ratio of O1s peak height above background to peak-to-peak noise) of 100, a normal survey scan will indicate no Al or Si to be present. Indeed the peak heights will be roughly 1/4 the size of the noise. To be confident of confirming the presence of the Al or Si would mean improving S/N 12-fold. Statistically, for particle counting, the S/N improves as the square root of the number of counts. Hence the same survey scan would take 144 times longer to acquire, converting a 10 minute exercise into a 24 hour marathon (assuming that specimen contamination and damage did not occur). As this is impractical the best approach is to be aware of the elements of low sensitivity for the given technique and the likely level at which they will be detected. If it is thought that they might be present in the sample, then narrow width scans at the appropriate locations can always be acquired at the necessary S/N. Note that with XPS neither surface-associated H nor He can be detected in practice due to low cross-sections and likely overlap with substrate valence level features. For Auger spectroscopy neither H nor He can emit a characteristic electron due to an insufficient number of electrons.

1.5.3 Practical Detection Limits

Analytical detection limits are typically quoted as percent, ppm or ppb with no due care as to whether the units are in terms of moles or by mass. In surface analysis this is further complicated since it is often not stated whether only the outer layer is being referred to or indeed the detectable selvedge. A true inter-technique comparison is difficult to assess due to the wide variety of the probes and the wide variations in energy densities that can be applied. In Fig. 1.11 an approximate detection limit (under "typical" conditions) in terms of a monolayer fraction of the outermost surface layer is indicated for XPS. Detectability assumes a S/N of at least 2. Detection limits for the outermost surface layer and the selvedge probed are quite different as the quantity of material in the latter may be 10-fold that of the former. The stated limits are really only a guide and with due care a factor of 10 improvement ought to be possible in most circumstances. SIMS has a particularly low detection limit, comparable with that achieved in thermal desorption spectrometry (TDS) in which a mass spectrometer is also used for detection, but the reliance on thermal excitation greatly restricts the variety of molecules which will desorb.

As noted in the prior section, a comparison of SIMS "practical" detection limits has limited meaning at best. Some quoted limits can vary from 4×10^{12} atoms cm^{-3} (Mg in Si with O_2 source) to 3×10^{16} atoms cm^{-3} (C in Si with Cs source) [1.43]. Based on an atomic density in Si of 4.99×10^{22} atoms cm^{-3}, these limits translate to 0.1 ppb to 1 ppm (mole fractions). On average the density of atoms in a Si monolayer is 1.36×10^{15} atoms cm^{-2}. Since the optimum detection limits are based on multilayer analysis, converting from, say, a 50 Å selvedge to a single monolayer, then a feasible detection range would be $\sim 2.5 \times 10^6$ to $\sim 2.5 \times 10^{10}$ atoms cm^{-2} i.e. at the very least $< 10^{-4}$ of a monolayer. As pointed out by *Magee* [1.41], Rutherford backscattering spectroscopy (RBS) and nuclear reaction analysis (NRA – see Sect. 1.9 and Chap. 9) constitute with SIMS a complementary trio. Although detection sensitivities of RBS and NRA are inferior to SIMS e.g. RBS under favourable conditions can at best detect 10^{-4} of a monolayer of heavy atoms on silicon and NRA 3×10^{-2} of a monolayer of H atoms on silicon [1.41], they can provide superior quantitative capabilities. This is especially true of RBS, e.g. analysis of anodized films on aluminium [1.44] indicated the inner region to be composed of Al_2O_3 with an outer region of aluminium-deficient alumina incorporating electrolyte anions. The fraction of film thickness with this incorporation typically varied from 0.2 with molybdate at 0.61 ± 0.03 atom % to 0.7 with phosphate at 5.53 ± 0.24 atom %. Moreover RBS could determine the anion concentrations (as atomic % of the characteristic anions and Al) to an accuracy of a few percent.

The total quantity of material being probed can vary from a single atom as in STM to up to 10^{16} atoms in a broad area spectroscopy such as XPS. In order to detect a single element, XPS would need the presence of up to 10^{13}–10^{14} atoms so that STM would win in the adjusted sensitivity competition. The advantage in XPS is the wealth of chemical information provided. By comparison a destructive bulk analytical method such as atomic absorption spectroscopy may need as few as 10^{10} atoms for detection [1.45]. Despite their many advantages neither XPS nor AES display sufficiently high sensitivities for use in areas such as semiconductor doping. Here the highest impurity levels which affect performance can be below their detection limits. SIMS is generally used for this reason. In an alternative area, XPS has been applied to trace analysis by *Hercules* et al. [1.46]. Utilizing chelating glass surfaces it was found that heavy metals in solution could be easily analyzed down to 10 ppb, without any optimization of the method.

1.5.4 Practical Spatial Limits

The limit of spatial resolution for the selvedge volume element $\Delta x \Delta y \Delta z$ is invariably controlled by the lateral dimensions Δx and Δy. For many of the techniques considered in this book the analysed depth is a few atomic layers, which, with the exception of STM, is much less than the possible lateral dimensions, which can be up to mm. Improvements in lateral spatial resolution can be approached in two ways: either by controlling the dimensions of the incident probe beam or by adjusting the input optics of the analyser so that it views only a selected region of the irradiated surface. The first approach is the preferred method as it maintains high incident fluxes and consequently good S/N characteristics, although this can often be at the expense of considerable radiation damage. Generally this approach is only feasible for charged particle beams. The second method of selected area analysis is utilized when the incident probe cannot be easily focussed. In this case, generally poor S/N has to be accepted and, unless non-selected areas are protected, the whole surface is radiation exposed even if it is not contributing to the useful signal.

Using conventional thermionic emitters and simple electrostatic optics, electron beams can be relatively easily focussed down to $2000\,\text{Å}$ spots with currents of up to $2\,\text{nA}$ at $10\,\text{keV}$. Such electron guns are commerically available [1.42]. Importantly however, even at that resolution the current density involved is $6.4\,\text{A}\,\text{cm}^{-2}$. This equates to 4×10^4 electrons per surface site per second. This is a substantial radiation level considering the energy of the beam. Placing it into perspective the energy density is $6.4 \times 10^4\,\text{W}\,\text{cm}^{-2}$ or 4.7×10^5 times the solar radiation constant [1.47]. Finer electron beams down to $200\,\text{Å}$ have been achieved using high brightness LaB_6 sources or field emission tips in combination with magnetic optics. The temptation to use beam energies higher than $30\,\text{keV}$ for more precise focussing provides diminishing returns because of the reduced cross-sections for Auger electron production, especially for the lighter elements. Most electron excited Auger work at high spatial resolution is carried out in the 5–$10\,\text{keV}$ range. As noted in Chap. 6, the lateral resolution limit in SAM has not been reached and $30\,\text{Å}$, i.e. the IMFP limit, is not considered impossible. Currently, in terms of practical surface analysis in materials science, the "limit" is about $500\,\text{Å}$, see e.g. [1.48].

As with electrons, incident ion beams can also be focussed into fine beams. The popular technique of static SIMS has a scanning variation akin to SAM's relationship to conventional AES. A variety of means exist for the production of ion beams, but for microfocussing the most effective is the liquid metal field emission ion source. A liquid metal film, such as Ga, wets a solid needle and is drawn out into a fine tip at the needle's end. By the application of a high field, positive metal ions are extracted by field emission [1.49]. Such beams can be focussed down to $500\,\text{Å}$.

X-rays, in contrast to charged particles, are not easily focussed into small spots. A Fresnel phase plate lens has been constructed which can focus Al K_α radiation with $4000\,\text{Å}$ spatial resolution [1.50]. However, this is yet to be applied in surface analysis. Less successful methods of localizing X-rays, such as the use of crystal optics to focus the beam down to $0.15\,\text{mm}$ or the proximity method (generating the X-rays at a fine point, adjacent to a thin film to be analysed) down to $20\,\mu\text{m}$, have found some use to date [1.51]. Other than X-ray localization, several methods do exist to achieve selected area XPS (SAXPS), or the more general effect of photoelectron microscopy (PEM). These are considered by *Drummond* et al. [1.51]. The approach which is easiest to implement, because of its compatability with existing spectrometers, is the use of a transfer lens system to magnify the surface area projected onto the spectrometer entrance aperture [1.52]. Selection of a small entrance aperture then means the spectrometer is only viewing part of the surface. This can be further restricted by an additional aperture prior to the transfer lens. Selected areas down to $150\,\mu\text{m}$ in diameter are available commercially [1.53]. Whilst relatively easy to implement, this approach does suffer from reduced sensitivity with the photoelectron count rate decreasing in proportion to the area being sampled. In practice this usually necessitates

Table 1.7. Comparison of the possible different methods of selected area surface compositional analysis

Method	Information available	Spatial resolution	Advantages and disadvantages
Selected area XPS (SAXPS)	Surface elemental composition and chemical state analysis	20–150 μm depending upon method selected (see text)	Chemical shift information obtainable with a minimum of radiation damage. Physical movement of specimen may be required to obtain an image. Generally suffers from inferior signal to noise.
X-ray photoelectron microscope	Surface elemental composition and chemical state analysis	10 μm	Chemical shift information obtainable with a minimum of radiation damage. Image obtained by control of electron optics. Signal to noise superior to SAXPS.
Scanning Auger microscopy (SAM)	Surface elemental composition and some chemical state analysis via "fingerprinting"	30–500 nm	Chemical state information is difficult to obtain although phase identification is possible from a multielement correlation diagram analysis.
Secondary ion imaging mass spectrometry (SIIMS)	Surface elemental composition and isotope distributions	50 nm	Excellent elemental sensitivity. Based on fragments, some chemical state information is possible although the interpretation can be difficult.

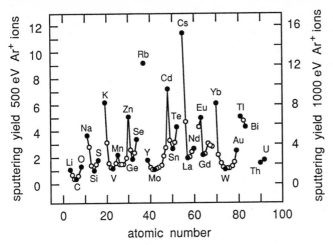

Fig. 1.13. Elemental sputtering yield across most of the periodic table calculated for Ar ions of energies 500 and 1000 eV [1.41]

operating at reduced resolution and scan widths in order to obtain information in a viable time interval. By an extension of the detection electron optics it is feasible to produce an energy filtered photoelectron image of the surface with a resolution of 10 μm [1.54]. Note that, unlike SAM or imaging SIMS, the input beam is not rastered over the surface, and only a single selected area is analysed at a given time. A comparison of the possible different methods of selected area surface compositional analysis is given in Table 1.7.

Although in principle the depth, Δz, offers the highest spatial resolution (i.e. layer-by-layer) of the volume element dimensions, the achievable resolution with depth will depend upon the profiling approach. The commonly-employed method of sputter profiling can be fraught with pitfalls as outlined in Chap. 4. Of most significance is the variation of sputtering yields e.g. Fig. 1.13, leading to preferential sputtering which can dramatically alter relative selvedge compositions. Two alternative approaches exist to extract compositional information with depth. The first, principally applicable to the electron spectroscopies, simply relies upon altering the surface sensitivity by rotating the specimen such that the analyser (of narrow acceptance angle) receives electrons from increasing angles θ off-normal. As the IMFP does not alter, the analysed depth alters by cos θ and the technique thus becomes more surface sensitive at high θ. Angle-resolved XPS is explained in more detail in Chaps. 7 and 16. A typical example of this is illustrated in Fig. 1.14. It is the marked alteration in surface sensitivity which enables the detection of the variety of altered (oxidised) surface groups on plasma-treated polystyrene.

The second approach, which is especially useful for deeper interfaces, involves a mechanical lapping through the selvedge layer at some angle to the original surface i.e. bevelling. Analysis is then achieved by scanning across the lapped face, either mechanically or by beam raster. Depth resolution is controlled

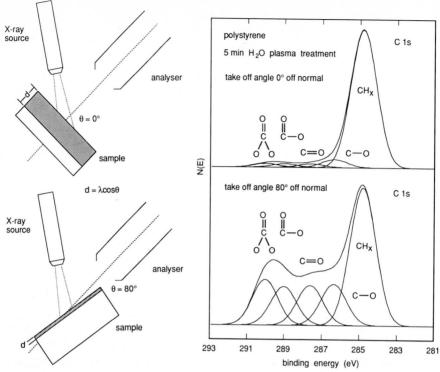

Fig. 1.14. Method of enhancing the surface sensitivity of the electron spectroscopies by increasing the take-off angle at which the analyser views the surface. Here the surface functionalities created by plasma treatment of a polystyrene surface are more readily detected. Adapted from *Evans* et al. [1.55]

by the angle of the lap and the lateral resolution of the probe. *Tarng* and *Fischer* [1.56] point out that an ultimate depth resolution of $\sim 35\,\text{Å}$ would be possible with AES using a 0.2° lap and a beam and manipulator of 0.5 μm resolution. A variation on straight lapping is ball cratering [1.57], in which a shallow spherical pit is ground into the surface and the spatially resolved probe is moved across the crater and thus to greater depths (Fig. 1.15). Craters are more rapidly produced than an angled lap, but both methods have a drawback in that the surfaces still have to be ion etched prior to analysis due to contamination from the mechanical grinding. The latter can also smear out phase constituents if excessive pressures are applied.

One technique in which a relatively deep selvedge is probed is Fourier transform infrared spectroscopy (see Chap. 8). Absorption of the infrared radiation by the characteristic vibrations of the surface species usually provides direct evidence for their structure and bonding. Both FTIR and its electron analogue, electron energy loss spectroscopy (EELS), can provide very detailed structural information on atomic, molecular and multi-atomic ionic species. Surface sensitivity in FTIR is usually achieved using high specific surface area samples,

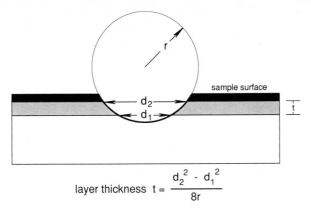

$$\text{layer thickness } t = \frac{d_2^{\ 2} - d_1^{\ 2}}{8r}$$

depth of crater is $\dfrac{d_c}{8r}$ where d_c is crater diameter

Fig. 1.15. Schematic cross-section of a ball cratered surface selvedge. Rastering the incident probe across the crater enables a profile analysis over a considerable depth

diffuse, specular and attenuated total reflectance, or reflection-absorption methods, but the analysed depth is still roughly 1/10 to 1/4 of a wavelength (i.e. 0.1-5 μm) depending on the material, i.e. relatively deep compared with other surface techniques discussed here. Nevertheless, microprocessor-controlled signal averaging has allowed good spectra of submonolayer films to be recorded. FTIR has been extensively applied to studies of catalytic reactions, polymer and glass surfaces [1.58], mineral processing [1.59], modification of oxide surfaces [1.60] and studies of most natural materials. It is an easy technique to use requiring little sample preparation and can be applied to any material in-situ i.e. in media other than gas or vacuum. It is also readily accessible in most laboratories.

Of course the extent to which a given technique probes the selvedge must be borne in mind when comparing results from the various techniques. For example the initial stages of 3-dimensional oxidation of Al(111) were examined by *Ocal* et al. [1.16] utilizing both ISS and XPS. With two slightly different oxidation schemes the outer layer of the oxide could be produced to be stoichiometric or reduced as determined by ISS. By contrast, XPS (probing well below the outer oxide layer, e.g. Tables 1.3 and 1.4) could not distinguish between the oxides, as the reduced outer layer only made a small contribution to the signal.

1.5.5 Chemical State Information

As mentioned at the beginning of this chapter, simply determining the elements present in the surface selvedge or even evaluating their mole proportions, does not guarantee phase identification. In any broad area probe the likelihood of multiple phases being present will confuse the result obtained. Stoichiometric errors aside, the "phase" will merely be an average of those actually present. The situation is improved at higher spatial resolutions when the physical dimensions of the phase become larger than the probe beam or the selected area. This is dramatically illustrated with the use of ratioed scatter diagrams in SAM ([1.49], see also Chap. 6) where accurate phase assessments and their distribution can be evaluated on the basis of stoichiometry.

Chemical state information greatly simplifies the identification of the phase or phases present, and is essential with multiple phases if a broad area is being analysed. Chemical state information in the technique response arises either from the indirect influence of chemical bonding on the core levels of the atoms concerned or from its more direct influence on the valence or molecular levels of those atoms or molecules. The photoelectron spectroscopies probe the levels directly, in the case of XPS the core or near-valence core levels, and in UPS the occupied valence states. UPS (Chap. 14), uses excitation sources in the UV to produce photoionisation and is more sensitive to changes in valence band structure. For the core levels in XPS, chemical state interpretation is very easy to a first approximation. Core electrons ejected from an increasingly oxidised atom (i.e. a more positive atom) have a lower kinetic energy (and hence a higher "initial state" binding energy). For instance, the binding energy of the $S\,2p$ alters from sulphide (S^{2-}) through elemental S^0 to sulphate (S^{6+}) from 161 eV to 169 eV. Numerous examples of this chemical shift illustrate the original publication of *Siegbahn* et al. [1.61], which justly served to promote the value of electron spectroscopy for chemical analysis (ESCA). Although the emphasis throughout this book is on the current practice of surface analysis in materials science, Seigbahn's early examples eloquently argue the power of XPS. Reproduced in Fig. 1.16 are the $C\,1s$ lines from the sodium salts of the first four fatty acids. The carboxylic carbon attached to electronegative oxygen atoms appears at higher binding energy, well separated from the alkane carbons. Note the expected carbon atom ratios for each of the acids. Of course, it is immediately evident that a surface upon which fatty acids have adsorbed may not be amenable to a simple XPS analysis if several of the different acids are coexisting, thereby causing considerable $C\,1s$ peak overlap from different surface species.

The chemical information obtained from UPS can be less obvious. Although the occupied valence levels are probed directly, the valence region of the spectrum is narrow in energy terms, and the number of molecular levels increases with the complexity of the molecular unit. The final state of the UPS photoelectron is also perturbed by the unoccupied valence states. Except in special circumstances e.g. weakly adsorbed molecular oxygen on W(110) at 26 K [1.62], the vibrational level structure is completely smeared out for the selvedge constituents and a broadened multipeak spectrum is usually the result. In specific circumstances e.g. occupied valence band density of states measurements or simple molecular processes on well defined surfaces, UPS can be extremely valuable, but it is best utilized as a complementary technique. One unusual application of UPS has been to use adsorbed Xe atoms as an atomic size probe of the local surface work functions by following the $5p_{3/2,1/2}$ BEs. This in turn enables the evaluation of the microfacets that exist on polycrystalline metal surfaces [1.63]. UPS is considered further in Chap. 14.

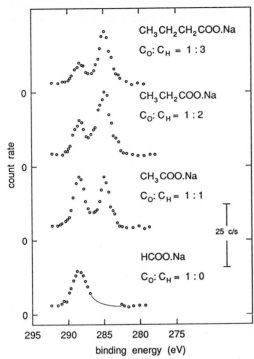

Fig. 1.16. C $1s$ photoelectron peaks for the sodium salts of the first four fatty acids. Adapted from the original work on ESCA by *Siegbahn* et al. [1.61]. The oxygen-attached carboxyl carbon appears at the higher binding energy

As might be expected the Auger electron also contains some chemical information. But since it results from a three-electron process (Chap. 6) that information is subtly convoluted into the spectral feature, which, in the differential mode of data acquisition (still commonly employed with AES), can nonetheless lead to striking differences in peak shapes and energies. Even if the reasons for the differences are unclear, they can be usefully employed on a "fingerprint" basis for the comparison of known and unknown phases. As illustrated by *Madden* [1.64], for even a "simple" process such as oxidation, the core-core-valence and core-valence-valence Auger peaks can display energy shifts from -15 to $+10\,\mathrm{eV}$ between the metal and its oxidised surface. (By comparison, in XPS, the shift would be a kinetic energy decrease of a few eV). An example of Auger chemical information is given in Fig. 1.17 (from [1.64]). The Si $L_{23}VV$ peak is seen to shift to lower kinetic energies with increasing electronegativity difference in the series Si-X. Noteworthy is the influence of H, as it is itself invisible to Auger spectroscopy.

SIMS does not represent a simple case with respect to the extraction of chemical information (see Chap. 5). In order for useful chemical information to be obtained, molecular fragments need to be collected, with those fragments representing the character of the selvedge i.e. without extensive molecular rearrangements being prompted by the act of sputtering. A variant of static SIMS

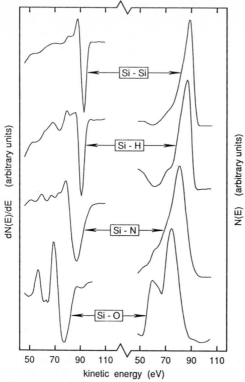

Fig. 1.17. Si $L_{23}VV$ Auger peaks [in differential $dN(E)/dE$ and integral $N(E)$ modes] from elemental silicon (Si-Si), chemisorbed hydrogen on Si(100) (Si-H), silicon nitride film (Si-N) and silicon dioxide film (Si-O) [1.64]

which is capable of this is fast atom bombardment mass spectrometry (FABMS). This approach reduces the charging effects, ionic migration and radiation damage normally associated with ion bombardment, especially of insulators. FABMS was originally developed as a structural diagnostic for high molecular weight, nominally involatile, compounds [1.65].

Fourier transform infrared (FTIR) is quite different from the other analytical techniques in that it does not detect individual atoms, but rather bond vibrations due to the resonance absorption of infrared radiation at frequencies specific to the characteristic vibrations. In ascertaining chemical states at interfaces it can thus be a very powerful probe. Infrared techniques have been applied to surface studies for several decades and predate the electron and ion spectroscopies in extensive use. The generally poor signal strengths of the past have now been largely overcome with the improved sensitivity of FTIR over earlier dispersive instruments. In the attenuated total reflectance (ATR) mode it is particularly valuable for *in situ* solution studies of the solid-liquid interface. (This, in particular, will be considered in further detail in Chap. 8). This contrasts with most modern surface analytical methods where the studies are generally carried out *ex situ*.

1.5.6 Laboratory Standards

In spite of the present state of technological advancement of surface analysis, with system costs ranging up to millions of dollars, there is no guarantee of accuracy or precision. Problems still remain in the comparison of results between different laboratories and hence all data needs to be critically assessed. This is even more important for the non-surface specialist commissioning work performed elsewhere on a contract basis. Reliability of literature data is limited so that laboratory independence is still required and laboratory standards are indispensable. This is not to say that the wealth of literature data on photoelectron binding energies and cross-sections or modified Auger parameters etc., is inaccurate, but that, in practice, other factors can modify spectra. If only for reasons of different referencing methods, experimental set-ups and spectrometer transmission functions, there is no substitute for examining comparative standards on the same instrument and under the same conditions that the problem specimen is being examined. This is especially true for quantitative analyses. Whilst this can considerably lengthen the time taken to achieve a result, the correct interpretation is more likely to be made.

A fundamental problem in the choice of a standard is the selection of a material with known bulk composition which translates right through the surface selvedge. As pointed out already, very often the phase composition of the selvedge bears little resemblance to the bulk. The more surface sensitive the technique the more severe this problem. Those materials with greatest thermodynamic and kinetic stability, exhibiting the least tendency toward any environmental degradation are usually the most reliable standards and also the easiest to handle. If the material has a known susceptibility toward oxidation or hydrolysis, then in-situ fracture or glove box preparation is warranted. As a general rule standard materials are best obtained in the most pure state. However, lower qualities are quite acceptable if the impurities are evenly distributed throughout the bulk e.g. a 99% pure compound in this category would be far more acceptable than a 99.999% sample in which the remaining 0.001% was concentrated entirely in the surface.

1.5.7 Inter-laboratory Errors

The ASTM Committee E-42 on surface analysis has a variety of sub-committees covering AES, XPS, ISS, SIMS, ion beam sputtering, standard reference materials and standard reference data. Joint round-robin investigations sponsored by E-42 have been conducted for XPS [1.66] and AES [1.67] to compare interlaboratory performance. The XPS round robin compared results from Ni, Cu and Au foils on 38 different instruments manufactured by 8 companies and found a spread in reported binding energy (BE) values typically greater than 2 eV, whereas the individual imprecisions in measurements were believed to be less than or equal to 0.1 eV. Reported intensity ratios of photoelectron peaks for Cu and Au spectra (for clean surfaces) spread over a factor of ten, although reported imprecision was less than 10%. The subsequent AES round robin compared results from Cu and Au foils on 28 different instruments manufactured by 4 companies. Here the spread in reported kinetic energy (KE) as a function of kinetic energy varied from 7 eV at a KE of 60 eV, to 32 eV at a KE of 2 keV, with an imprecision range of $\sim 1-3$ eV. For an incident energy of 3 keV, the reported intensity ratios varied with a factor of ~ 38 for Cu to one of ~ 120 for Au, whilst again the individual imprecision was less than 10%.

At least part of the observed spreads could be ascribed to erratic instrumental performance, while in other cases it was believed that operator error was responsible. Both round-robins clearly demonstrated the need for improved calibration methods and operating procedures if these surface techniques are to produce accurate results on a routine basis. In the interim the only reliable approach is to utilize internal laboratory standards to aid the interpretation of the work performed in that laboratory.

1.6 Defect and Reaction Sites at Surfaces

The reactivity of a surface is nearly always increased if the concentration of surface defects is increased. In most cases, point defects, as in cation, anion (or atomic) vacancies and "kink" sites, provide sites with altered electronic structures permitting enhanced electron-transfer reactions not possible on the perfect surface. Extensive theoretical studies have successfully modelled the energies of vacancy and kink sites [1.11, 14, 15, 68, 69]. For instance, on ionic oxide surfaces with rock salt structure (e.g. MgO, NiO), when an oxide ion is missing from a (100) surface, the screening of the cations from each other by the large, polarizable oxide ions is much reduced. This produces changes in the electron energy levels of the surrounding, now 4-coordinate, cations and anions of the top layer such that molecular hydrogen can be dissociated at this site where it would not on a perfect 5-coordinate surface. For kink sites at steps and edges, coordination is reduced to four and, at corner sites, to three. The model also predicts substantial ionic displacements, tending to "smooth" the electron density contour across the step, as well as significantly altered energy levels for reaction.

Examples of changes in concentrations of such sites are myriad in materials science. Figure 1.18 presents TEM images using phase contrast from NiO subjected to increasing annealing temperatures from 700°C to 1450°C. The initially rounded 700°C-formed particles exhibit relatively high concentrations of surface (and bulk) defects "frozen" into the structure during decomposition. The surface sites are disordered with variable coordination and high reactivity (e.g. in acidic dissolution kinetics). The 1450°C-formed oxide has grown quite extensive flat (100) faces with well-developed steps and ledges. The majority of the surface ions are in 5-fold coordination and the reactivity is accordingly reduced roughly ten-fold. These vacancy and kink sites in NiO surfaces can be directly correlated

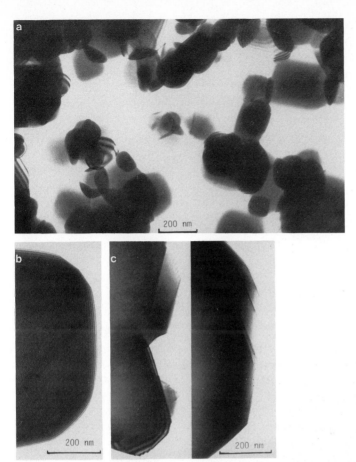

Fig. 1.18. Transmission electron micrographs of NiO crystals annealed in air for 4 *h* at (a) 700°C, (b) 1100°C, and (c) 1450°C (2 regions). The transition from defective, equiaxed crystals with rounded, relatively smooth surfaces to relatively perfect crystals with basically flat steps and ledges can be seen. [Reprinted with permission from Figs. 1b,c, C.F. Jones, R.L. Segall, R.St.C. Smart, and P.S. Turner,. J. Chem. Soc. Faraday Trans. I **73**, 1710 (1977)]

with the concentration of sites with Ni^{3+} and O^- electronic structures derived from XPS studies, a result in agreement with the theoretical model predictions [1.70].

Hence, three of the techniques most suited to studying defect sites to define their atomic positions and their electronic structure are high resolution, phase contrast TEM with XPS and/or UPS. TEM has been discussed in Sect. 1.4 above and XPS in Sect. 1.5. UPS has been used [1.71] to show that, although nearly perfect (110) TiO_2 surfaces do not dissociate H_2O on adsorption, oxygen vacancies interact strongly with H_2O dissociating the molecule to produce OH^- species [1.72]. Even more subtly, UPS shows that H_2O does not dissociate on the (100) NiO face and only dissociates to a very limited degree on the defective surface.

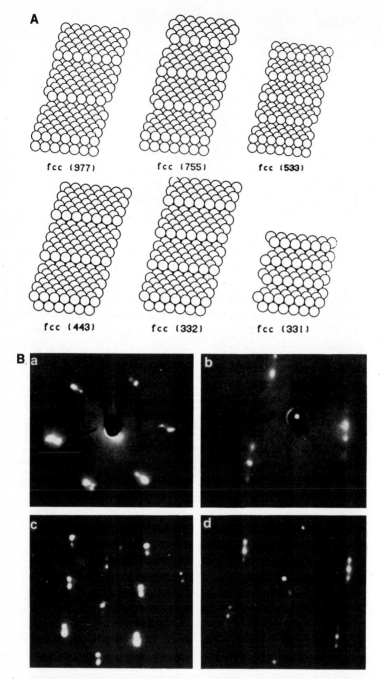

Fig. 1.19. (A) Structure of several high-index stepped surfaces with different terrace widths and step orientations. (B) LEED patterns of the (a) Pt(755), (b) Pt(679), (c) Pt(544), and (d) Pt(533) stepped surfaces. [Reprinted from Gabor A. Somorjai: Chemistry in Two Dimensions: Surfaces. Copyright (c) 1981 by Cornell University. Used by permission of the publisher, Cornell University Press]

Prior adsorption of O_2 however, giving O atoms at the defect sites, results in the immediate appearance of OH^- ions after H_2O admission. The role of oxygen in promoting H_2O dissociation through defect interaction is of major importance in catalysis [1.73] and energy conversions [1.68].

Two other techniques, already discussed in Sect. 1.4, are also of great value in defining defect sites on surfaces. The STM can image vacancies, low-coordinate kink sites at steps and ledges, and atomic displacements resulting from these defect structures. LEED has been used to study low-coordination defect sites through an approach based on stepped, high Miller index surfaces. The high index surfaces are formed on stable, low index surfaces, by atoms breaking away prior to desorption or migrating prior to condensation. In either case, kink sites result with different binding energies and reactivity [1.10, 74]. In LEED patterns, multiple diffraction spots characterise the formation of steps with different orientations, and hence, high Miller indices as in Fig. 1.19. On the ZnO (0001) face, annealing at 900 K induced regular step arrays oriented in a single direction as evidenced by additional spots in the pattern which changed with changing electron energy [1.75]. The step height could be determined to be a single lattice unit (or two Zn-O layers). On the polar (000$\bar{1}$) face, step arrays oriented in three different directions were found and the streaky, diffuse form of the additional features in the pattern showed that the arrays were distinctly irregular [1.76]. The concentration of defect kink sites can be estimated from such observations.

FEM and FIM can, as with physical imaging, image defects and high Miller index planes on surfaces but the information is confined to the limited class of materials that can withstand the strong electric field (i.e. $\sim 10^{10}$ V m^{-1}) at the sample tip without desorption, evaporation or destruction of the surface. The two techniques have been immensely useful in studying oxide structures and particularly stable atomic defect positions on refractory metals like W, Ta, Ir, Re (e.g. [1.77]) and some oxides like ZnO [1.78]. The work functions of individual planes can be determined as well as field migration of defects.

1.7 Electronic Structure at Surfaces

The electronic structure of surfaces embraces a variety of concepts such as:

- the electronic barrier at the interface of a metal or semiconductor with another phase (vacuum, gas, liquid or solid);
- the space-charge (or diffuse) layer in semiconductor materials (including oxides, minerals and some ceramics);
- the electronic states (e.g. electron densities, valencies, bonding etc.) of surface atoms;
- defect sites (or states), active in adsorption and reaction, arising from the geometrical structure of the surface (e.g. faceting, vacancies, dislocations etc.);

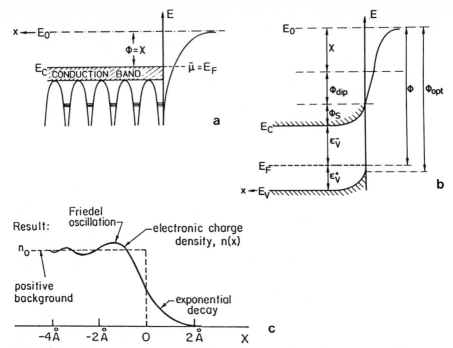

Fig. 1.20. (a) Schematic illustration of the band model of metals (without electric fields). The work function ϕ is defined relative to E_0. (b) The band model of a p-type semiconductor surface involving both the space charge (ϕ_s) and dipole moment (ϕ_{dip}) components. [Reprinted with permission from Figs. 7.1, 7.3, J. Nowotny and M. Sloma in "Surface and Near-Surface Chemistry of Oxide Materials", ed. by L.-C. Dufour and J. Nowotny, Ch. 7 (Elsevier, Amsterdam 1988) p. 281]. (c) Charge density oscillations and redistribution at a metal-vacuum interface. [Reprinted from Gabor A. Somorjai Chemistry in Two Dimensions: Surfaces. Copyright (c) 1981 by Cornell University. Used with permission of the publisher, Cornell University Press]

   – impurity segregation into or out of surfaces and consequent reactive surface sites;

   – surface conduction and surface charging consequent upon any or all of the above considerations.

We have already described techniques to study the electronic structure of defect sites in the previous section. Information on the electronic (chemical) states of surface atoms, largely obtained from XPS, UPS, EELS and AES, has been discussed in Sect. 1.5.4 above. Impurity segregation and doped reactive sites will be covered in Sect. 1.8 below. Hence, we will limit ourselves here to techniques relevant to studies of the electronic surface barrier, space-charge layers and surface conduction or charging.

Schematically, the potential energy at the surface of a metal and a semiconductor in vacuum can be represented as in Fig. 1.20a and b, respectively showing the work function ϕ or potential that an electron at the Fermi level (average energy of the last electron added) must exceed in order to achieve complete ionisation with zero kinetic energy in the vacuum. The magnitude of this work function is related to interaction of the electron with the charge distribution at the surface (i.e. electrostatic, exchange and correlation energies) and changes as the surface charge is redistributed. Redistribution of this kind can be caused by changes in surface geometry (e.g. faceting, reconstruction, relaxation defects) surface composition and adsorption. In practice, measurements of $\Delta\phi$, the change in work function, are of most value. Dynamic or vibrating condenser methods, as in the Kelvin probe, are normally used to determine $\Delta\phi$ [1.79]. Uncertainties in value of the reference levels in the $\Delta\phi$ term, and relatively complicated charge transfer processes at surfaces have limited interpretation [1.10] and application [1.79]. Nevertheless, the method has been used to monitor adsorption in simple gas/solid systems where the solid is either a metal or semiconductor, oxidation of metals, oxygen interactions with oxides and ceramics, defect equilibria, phase transitions, surface segregation and catalysis.

LEED has been extensively applied to studies of the shape, extent and magnitude of the surface barrier, i.e. to the definition of the barrier itself, using the fine structure of intensities of the diffracted beams as a function of beam energy and direction (Chap. 13). It has elucidated barrier structures on metals and semiconductors in vacuum. The major limitation is that it can only be applied to extensive, single crystal surfaces in UHV conditions. It cannot be easily applied to granular (e.g. ceramics), multiphase or powdered material.

Another way of describing the surface barrier is associated with the concept of surface dipoles. The surface asymmetry (i.e. lattice termination) produces a redistribution of charge beyond the surface itself, into the vacuum (or other media) inducing a surface dipole which, in turn, modifies the barrier potential. This induced dipole varies in magnitude at different sites on the surface. On ionic surfaces, this variation is extreme, giving a net static charge separation. On the surfaces of semiconductors and insulators, separation of positive and negative charges results in a space-charge region and this partly controls the transport of charge (as electrons, holes, ions) and atomic species across the surface. These species, and adsorbed species, experience different dipoles (and hence, polarisation) as they move from site to site leading to different tendencies to bond formation. Techniques for monitoring these space-charge effects, conduction and charging, must contribute considerably to our understanding of the surface of any material. Descriptions of the older techniques, including work function measurements, can be found in references [1.80, 81] but will not be explained in detail here.

The relatively new technique of STM allows an associated spectroscopy, scanning tunneling spectroscopy (STS), in which the voltage dependence of the tunneling current is used (Chap. 10). The densities of electronic states, surface band structures and band bending in the space charge region can be obtained from this dependence. Local work functions (or barrier heights), with atomic resolution, can be measured using the dependence of tunneling current (I_t) on tip distance (d) from the surface through the (approximate) relationship

$I_t \propto \exp(-\phi^{0.5}d)$. The STM/STS technique promises, with further development, images of the spatial distribution of both local work functions and electronic states including space-charge effects.

The high-field analogue of STS, field emission spectroscopy (FES), is of some value in defining electronic states of single-atom adsorbates (e.g. [1.82]) but, as with FEM, its application is limited to refractory metals and oxides. The other high-field technique, field ionisation spectroscopy (FIS), gives information, from the kinetic energy distribution of the ions, on the density of unoccupied states with similar limitations to those applying to FES, FEM and FIM.

Ion neutralisation spectroscopy (INS) is another electron tunneling technique in which electrons neutralise He^+ ions directed at the surface inducing an Auger transition in which the second (Auger) electron is analysed for its kinetic energy [1.83, 84]. A focussed ion beam gives lateral resolution to $< 100 \, \mu$m in a spectrometer set up for AES, XPS or UPS. INS has been used, for instance, to show that oxygen diffuses below the surface of Ni(100), a result which is supported by work function measurements of an inverted dipole and by UPS results. The electronic state of a surface space-charge layer is reflected in the tunneling characteristics of the surface and, hence, in all of the tunneling spectroscopies.

The other techniques that have been applied to studies of electronic states are not included in this compilation because the applications have so far been mainly in fundamental research. Surface photovoltage measures surface conductivity induced by transitions between electronic states (bands, polarons and discrete levels) under photoillumination. It can, particularly with elemental semiconductors, semiconducting compounds, oxides and some minerals, define these states on both single crystal and powder surfaces [1.85, 86]. Photoluminescence, and related UV/visible optical spectroscopies, have produced new results correlating surface electronic states with changes in the coordination of oxide ions in metal oxide surfaces [1.87–89]. This work is paralleled by extensive studies using electron spin resonance (ESR) on the same topic [1.89]. The theoretical modelling of these surfaces referred to in Sect. 1.6, with these optical and ESR studies, appears to point the way to new methods of correlating defect structure and electronic states which may be applicable to a range of materials before long.

Surface conduction and surface charging are both considerations of singular importance in many of these measurements, and indeed, in materials science generally. Non-conducting or poorly conducting surfaces charge heavily under photoemission, electron emission or tunneling, ion beams and photovoltage measurements, in many cases severely limiting the data available. The level of charging can be used, however, to estimate surface conductivity and to infer information on surface electronic structure. In XPS, this approach has been used to show that an increased surface concentration of defects in NiO, as Ni^{3+} and O^-, leads to higher surface conductivity. For powdered samples the expression for surface charge:

$$V_{ch} = \frac{K_1 - K_3}{B(\sigma_0)} \exp(eV_s/kT) \tag{1.8}$$

(where K_1 and K_3 are constants, V_s is the band bending and $B(\sigma_0)$ is a function of the sample conductivity) has been tested [1.90].

This is in accord with surface photovoltage results on the same surfaces correlating these defects with specific electronic states in the band structure [1.86]. The combination of XPS and photovoltage measurements is a useful approach to this problem. Information on surface conduction can also be found using the STM, SAM, SEELFS, UPS, INS and EELS techniques.

1.8 Structures of Adsorbed Layers

In addition to the structure of the surface of the solid material, it is obvious that we would like to know as much as possible about the structure of adsorbed or reacted layers in order to understand properties such as:

- coverage of the surface (continuous or discontinuous);
- degree of ordering in the layer(s) and effects of reaction conditions (particularly temperature) on this ordering;
- effects of the solid surface (e.g. defects, facets) on the adsorbed layer,
- local structure and geometry of adsorbed atoms, molecules or ions;
- multilayer formation and reaction profiles into the surface

The importance of this information is obvious in relation to contamination of surfaces, particularly by carbon, before reaction or coating. It is also central to materials technology in areas like corrosion and passivation, adhesion, minerals beneficiation, joining and plating technologies, catalysis, thin film technology, surface modication of materials for composite compatibility, slip (frictional) properties, wear resistance and reaction resistance. A good deal of materials research and development is concerned with the verification (or failure) of surface treatments of this kind.

Classical adsorption isotherms (e.g. Langmuir, BET) can be used to estimate the surface coverage and give some indirect information, from isotherm shapes, on monolayer or multilayer formation [1.91]. More directly, STM can image discontinuous carbon layers on surfaces, as in carbon islands and thread-like structures seen on the Au (100) surface (Fig. 1.7), as well as local structure and geometry of atoms, molecules and ions [1.92]. SAM is also directly applicable to studies of surface coverage, localised adsorption and reaction and changes in the structure of these layers with altered reaction conditions. It lacks resolution for structure of individual species as does SEM. Despite the high resolution of TEM, it is not much used to study adsorbed layers because the less stringent vacuum conditions in most instruments leads to surface carbon contamination and contrast differentiation from single, discontinuous layers of adsorbed species is not usually sufficient. TEM, however, gives direct imaging of surface structural alterations caused by adsorption and reaction as in the formation of extensive facetting, recrystallised overlayers and porosity induced by etching, oxidation or reduction on ZnO surfaces [1.93]. FEM and FIM, on the limited materials applicable to these techniques, give detailed images of adsorbed layer coverage, ordering, surface facets, effects of surface defects, and specific atom sites on the surface (e.g. [1.94]) but not on the structure of individual adsorbed species. ABS, using He beams, has been extensively used to study coverage, order (via diffraction of the beam) and disorder, atomic defects, and surface site configurations [1.30]. Ion scattering (LEIS, MEIS and HEIS) techniques are also directly applicable to these studies. LEIS has been used [1.95] to identify specific sites (and depths below the surface) for oxygen on mineral surfaces. HEIS (or RBS) is particularly suited to identification of the positions of adsorbates dissolving or reacting into the surface (Chap. 9).

Ordering beyond the atomic scale, i.e. surface crystallography in adsorbed or reacted layers, is generally studied using LEED. These applications of LEED are similar to those described in Sect. 1.4.2. Order and disorder are represented in the diffraction patterns by new spots, multiple splitting of spots, changes of intensity and alterations in the pattern as a function of beam energy and entrance (or exit) angles. As mentioned above, ABS and ion beam scattering give data indicative of surface ordering. SEXAFS, although less accessible since it requires synchroton radiation, and SEELFS should be added to this list because both can identify atomic positions and ordering of surface layers.

SIMS is also capable of producing useful information on the structure and composition of adsorbed layers. Static SIMS (Chap. 5), with its very slow sputter rate, can be used to determine: whether the surface is fully covered by an adsorbed or reacted layer, whether multilayer formation has occurred; the structure of molecular and ionic species in each layer; and the structure of reaction profiles in the surface [1.96, 97].

For instance, a static SIMS study of an oxidised copper-zinc alloy [1.98] showed that the top layer only consisted of oxidised copper yet zinc was preferentially segregated into the first few surface layers below the top layer (as seen by XPS). The ability to see the structure layer-by-layer is a major feature of static SIMS applications. It is a relatively straightforward technique which can be applied to most materials although many polymers are decomposed by the beam and insulators are difficult to handle, requiring special discharging techniques. Its main disadvantage is that it is not reliably quantitative for most materials because sputter yields are radically altered by changes in the bonding and composition of the matrix. Multilayer formation and reaction profiles will be further considered in the next section where the use of HEIS (RBS), nuclear reaction analysis (NRA) and ellipsometry will be discussed.

Finally, EELS, like SEELFS, measures inelastic losses in the scattering of electrons from surfaces. These losses encompass: core level ionisation, valence electron excitation and ionisation, plasmon (collective) excitation of valence electrons and vibrational excitations at the surface. The last of these, like FTIR, can give molecular structure but very high resolution energy measurement of the electron loss is required. This technique is instrumentally sophisticated and not readily available. It is sometimes coupled with LEED apparatus.

1.9 Structure in Depth Profiles Through Surfaces

Very often in materials science, it is not the top atomic or molecular surface layer that we are concerned with but a near-surface region from 2–100 nm thick. Some examples of surface phenomena in this category are:

- segregation of particular elements of the solid or impurities (e.g. dopants) preferentially into or out of the surface region;
- thin, deposited films as in optical and microelectronic applications;
- reprecipitated (from solution) or recrystallised (in-situ) layers (for nucleated crystallites, see Sect. 1.10.4 below);
- reaction layers of different composition (e.g. hydrolysed, oxidised, reduced);
- leaching or dissolution profiles after aqueous attack;
- applied passivation layers (often multilayers).

The range of surface concentrations in these examples may vary from major matrix elements to ppm (as in n-type or p-type semiconductor dopants).

In some materials (e.g. oxides, polymers, glasses), the immediate surface can be dominated by one particular functional group of the matrix (e.g. hydroxyls, alkyl groups or alkali metal cations) while excluding others (e.g. metal ions, amine groups or other cations respectively). This segregation in the first few layers can be studied non-destructively (i.e. without ion beam etching) using angle resolved XPS (ARXPS) as discussed above. Photoelectrons emitted from the surface at angles close to grazing exit angle (i.e. 80–85°) come almost entirely from the top monolayer. At 0° exit angle, the depth penetration is at a maximum corresponding to the IMFP in the solid. Comparison of spectra from 0°, 45° and 80° reveals structure of this type.

Segregation of impurities, positive or negative as in the Gibbs surface excess, is a characteristic of most materials and they are known to play crucial roles in surface reactivity and other properties. In ceramics, minerals and electronic materials this can be the single most important factor in their behaviour in application. Theoretical calculations similar to those used to model defects in oxide and ceramic surfaces [1.27, 68], have been applied to impurity segregation [1.15]. Energetically preferred sites emerge from these studies. For instance, Ca^{2+} is found to preferentially incorporate into an MgO surface while Li^+ segregates out of it. Experimental verification of this kind of structure has come from ARXPS, AES and LEIS. Static SIMS is also useful particularly for trace concentrations of impurities but, like LEIS, is more damaging to the surface. If impurity segregation results in ordered overlayer structures, as in K diffusing to ZnO surfaces [1.99, 100], LEED patterns reveal this as new spots.

Depth profiles are most commonly generated using XPS, AES or SIMS with an ion beam to erode the surface under controlled conditions. The analysis may be recorded either intermittently by interrupting the sputtering or simultaneously. The rate of removal of surface material by the ion beam can vary from the equivalent of < 0.1 to > 35 nm min^{-1} so that quite different surface effects are found

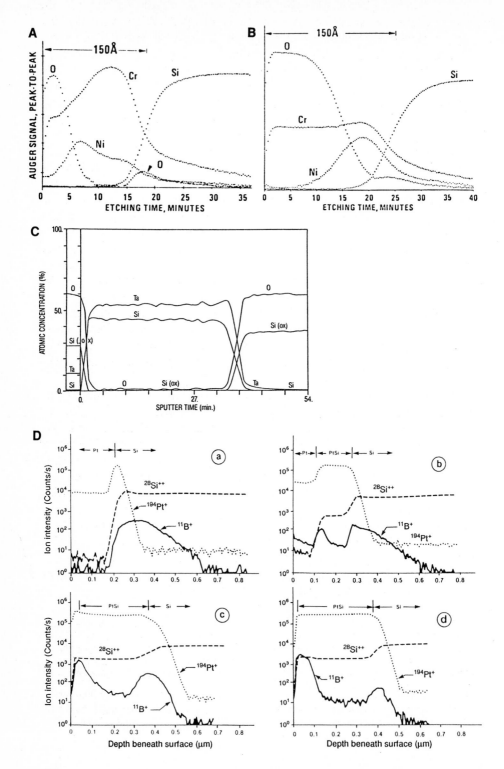

50

at the extremes of this range. At high beam currents the following difficulties apply: preferential sputtering giving compositional variation; enhanced diffusion, reconstruction and segregation; decomposition and loss of specific elements (e.g. F, S, Cl, O); spatial mixing distorting layer separation; and surface roughness (e.g. formation of cones, pyramidal pits and other specific structures). Removal is very often non-uniform. Despite all of these problems, under controlled conditions very good results can be obtained as illustrated in Fig. 1.21. A full discussion of ion beam effects and profiling is in Chap. 4.

These films and deposited layers can be very usefully analysed with RHEED, HEIS (RBS) and ellipsometry. As explained in Sect. 1.7 and Chap. 12, RHEED is particularly suited to the study of near-surface layers in thin films and reaction profiles. The use of grazing angles of incidence and emergence, in order to minimise the depth of analysis in RHEED, places requirements on surface planarity of the samples that are often difficult to meet so that it is primarily a near-surface analysis technique. Changing these angles, however, generates data on segregation, corrosion, wear hardening, passivation and other systems where a concentration gradient through the surface applies. Alternatively, HEIS or RBS has some significant advantages for structure in depth profiles such as: cross sections and scattered ion yield proportional to Z^2; no sputter damage; and correlation of the energy of the backscattered ions with penetration depth. Thus, non-destructive profiles can be generated giving both thickness and composition of each layer. It is therefore applied not only to segregation of impurities but also to thin film determination.

Ellipsometry has not yet been introduced here but it is also valuable for studies of thin films, reprecipitated, recrystallised or reacted layers, passivation layers and leached layers. The amplitude and phase of polarised light reflected from a surface change when one or other of these layers is present. There is a linear relationship between these changes and optical thickness which, in turn, is related to the average thickness of the layer. Changes in the polarisation (i.e. parallel and perpendicular electric vectors) of the reflected light can be measured to give the refractive index and the absorption index of the film, values which can ideally be matched to those predicted for the film structure and composition. The technique is relatively simple to set up and use but requires surfaces with good reflectivity, i.e. fairly flat, and gives averaged values in multiphase systems. It is mostly applied to deposited (e.g. from vacuum, solution or electrochemical) films on extensive substrates before and after reaction.

Fig. 1.21. (A) Auger (AES) depth profiles of 15 nm thick nichrome film deposited on a silicon substrate recorded immediately after deposition and (B) after heating in air at 450°C for 30 s. The heat treatment has produced both oxidation and migration of the Cr and Ni. (C) XPS depth profiles of a tantalum silicate film on SiO_2 substrate. The Si(ox) signal used was the Si($2p$), 103 eV binding energy emission; the elemental Si is the Si($2p$) intensity at 99 eV binding energy. (D) SIMS depth profiles showing redistribution of B implanted in Si during platinum silicate growth; (a) as-deposited; (b) after annealing at 400°C for 5 min; (c) after annealing at 400°C for 30 min; (d) after 600°C annealing for 30 min. [Reprinted with permission from Figs. 7, 8, 12, K. Bomben and W.F. Stickle, Surface Analysis Characterisation of Thin Films in "Microelectronic Manufacturing and Testing" (Lake Publishing, Libertyville, IL, USA 1987)]

In relation to depth profiling, we should also introduce another ion beam technique mentioned in Sect. 1.5.2 but not previously discussed, namely nuclear reaction analysis (NRA) (see Chap. 9). This, like HEIS, uses very energetic (i.e. > 1 MeV) ion beams from an accelerator and is not widely available. In NRA, it is the products of nuclear disintegration, i.e. γ-rays, α- or β-particles and nuclear fragments, that are analysed not the incident ions. These nuclear reactions occur only when the energy of the beam, which progressively attenuates as it penetrates the solid, resonates with the intranuclear process. Hence, by changing the beam energy, resonance occurs at different depths in the sample generating a profile of the material. It has been applied particularly to hydrogen impurity profiles in amorphous semiconductors.

1.10 Specific Structures

There are still a few surface structural features as yet not fully explored by the preceding sections. They tend to be specific or localised features on or in the surface and are largely associated with ceramic materials although, in all cases, they can be found in other materials as well. We will look at techniques for their examination.

1.10.1 Grain Structures, Phase Distributions and Inclusions

Ceramics, rocks, ores, and composites exhibit these features. They are of major importance in determining the surface properties of the material.

Large grain (i.e. > 1 μm) multiphase materials can be imaged and chemically mapped using SEM/EDS and backscattered electron images (see Sect. 1.4.1). The analysis depth is relatively large (i.e. 0.1–1 μm) in these modes. For more surface-sensitive structural differentiation (i.e. 1–10 nm) SAM can map features down to 50 nm. An example is shown in Fig. 1.22. If the structural differentiation is even finer than this, TEM or STEM, with EDS or EELS for analyses, must be used. These techniques can distinguish grain structures, phase distributions and inclusions down to unit cell level (i.e. 1–2 Å) but do require electron-transparent samples. STM, with atomic resolution, would rarely be justified for this application unless there is a need to image and analyse (via STS) sub-structure within one of these regions.

All of these techniques can be used on the majority of materials as listed in the beginning of this book. Extensive use of these techniques can be found in the literature of metals, alloys, catalysts, minerals (i.e. in rock-forming) and ceramics. A particularly valuable application is in the definition of regions of different structure in blended polymers. These can be: elastomeric inclusions; phase-separated regions; segregated layers or regions (e.g. silicones in polyester matrices); unreacted monomer inclusions; and regions containing oxidised products or unsaturated bonding (e.g. Fig. 1.23). Similar applications can be found for composites and many natural materials.

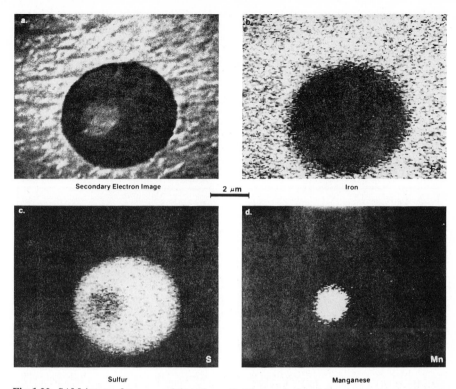

Secondary Electron Image   2 μm   Iron

Sulfur   Manganese

Fig. 1.22. SAM images from a precipitate in an Fe/Mn metal alloy (**a**) secondary electron image showing substructure in the precipitate; (**b**) Fe distribution showing absence of iron in the precipitate region; (**c**) S distribution with high concentration except in the substructure and; (**d**) Mn distribution showing separation of Mn into substructure of the precipitates. [Reprinted with permission from Applications Note No. 7903 (Perkin Elmer Physical Electronics Division, MN, USA)]

1.10.2 Fracture Faces and Intergranular Regions

Materials generally fracture along regions of least resistance, i.e. those across which bonding at the boundary is weakest. In single-phase materials, this may occur predominantly along grain boundaries (although trans-granular fracture also occurs under high-energy deposition). Surface analysis of the presented grain boundaries, using SAM, XPS and static SIMS, has revealed segregation of large impurity atoms into the atomc-width boundaries. The detailed structure of displaced atoms can be seen using TEM, STM, LEIS and ABS.

In multiphase materials, particularly multiphase ceramics, rocks and minerals, the region between grains of different phase is usually wider than a single atomic layer and has a composition different from that of either of the two adjacent phases. This integranular region can vary in width from 2 nm, as in the ceramic Synroc [1.101], to several micrometres, as in alumina-based ceramics. The structure of these intergranular films is often amorphous. Fracture in these materials is predominantly intergranular with the layer tending to "peel" off with

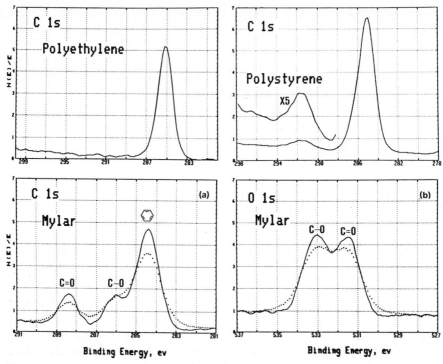

Fig. 1.23. C1s XPS spectra from polyethylene and polystyrene. Note the $\pi \rightarrow \pi^*$ shake-up satellite in the polystyrene spectrum, at ca. 6.7 eV higher binding energy than the new signal, indicative of the phenyl groups. C1s and O1s XPS spectra from Mylar (polyethylene terephthalate) surface before (*dotted*) and after (*continuous line*) removal of the X-ray line width, i.e. deconvolution. Note the contributions from C–O and C = O groups in the structure. [Reprinted with permission from Figs. 1, 2, 4, Applications Note No. 7905 (Perkin Elmer Physical Electronics Division, MN, USA]

one grain or the other leaving only a monolayer or two on the other surface. The material in these regions, the bonding and structure, can dominate such properties as mechanical strength, heat transfer, physical and chemical durability and mineral separation (i.e. processing).

Figure 1.24a illustrates the use of SAM to study fracture faces and intergranular films from a ceramic material. A surface showing enhancement of alkali metal cations [e.g. (Na, Cs), Al and O is found, removable by etching with an ion beam to a depth of > 2 nm. These surface enhancements are confirmed by XPS and static SIMS as in Fig. 1.24b and c. TEM can be used to image the width of .the film (Fig. 1.24d)]. A considerable body of work now exists on the study of intergranular films in ceramics using these techniques (e.g. [1.102]).

1.10.3 Pore Structures

In ceramics, rocks, polymers, composites and some natural materials, porous structures can intersect the surface in fracture faces or cut sections. The concentration, size and structure of these pores is important in determining the mechanical properties of these materials. Additionally, they may contain specific elements or compounds trapped during their formation. These species are often soluble or reactive leading to undesirable leaching and reactivity properties. Trapped (i.e. closed) pores below the surface can also affect these properties by diffusion down intergranular films to the surface.

SEM and SAM can image these structures since they are usually larger than 500 nm in major dimensions. SAM can also analyse condensed layers on their surfaces. An illustration of this application can be seen in Fig. 1.24a where a pore intersects the surface of a ceramic with a monolayer of volatile Cs condensed during pore formation and subsequent cooling.

1.10.4 Precipitates, Reaction Products and Recrystallised Particles on Surfaces

Separate particles on surfaces can be formed by: precipitation from saturated solution; reaction to form new phases; and leaching, dissolution or hydrolysis of a surface layer (i.e. loss of specific elements) followed by recrystallisation in situ to form new phases. In all three cases, nucleation occurs at specific sites on the surface rather than as a relatively uniform surface layer. Precipitation of sequentially supersaturated phases can sometimes be found in successive layers of the same crystallite at the same nucleation site (i.e. changing composition with depth through the crystallite). SAM can reveal these layers by imaging individual crystallites and using ion beam sputtering to develop depth profiles as the precipitate is removed. The spatial resolution and surface specificity of the technique are both essential to the structural analysis. Reaction at a surface can lead to extensive surface diffusion, facilitated by local exothermic processes, and structural rearrangement of its products. SAM, or in many cases SEM when the crystallites are $> 1\,\mu m$, can provide the required information on these new phases. The newest generation of XPS analysers can image regions $< 10\,\mu m$ in diameter. (Recent instrumental advances have also produced scanning XPS with resolution down to 500 nm but these instruments are as yet only available for fundamental research). Hence, larger crystals or collections of crystallites can be analysed in this field.

Sometimes, compositional information is not unequivocal and it is essential for structural identification to have diffraction patterns. TEM of thin sections with crystallite formation can provide this using selected area electron diffraction. An example is illustrated in Fig. 1.25 where TiO_2 crystallites in two different phases, as brookite and anatase, are found recrystallised in situ after hydrolysis of the surface of a $CaTiO_3$ perovskite.

A

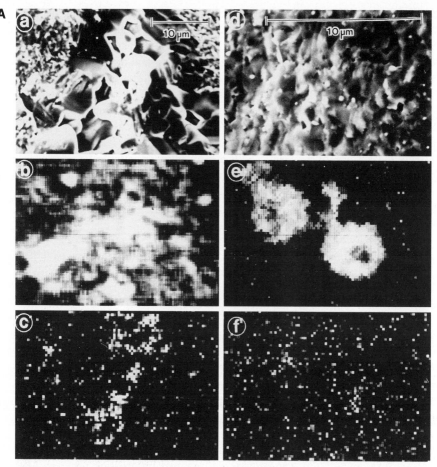

Fig. 1.24. (A) Secondary electron image (*a*) of a pore in a fracture face of the ceramic Synroc C with Cs maps of the same area (*b*) before and (*c*) after removal of ca. 15 nm by ion etching. Similar images (*d,e,f*) of a Cs-rich area (*e*) before and after (*f*) removal of ca. 375 nm. The fracture face exposes closed porosity and thin (< 2 nm) remnants of intergranular films in the ceramic. (**B**) XPS depth profiles from a Synroc C fracture face before (*full line*) and after (*dashed line*) immersion in doubly distilled water for ca. 30 s. An enhancement of Cs, Al and, possibly Mo in the fresh fracture face is reduced by the water. The numbers on the right indicate averaged atomic % of each element in the bulk of the material. (**C**) Static SIMS depth profile of a fresh fracture face of Synroc C demonstrating enchancement of Cs and Na, but not Ca and Ti, in the intergranular region. (**D**) High resolution TEM of an intergranular film at the interface between grains of perovskite and magnetoplumbite. Cleavage has been initiated, with the crack (arrowed) running between grains, the intergranular film adhering to the pervoskite grain. [Reprinted with permission from Figs. 1,4,5,6; J.A. Cooper, D.R. Cousens, J.A. Hanna, R.A. Lewis, S. Myhra, R.L. Segall, R.St.C. Smart, P.S. Turner, and T.J. White, J. Amer. Ceram. Soc. **69**(4), 347–352 (1986)]

B

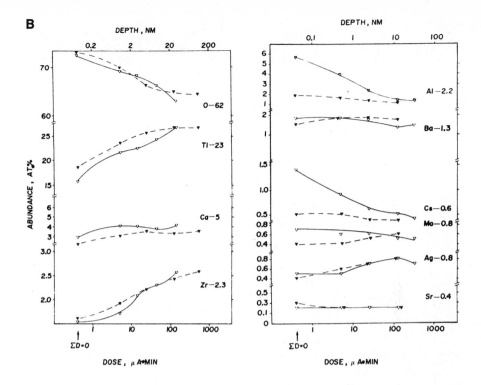

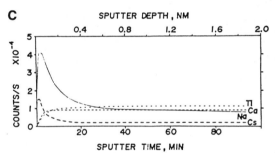

Fig. 1.24B–D. For Caption see opposite page

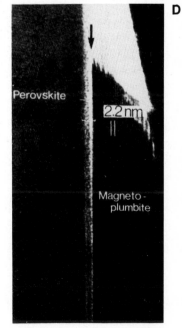

57

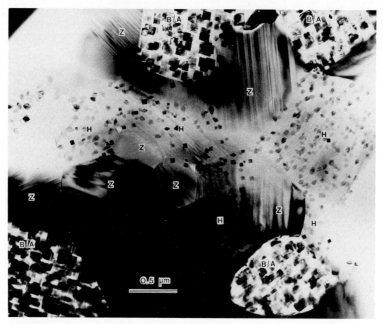

Fig. 1.25. TEM of an ion beam – thinned section of Synroc B (Synroc C without addition of simulated nuclear waste) after hydrothermal attack for 1 day. The perovskite grains have completely dissolved, recrystallising in-situ as brookite (B) and anatase (A). Hollandite (H) grains are slightly attacked with a few small TiO_2 crystallites on their surfaces. Zirconolite (Z) appears unaltered by this treatment. [Reprinted with permission from T. Kastrissios, M. Stephenson, P.S. Turner, and T.J. White, J. Amer. Ceram. Soc. **70**, 144–146 (1987)]

1.10.5 Magnetic Domains

Chapter 15 explains techniques for measuring and mapping different domains of magnetic fields in the surface of materials. The techniques are relatively specific to this use and examples will be left to that section of the book.

1.11 Technique-Induced Artifacts

Each of the techniques utilized in surface analysis in materials science will perturb the system being analysed to some extent. With methods such as SIMS, where the act of detection involves sputtering off layers of surface material, the initial perturbation can be obvious. However, more subtle secondary consequences may also arise. For example, in any ion beam technique, the preferential loss of one element and atomic (knock-on) mixing may induce a result difficult to relate to the original selvedge material. It is useful to consider two artifacts in particular, both commonly encountered, which can obscure useful information.

1.11.1 Radiation Damage

A chemical bond strength of $100 \, kJ \, mol^{-1}$ translates to a mere $1.036 \, eV$. As can be seen from Table 1.4 incident probes can have energies up to MeV. Hence it is not surprising that chemical bonds will be disrupted. As a general rule of thumb the higher the energy of the incident probe the greater the total damage, with photons being more benign than charged particles. In fact, the probability of local damage at a particular site is a function of the product of energy and cross section. For medium and high energy particles, this function is related to $E \times E^{-2}$ or E^{-1} so that, while the total damage increases, the local damage decreases with increasing energy. Logically, if a sample were to be analysed with a sophisticated multi-technique instrument, offering say, scanning Auger microscopy (SAM), XPS, SIMS and ultraviolet photoelectron spectroscopy (UPS), the rule of thumb would dictate performing the sequence of analyses: UPS, XPS, SAM and finally SIMS, i.e. in the order of increasing radiation damage.

The destructive technique of SIMS has historically developed along two pathways [1.103]. Originally intended as a non-surface technique for in-depth and microanalysis in materials science, the method of *dynamic* SIMS utilizes high incident ion fluxes ($> 1 \, \mu A \, cm^{-2}$) with consequently large sputtering rates ($> 50 \, \text{Å} \, min^{-1}$). The surface oriented method of *static* SIMS with considerably lower incident fluxes ($< 1 \, nA \, cm^{-2}$) has sub-monolayer sensitivity and results in much less damage than dynamic SIMS. To increase the signal level in static SIMS the area of analysis is usually increased, either by a broad beam or via rastering. The rastering can be synchronized with an analyser gate for a particular signal to provide the variant known as *imaging* SIMS. Although the atomic mixing in static SIMS can be reduced to submonolayer levels, it remains larger than the damage induced with incident electrons at the same current density. Dynamic SIMS has the disadvantages of rapid loss of surface layers, lateral and vertical (recoil) mixing of species, preferential sputtering of particular elemental species and sputter-induced cratering/protusions. By contrast, static SIMS removes a single surface layer over periods of hours with much less damage or mixing effects.

Considering electrons as the incident probe and an incident current density of $1 \, mA \, cm^{-2}$ or $10 \, \mu A \, mm^{-2}$ (this equates to about 6 electrons per surface site per second into the outermost layer), Table 1.8 shows, for a number of materials and electron energies, the accumulated threshold charge densities for detectable damage to occur [1.104]. Also indicated are the times that the current density can be maintained without detectable damage and the lateral spatial resolution that may be afforded in an Auger analysis (based on a beam of $100 \, nA$ residing on a spot for 5 minutes [1.105]). Certainly for a number of materials it is not viable to analyse them via electron excited Auger as damage will occur well before any meaningful data can be acquired. The radiation damage can occur as electron stimulated diffusion, reaction or desorption of adsorbed species, all leading to an altered selvedge. The action of electron-stimulated desorption (ESD) can also be deliberately employed as an analytical method, and, in its more sophisticated

Table 1.8. Accumulated threshold charge densities D_c for detectable electron radiation damage to occur for a variety of materials and electron energies. The threshold levels relate to maximum exposure times and optimal spatial resolutions possible without damage occurring

Material	Energy (keV)	D_c (C cm^{-2})	Time at 1 mA cm^{-2}	Spatial resolution (mm)
Si$_3$N$_4$	2	Stable	–	Small
Al$_2$O$_3$	5	10	3 h	0.02
Cu, Fe phthalocyanines	1	> 1	> 15 min	0.06
SiO$_2$	2	0.6	10 min	0.08
Li$_2$WO$_4$	1	0.5	8 min	0.09
NaF, LiF	0.1	0.06	60 s	0.25
LiNO$_3$, Li$_2$SO$_4$	1	0.05	50 s	0.28
KCl	1.5	0.03	30 s	0.36
TeO$_2$	2	0.02	20 s	0.44
H$_2$O$_{(s)}$	1.5	0.01	10 s	0.62
Native oxides	5	2×10^{-3}	2 s	1.4
C$_6$H$_{12(s)}$	0.1	3×10^{-4}	0.3 s	3.6
Na$_3$AlF$_6$	3	$10^{-4} - 10^{-3}$	0.1 s	2–6
CH$_3$OH$_{(s)}$	1.5	2.5×10^{-4}	0.3 s	3.9
Formvar	75	2×10^{-3}	0.2 s	1.4

form of electron stimulated desorption ion angular distribution (ESDIAD), yield information on bond geometries of adsorbed species. Complementary to ESD is the subtler form of damage known as electron stimulated adsorption [1.104]. The sticking probability of a vacuum residual is enhanced by radiation placing it in an excited state or causing its fragmentation. Note that the concept of electron beam radiation damage is equally applicable to the diffraction techniques of low energy electron diffraction (LEED) and reflection high energy electron diffraction (RHEED) and the electron microscopies. Scanning tunnelling microscopy (STM) is somewhat of a different case. Although the total tunnelling current in STM is very small (see Chap. 10), say 1 nA, this tunnels into about a 5 Å diameter area and represents a massive current density of about 5×10^5 A cm^{-2} or 3×10^9 electrons per surface site per second (compare this with 6!). The fact that apparently little damage occurs even after lengthy exposure can only be a consequence of the small tunnelling bias, typically 0.1 V, well below the threshold energy involved in a bond. Indeed the most severe damage from STM results from physical tip crashing.

As an incident photon method, XPS has the general reputation of being fairly benign. However, even beyond the photoionization of core levels and subsequent relaxation processes which may cause chemical alterations, the exiting photoelectrons and associated secondaries represent a definite, though small, electron flux. Due to small subtended solid angles, band pass electron spectrometers generally have small collection efficiencies. Based on detected count rates the actual total electron flux passing through the irradiated surface area is typically up to 10^{10} electrons s^{-1}. Over a 5 mm diameter circle that converts to about 5×10^{-5} electrons per surface site per second. Purely on the grounds of electron flux that makes XPS a factor of 10^5 less likely to produce radiation damage than conventional electron excited Auger. Nevertheless radiation damage can be observed with XPS e.g. alkali halides readily form colour centres, the damage being macroscopically visible after just a few minutes X-ray exposure. The Au(I) state in $KAu(CN)_2$ readily disproportionates under X-ray irradiation to Au(0) and Au(III), thereby forming a mixture of three states of gold. With extended exposure the quantity of metallic gold formed is also macroscopically visible [1.106].

In systems prone to radiation damage the effect cannot be entirely eliminated but the extent of damage can be minimized [1.107]. The occurrence of damage is readily established by sequential spectral acquisition during continuous irradiation, thereby enabling the threshold dose at a given flux to be estimated. It may then be possible not to exceed the dose by minimizing the incident flux (even at the expense of spatial resolution), turning off the radiation source except when data is being acquired and scanning over narrow energy ranges. In some instances it can also help to cool the specimen to liquid nitrogen temperatures since some radiation damage can be mass diffusion dependent. Cooling can also reduce the secondary effects of possible specimen heating, especially in the case of poor thermal conductors.

1.11.2 Electrostatic Charging

After radiation damage probably the most disruptive artifact is that of induced electrostatic charging of the specimen surface. Except for techniques utilizing neutral incident and exit probes, charging will always be experienced to some extent with specimens of low electrical conductivity, or with insulating selvedge layers e.g. oxide films on semiconductors (or even conductors). In the case of incident electrons, when the secondary electron yield is less than unity i.e. less than the nett incident current, the surface will accumulate negative charge. This accumulation and the resultant negative surface potential increases with time, often leading to values which induce dielectric breakdown. Influences may range from deflection of the incident beam to gross distortions in the energy of the characteristic electrons produced. Under these unstable conditions meaningful spectra cannot always be obtained. It is possible to reduce this type of charging by judicious sample mounting (e.g. in In foil) and by careful choice of both electron beam energy and angle of incidence, thereby altering the value of the secondary electron yield. These high surface potentials may also be capable of causing field-induced migration of mobile ions, thereby becoming another source of radiation damage [1.105].

Under conditions where the secondary electron yield is greater than unity the surface charge becomes positive rather than negative. For example, this will always be the case for XPS applied to insulators since photons arrive and electrons leave. The induced positive charge prevents the lowest kinetic energy secondary electrons from leaving, thereby rapidly establishing a stable equilibrium situation. In the case of Auger or XPS spectra the positive surface potential will shift the entire spectrum uniformly to lower kinetic energies, typically 2–15 eV. Since for a given specimen and given incident flux the static charge shift is constant, careful energy referencing, to a known spectral feature will enable a correction to be applied. Such energy referencing is particularly important in evaluating binding energies in XPS and considerable effort has gone into the problem. This will be considered in more detail in Chap. 7. In an alternative to referencing, the surface may be flooded with an adjustable flux of additional low energy electrons in order to balance out the positive charge.

References

1.1   D. Tabor: Surf. Sci. **89**, 1 (1979)
1.2   P.H. Abelson: Science **234**, 257 (1986)
1.3   W.A. Grant, R.P.M. Procter, J.L. Whitton (eds.): *Surface Modification of Metals by Ion Beams* (Elsevier Sequoia, Lausanne 1987)
1.4   J.W. Rabalais, S. Kasi: Science **239**, 623 (1988)
1.5   P. Danielson: In *Kirk-Othmer Encyclopedia of Chemical Technology*, Vol. 20 (Wiley, New York 1982) pp. 811, 812
1.6   F. Jona: J. Phys. C **11**, 4271 (1978)
1.7   J. Verhoeven: J. Environ. Sci. **22**, 24 (1979)
1.8   R.G. Musket, W. McLean, C.A. Colmenares, D.M. Makowiecki, W.J. Siekhaus: Appl. Surf. Sci. **10**, 143 (1982)
1.9   G.A. Somorjai, M.A. Van Hove: Structure and Bonding **38**, 1 (1979)
1.10 G.A. Somorjai: *Chemistry in Two Dimensions: Surfaces* (Cornell University Press, Ithaca 1981)
1.11 V.M. Bermudez: Electronic structure of point defects on insulator surfaces. Prog. Surf. Sci. **11**, 1 (1981)
1.12 E.A. Colbourn, W.C. Mackrodt: Theoretical aspects of H_2 and CO chemicsorption on MgO surfaces. Surf. Sci. **117**, 571 (1982)
1.13 C.F. Jones, R.A. Reeve, R. Rigg, R.L. Segall, R.St.C. Smart, P.S. Turner: Surface area and the mechanism of hydroxylation of ionic oxide surface. J. Chem. Soc. Faraday I, **80**, 2609 (1984)
1.14 E.A. Colbourn, W.C. Mackrodt: Irregularities at the (001) surface of MgO: topography and other aspects. Solid State Ionics **8**, 221 (1983)
1.15 P.W. Tasker, E.A. Colbourn, W.C. Mackrodt: The segregation of isovalent impurity cations at the surfaces of MgO and CaO. J. Am. Ceram. Soc. **68**, 74 (1985)
1.16 C. Ocal, B. Basurco, S. Ferrer: Surf. Sci. **157**, 233 (1985)
1.17 J.E. Crowell, J.G. Chen, J.T. Yates Jr.: Surf. Sci. **165**, 37 (1986)
1.18 C.F. McConville, D.L. Seymour, D.F. Woodruff, S. Bao: Surf. Sci. **188**, 1 (1987)
1.19 I.P. Batra: J. Electron. Spectrosc. Relat. Phen. **33**, 175 (1984)
1.20 W.A. Anderson, W.E. Haupin: In *Kirk-Othmer Encyclopedia of Chemical Technology*, 3rd ed., ed. by M. Grayson, D. Eckroth (Wiley, New York 1978) Vol. 2, pp. 181–183
1.21 W.W. Binger, E.H. Hollingsworth, D.O. Sprowls: In *Aluminum, Properties, Physical Metallurgy and Phase Diagrams*, Vol. I, ed. by K.R. Van Horn (American Society for Metals, Metals Park, Ohio 1967) pp. 226–235
1.22 L.F. Mondolfo: *Aluminum Alloys: Structure and Properties* (Butterworths, London 1976) pp. 123, 148
1.23 K. Nisancioglu, O. Lunder, H. Holtan: Corrosion **41**, 247 (1985)
1.24 E.W. Müller, T.T. Tsong: *Field Ion Microscopy* (Elsevier, New York 1969)
1.25 E.W. Müller: Ann. Rev. Phys. Chem. **18**, 35 (1967)
1.26 E. Bauer: Surf. Sci. **162**, 163 (1986)
1.27 A.M. Stoneham, P.W. Tasker: The Theory of Ceramic Surfaces, in *Surface and Near-Surface Chemistry of Oxide Materials*, ed. by J. Nowotny, L.-C. Dufour (Elsevier, Amsterdam 1988) pp. 1–22
1.28 K. Heinz, K. Müller: LEED-intensities – experimental progress, and new possibilities of surface structure determination, In *Structural Studies of Surfaces*, Springer Tracts Mod. Phys. 91 (Springer, Berlin, Heidelberg 1982) pp. 1–54

1.29   I.P. Batra, T. Engel, K.H. Rieder: In *The Structure of Surfaces*, Springer Ser. Surf. Sci., Vol. 2, ed. by M.A. van Hove, S.Y. Tong (Springer, Berlin, Heidelberg 1985) p. 251

1.30   T. Engel, K.H. Rieder: Structural studies of surfaces with atomic and molecular beam diffraction, In *Structural Studies of Surfaces*, Tracts Mod. Phys. 91 (Springer, Berlin, Heidelberg 1982) pp. 55–180

1.31   J. Stöhr: Surface crystallography by SEXAFS and NEXAFS, In *Chemistry and Physics of Solid Surfaces V*, ed. by R. Vanselow, R. Howe, Springer Ser. Chem. Phys. Vol. 35 (Springer, Berlin, Heidelberg 1984) pp. 231–256

1.32   W.M. Gibson: Determination by ion scattering of atomic positions at surfaces and interfaces, In *Chemistry and Physics of Solid Surfaces V*, ed. by R. Vanselow, R. Howe, Springer Ser. Chem. Phys. Vol. 35 (Springer, Berlin, Heidelberg 1984) pp. 427–454

1.33   B.K. Teo, D.C. Joy (Eds.): *EXAFS Spectroscopy – Techniques and Applications* (Plenum, New York 1981)

1.34   Periodic Table of the Elements, Sargent-Welch Scientific Company, Skokie, Illinois, Catalogue Number S-18806. Based upon National Standard Reference Data System material

1.35   Table 1.4 is drawn from a variety of sources including tables in the references listed below, manufacturers specifications and the contributions of other authors in this book. Due to the limited scope in presenting information in such tables they should be utilized only as a guide.
        (a) H. Fellner-Feldegg, U. Gelius, B. Wannberg, A.G. Nilsson, E. Basilier, K. Seigbahn: J. Electron Spectrosc. Relat. Phen. **5**, 643 (1974)
        (b) D. Roy, J.D. Carette: In *Electron Spectroscopy for Surface Analysis*, ed. by H. Ibach (Springer, Berlin, Heidelberg 1977) pp. 14, 15
        (c) M.W. Roberts, C.S. McKee: *Chemistry of the Metal-Gas Interface* (Clarendon, Oxford 1978) p. 207
        (d) M.P. Seah, D. Briggs: In *Practical Surface Analysis by Auger and X-ray Photoelectron Spectroscopy*, ed. by D. Briggs, M.P. Seah (Wiley, Chichester 1983) p. 12
        (e) C.W. Magee: Nucl. Instrum. Meth. **191**, 297 (1981)

1.36   M.P. Seah, W.A. Dench: Surf. Interface Anal. **1**, 2 (1979)

1.37   R.E. Ballard: J. Electron Spectrosc. Relat. Phen. **25**, 75 (1982)

1.38   S. Tanuma, C.J. Powell, D.R. Penn: Surf. Sci. **192**, L849 (1987)

1.39   C.D. Wagner, W.M. Riggs, L.E. Davis, J.F. Moulder, G.E. Muilenberg (eds.): *Handbook of X-ray Photoelectron Spectroscopy* (Perkin-Elmer, Eden Prairie, Minnesota 1978) pp. 42, 50

1.40   M.F. Ebel: J. Electron Spectrosc. Relat. Rhenom. **22**, 157 (1981)

1.41   C.W. Magee: Nucl. Instrum. Meth. **191**, 297 (1981)

1.42   D.P. Leta, G.H. Morrison: Anal. Chem. **52**, 514 (1980)

1.43   Riber MIQ 256 SIMS/Ion microprobe

1.44   P. Skeldon, K. Shimizu, G.E. Thompson, G.C. Wood: Thin Solid Films **123**, 127 (1985)

1.45   Based on an absorption sensitivity for iron of 0.15 μg/ml/1% Abs (p. 825) and a sample volume of 5 μl (p. 355), In *Instrumental Methods of Analysis*, H.H. Willard, L.L. Merritt Jr., J.A. Dean (eds.) (Van Nostrand, New York 1974)

1.46   D.M. Hercules, L.E. Cox, S. Osnisick, G.D. Nichols, J.C. Carver: Anal. Chem. **45**, 1973 (1973)

1.47   R.C. Weast, M.J. Astle (eds.): *CRC Handbook of Chemistry and Physics*, 63rd edn. (CRC, Boca Raton 1982) F-161

1.48   R. Browning: J. Vac. Sci. Technol. A **2**, 1453 (1984)

1.49   P.D. Prewett, D.K. Jeffries: J. Phys. D **13**, 1747 (1980)

1.50   N.M. Ceglio, A.M. Hawryluk, M. Schattenburg: J. Vac. Sci. Technol. B **1**, 1285 (1983)

1.51   I.W. Drummond, T.A. Cooper, F.J. Street: Spectrochimica Acta **40B**, 801 (1985)

1.52   K. Yates, R.H. West: Surf. Interface Anal. **5**, 217 (1983)

1.53   For example, VG Scientific ESCALAB 20-X

1.54   For example, VG Scientific ESCASCOPE

1.55   J.F. Evans, J.H. Gibson, J.F. Moulder, J.S. Hammond. The PHI Interface **7**, 1 (1984) (Perkin Elmer, Eden Prairie, USA)

1.56   M.L. Tarng, D.G. Fischer: J. Vac. Sci. Technol. **15**, 50 (1978)

1.57   V. Thompson, H.E. Hintermann, L. Chollet: Surf. Technol. **8**, 421 (1979)

1.58   D.R. Clark, L.L. Hench: An overview of the physical characterisation of leached surfaces, Nucl. Chem. Waste Manag. **2**, 93 (1981)

1.59   J.W. Strojek, J. Mielczarski: Spectroscopic investigations of the solid-liquid interface by the ATR technique, Adv. Colloid Interf. Sci. **19**, 309 (1983)

1.60 R.F. Willis (ed.): *Vibrational Spectroscopy of Adsorbates* (Springer, Berlin, Heidelberg 1980)
1.61 K. Siegbahn, C. Nordling, A. Fahlman, R. Nordberg, K. Hamrin, J. Hedman, G. Johansson, T. Bergmark, S.-E. Karlsson, I. Lindgren, B. Lindberg: *ESCA, Atomic, Molecular and Solid State Structure Studied by Means of Electron Spectroscopy* (Almqvist and Wiksells, Uppsala 1967) p. 79
1.62 R. Opila, R. Gomer: Surf. Sci. **105**, 41 (1981)
1.63 J. Hulse, J. Küppers, K. Wandelt, G. Ertl: Appl. Surf. Sci. **6**, 453 (1980)
1.64 H.H. Madden: J. Vac. Sci. Technol. **18**, 677 (1981)
1.65 M. Barber, R.S. Bordoli, G.J. Elliot, R.D. Sedgwick, A.N. Tyler: Anal. Chem. **54**, 645A (1982)
1.66 C.J. Powell, N.E. Erickson, T.E. Madey: J. Electron Spectrosc. Relat. Phen. **17**, 361 (1979)
1.67 C.J. Powell, N.E. Erickson, T.E. Madey: J. Electron Spectrosc. Relat. Phen. **25**, 87 (1982)
1.68 V.E. Henrich: Electronic and geometric structure of defects on oxides and their role in chemisorption, In *Surface and Near Surface Chemistry of Oxide Materials*, ed. by J. Nowotny, L.-C. Dufour (Elsevier, Amsterdam 1988) pp. 23–60
1.69 A.B. Kunz: Theoretical study of defects and chemisorption by oxide surfaces, In *External and Internal Surfaces in Metal Oxides*, ed. by L.-C. Dufour, J. Nowotny, Materials Science Forum (Trans. Tech. Publications, Claustal-Zellerfeld 1988) pp. 1–30
1.70 R.L. Segall, R.St.C. Smart, P.S. Turner: Oxide surfaces in solution, In *Surface and Near-Surface Chemistry of Oxide Materials*, ed. by J. Nowotny, L.-C. Dufour (Elsevier, Amsterdam 1988) pp. 527–576
1.71 V.E. Henrich: Ultraviolet photoemission studies of molecular adsorption on oxide surfaces, Prog. Surf. Sci. **9**, 143 (1979)
1.72 V.E. Henrich, G. Dresselhaus, H.J. Zelger: Chemisorbed phases of H_2O on TiO_2 and $SrTiO_3$, Solid State Commun. **24**, 623 (1977)
1.73 M.W. Roberts: Metal oxide overlayers and oxygen-induced chemical reactivity studied by photoelectron spectroscopy, In *Surface and Near-Surface Chemistry of Oxide Materials*, ed. by J. Nowotny, L.-C. Dufour (Elsevier, Amsterdam 1988) pp. 219–246
1.74 G.A. Somorjai: Adv. Catal. **26**, 1 (1979)
1.75 G. Heiland, H. Lüth: In *The Chemical Physics of Solid Surfaces and Heterogeneous Catalysis*, Vol. 3 , ed. by D.A. King, D.P. Woodruff (Elsevier, Amsterdam 1984) p. 147
1.76 S.C. Chang, P. Mark: Surf. Sci. **45**, 721 (1974); ibid. **46**, 293 (1974)
1.77 C.C. Schubert, C.L. Page, B. Ralph: Electrochim. Acta **18**, 33 (1973)
1.78 J. Marien: Phys. Status Solidi A **38**, 339, 513 (1976)
1.79 J. Nowotny, M. Sloma: Work function of oxide ceramic materials, In *Surface and Near-Surface Chemistry of Oxide Materials*, ed. by J. Nowotny, L.-C. Dufour (Elsevier, Amsterdam 1988) pp. 281–344
1.80 S.R. Morrison: *The Chemical Physics of Surfaces* (Plenum, New York 1977)
1.81 H. Wagner: *Physical and Chemical Properties of Stepped Surfaces*, Springer Tracts Mod. Phys. (Springer, Berlin, Heidelberg 1978)
1.82 H.E. Clark, R.D. Young: Surf. Sci. **12**, 385 (1968)
1.83 H.D. Hagstrum: Science **178**, 275 (1972)
1.84 H.D. Hagstrum: Phys. Rev. **150**, 495 (1966)
1.85 G. Heiland, H. Lüth: Adsorption on oxides, in *The Chemical Physics of Solid Surfaces and Heterogeneous Catalysis*, ed. by D.A. King, D.P. Woodruff (Elsevier, Amsterdam 1984) p. 156
1.86 N.S. Huck, R.St.C. Smart, S.M. Thurgate: Surface photovoltage and XPS studies of electronic structure in defective nickel oxide powders, Surf. Sci. **169**, L245 (1986)
1.87 E. Garrone, A. Zecchina, F.S. Stone: Philos. Mag. B **42**, 683 (1980)
1.88 W. Göpel: Prog. Surf. Sci. **20**, 9 (1985)
1.89 J. Cunningham: Photoeffects on metal oxide powders, in *Surface and Near-Surface Chemistry of Oxide materials*, ed. by J. Nowotny, L.-C. Dufour (Elsevier, Amsterdam 1988) pp. 345–412
1.90 M.W. Roberts, R.St.C. Smart: XPS determination of band bending in defective semiconducting oxide surfaces, Surf. Sci. **151**, 1 (1985)
1.91 A.W. Adamson: *Physical Chemistry of Surfaces* (Wiley, New York 1986)
1.92 R.J. Bohm, W. Höseler: Scanning tunneling microscopy – a review, in *Chemistry and Physics of Solid Surfaces VI*, ed. by R. Vanselow, R. Howe, Springer Ser. Surf. Sci. Vol. 5 (Springer, Berlin, Heidelberg 1986) pp. 361
1.93 W. Hirschwald: In *Current Topics in Materials Science*, ed. by E. Kaldis, Vol. 7 (1981) p. 143
1.94 C.C. Schubert, C.L. Page, B. Ralph: Electrochim. Acta **18**, 33 (1973)
1.95 W. Heiland, E. Taglauer: Surf. Sci. **68**, 96 (1977)

1.96   A. Benninghoven: Developments in secondary ion mass spectrometry and applications to surface studies, Surf. Sci. **53**, 596 (1975)

1.97   A. Brown, J.C. Vickerman: Static SIMS for applied surface analysis, Surf. Interface Anal. **6**, 1 (1984)

1.98   W. Hirschwald: Selected experimental methods on the characterization of oxide surfaces, In *Surface and Near-Surface Chemistry of Oxide Materials* ed. by J. Nowotny, L.-C. Dufour (Elsevier, Amsterdam 1988) pp. 140–141

1.99   M. Grunze, W. Hirschwald, D. Hoffman: J. Cryst. Growth **52**, 241 (1981)

1.100 W. Hirschwald, P. Bonasewicz, L. Ernst, M. Grade, D. Hopmann, S. Krebs, R. Littbarski, G. Neumann, M. Grunze, D. Kobb, H.J. Schultz: Zinc oxide, Current Topics Mater. Sci. **7**, 143 (1981)

1.101 J.A. Cooper, D.R. Cousens, J.A. Hanna, R.A. Lewis, S. Myhra, R.L. Segall, R.St.C. Smart, P.S. Turner, T.J. White: Intergranular films and pore surfaces in Synroc C: structure, composition and dissolution characteristics, J. Amer. Ceram. Soc. **69**, 347 (1986)

1.102 D.R. Clarke: Observation of microcracks and thin intergranular films in ceramics by transmission electron microscopy, J. Am. Ceram. Sco. **63**, 104 (1980)

1.103 N.H. Turner, B.I. Dunlap, R.J. Colton: Anal. Chem. **56**, 373R (1984)

1.104 C.G. Pantano, T.E. Madey: Appl. Surf. Sci. **7**, 115 (1981). Note error in Table 1 re. D_c value of Li_2WO_4

1.105 M.P. Seah: In *Surface Analysis of High Temperature Materials: Chemistry and Topography*, ed. by G. Kemeny (Elsevier, London 1984) p. 124

1.106 C. Klauber: Unpublished data

1.107 ASTM Standards on Surface Analysis (1986) ISBN 0-8031-0948-2

2. UHV Basics

C. Klauber

With 3 Figures

2.1 The Need for Ultrahigh Vacuum

Modern surface analytical methods over the last three decades have been dominated by those requiring ultrahigh vacuum (UHV) chambers in which to carry out the analyses. This is not a universal requirement for surface analysis and several of the techniques such as Fourier transform infrared (FTIR), scanning tunnelling microscopy (STM) and ellipsometry do not have a mandatory vacuum requirement. Vacuum is of course required by those techniques utilizing beams of particles and higher energy radiation so that the beams may be generated and travel undisturbed until intercepting the surface. The requirement for UHV or vacua of $\leq 10^{-10}$ mbar (10^{-8} Pa) is fundamental to surface analysis when those beams are employed. This arises due to the flux of residual gas molecules striking the surface i.e. the number of molecules per unit area per unit time, which is responsible for the pressure that those gas molecules exert upon the surface. By knowing the pressure the flux can be evaluated. From the kinetic theory of gases [2.1] the molecular flux Z is given by the Herz-Knudsen equation:

$$Z = \frac{Nc}{4V} \, , \tag{2.1}$$

where N/V is the number of molecules per unit volume and c is the average speed of the molecules. For a gas of molecular weight M at temperature T [2.1]:

$$c = \sqrt{\frac{8RT}{\pi M}} \, , \tag{2.2}$$

where R is the gas constant. Combining the above with the ideal gas equation $PV = nRT$ and $N = nN_A$ where N_A is Avogadro's number, gives:

$$Z = \frac{nN_A P \sqrt{(8RT/\pi M)}}{4nRT} \, ,$$

and therefore

$$Z = \frac{N_A P}{\sqrt{2\pi MRT}} \, . \tag{2.3}$$

For pressure P given in Pa (units $N\,m^{-2}$) and M (units $g\,mol^{-1}$), substitution of the relevant values for N_A, π and R gives:

$$Z = \frac{2.635 \times 10^{24}\,P}{\sqrt{MT}} \text{ collisions m}^{-2}\,s^{-1}.$$ (2.4)

Substitution of P by a pressure-time integral gives the integrated gas flux or fluence (dose and exposure are used synonomously, though exposure is not technically equivalent).

The rate of molecular impingement can thus be seen to be dependent upon the molecular weight, temperature and pressure of the gas involved. At atmospheric pressure (1.01325×10^5 Pa) the flux is extremely high, e.g. for nitrogen at room temperature $Z = 2.91 \times 10^{27}$ collisions $m^{-2}\,s^{-1}$. Since a typical surface might have an exposed atomic density $\sim 2 \times 10^{19}$ atoms m^{-2}, each surface atom would be struck by $\sim 1.5 \times 10^8$ molecules every second. Surfaces in an atmospheric environment thus reach an almost instantaneous pseudo-equilibrium with the reactive gases present such as oxygen and water. If we created a virginal surface of some material by fracturing the bulk, then to maintain the virginal state we must reduce the gas pressure. At a pressure of 10^{-8} Pa the average time interval between a surface atom being struck by a gas molecule would be about 7×10^4 s or 19 hours. This is a useful time interval during which a variety of surface characterizations can be carried out, hence the need for UHV. In practical surface analysis often the surface of interest has been removed from a non-vacuum environment and placed in the instrumental system. The necessity for UHV remains since whatever reacted selvedge has formed it is essential that the residual gases in the vacuum system do not form an additional overlayer. In poorer high vacuum systems such as typically found in electron microscopy, reacted carbonaceous overlayers rapidly form due to the interaction of the impinging electron beam with adsorbed residual gases. Figure 2.1 illustrates the relationship between gas pressure, surface contamination times and mean free path lengths [2.2].

- Contamination time is the time taken for a perfectly clean surface to acquire a monolayer of contaminant. This time is dependent not only on the molecular weight, temperature and pressure, but also on the reactivity of the system. The latter can be simply expressed as a sticking probability s i.e. the probability that an incident molecule will remain on the surface (in some form) after collision, $0 \leq s \leq 1$.
- Mean free path λ is the average distance travelled by a particle before it undergoes a collision with another particle. This collision may or may not be inelastic i.e. involve the loss of energy. The latter case of inelastic mean free path (IMFP) is particularly important in surface analysis, not only for particles through gas, but particularly particles through solids as it is the crux of a technique's surface sensitivity (see Sect. 1.5.1).

The comparison with the variation of atmospheric pressure with altitude is interesting since it places the environment of earth orbit space flight in perspective.

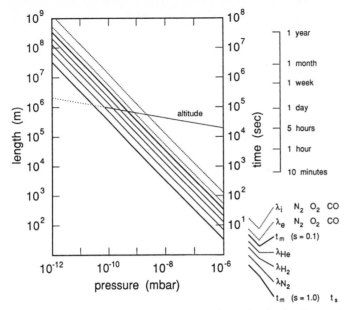

Fig. 2.1. Relationship between gas pressure, surface contamination times and mean free path lengths. The earth's atmospheric altitude variation is included for comparison. t_m is the time to form a monolayer of nitrogen with sticking probabilities $s = 0.1$ and 1.0, t_s is the average time before a surface atom is struck by a gas phase molecule. λ_e is the mean free path of a $100\,\text{eV}$ electron in N_2, O_2 or CO, λ_i is the mean free path for that collision to be ionizing, λ_x is the mean free path of a molecule of X in X [2.2]

UHV of 10^{-10} mbar is attained at an altitude of about $900\,\text{km}$ [2.3]. The American space shuttle has a maximum operational ceiling of about $1000\,\text{km}$.

Current practice is to measure pressure in millibar (mbar) units, mbar $\times 100$ yielding the SI derived unit of Pascals (Pa). This is convenient since mbar closely relates to the older metric, but non-SI unit, of the torr (1 torr = $1.333\,\text{mbar}$). The torr is still widely used, especially in the gas exposure unit of the Langmuir (L), where $1\,\text{L}$ represents a pressure-time integral of $10^{-6}\,\text{torr\,s}$ (approximately the time for the formation of a monolayer). In view of the inaccuracies associated with pressure-time integrals the substitution of $10^{-6}\,\text{mbar\,s}$ is equally useful, though Pa s is more often used. *Menzel* and *Fuggle* [2.4] suggested replacing the concept of gas exposure with measurements of true gas fluence, the unit of fluence to be the Ex, such that 1 Ex is equal to 10^{18} collisions m^{-2}. Though quite logical, this suggestion has not been widely adopted.

2.2 Achieving UHV

In order to practically achieve the UHV regime the vacuum system must not only be pumped adequately but it must be free of true and virtual leaks. The latter arise from gases, especially water, adsorbed on the internal chamber walls and instruments. At room temperature these will slowly desorb constituting a large virtual leak which can continue for years. These are eliminated in a matter of hours by heating or baking the whole vacuum system, typically to 150–200°C. UHV chambers can be constructed from materials such as glass and aluminium, but most commonly the material employed is non-magnetic stainless steel. Where electrical insulation is required, such as in feedthroughs and lens supports, ceramics are usually utilized. Great care must be taken to ensure that no oil or high vapour pressure materials are used in the system. Some polymers and elastomers such as teflon and viton can be used provided that they are not overheated especially whilst under compression or tension. A diagrammatic elevation and cross section of a typical multi-technique surface analytical vacuum rig is shown in Fig. 2.2. The various flanges are attached by a plethora of bolts, making such systems look somewhat like pressure vessels. These are necessary to compress steel knife edges within the flanges into softer gaskets (oxygen free high conductivity copper) to achieve an all-metal leak proof seal. The flange design along with that of all mechanical, electrical and fluid feedthroughs and internal components are designed to remain leak-proof with repeated thermal cycling up to 250°C. Any true gas leaks that do occur are easily detected via the use of a quadrupole mass spectrometer (usually a mandatory attachment) tuned to He in residual gas analysis mode. A jet of He is directed externally at the suspected area until an influx is observed. The technology required to fabricate such components is the principal cause of the high cost of UHV systems. Several methods for baking exist, the most common being heater bands around various parts of the system, heating elements contained within an oven-like shroud or the more recent practice of internal quartz encased radiation bars. Whatever approach is taken it is essential that even temperatures are reached over all internal surfaces, so as to avoid condensation in cooler regions. Once a system has been up to atmosphere, pumpdown to high vacuum normally takes several hours, baking a further 10–12 hours with a similar time for all internal components to cool. After UHV has been achieved an analysis of the residual gases making up that low pressure invariably indicates a composition that is not related to atmospheric constituents. The dominant residuals are usually H_2, H_2O, CO, CO_2 and a variety of hydrocarbons. These promote a reducing environment, which given time, will lead to carbonaceous deposits on all internal surfaces, including introduced specimens. Depending upon the preferential nature of the pumping, what pump oils and lubricants are used and the system's past history, the precise makeup of the residuals will change from system to system.

A variety of vacuum pumps exist which can pump down to the UHV regime, each type has its own virtues. The most popular main chamber pumps are diffusion, turbomolecular or ion pumps. Diffusion pumps using modern fluids such as polyphenyl ethers are excellent performers but liquid nitrogen cooled traps are mandatory. Turbomolecular pumps can be used without traps, although trapping improves their ultimate performance. Titanium sublimation pumps, utilizing chemical gettering, are useful for additional pumping. Cryogenic based sorption

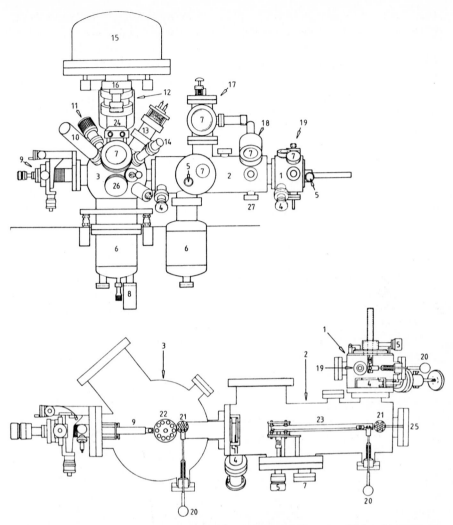

Fig. 2.2. Front elevation of a typical multitechnique surface analysis system (*top*) and a cross sectional view from above of specimen transporter mechanisms for such a system (*bottom*). Key to the diagrams: (*1*) Fast entry specimen insertion lock (stainless steel). (*2*) UHV specimen preparation chamber (stainless steel). (*3*) UHV experimental/analysis chamber (mu-metal). (*4*) Viton sealed gate valve. (*5*) Rotary drive to specimen transfer mechanism. (*6*) Titanium sublimation pump vessel (stainless steel). (*7*) Viewport. (*8*) Autocarousel motor drive. (*9*) High precision specimen translator (*X, Y, Z* translation and *θ* tilt). (*10*) Twin anode X-ray source (Al/Mg). (*11*) UV discharge source (UPS). (*12*) Monochromated X-ray source (Al/Mg). (*13*) 2000 Å electron source (AES, SEM, SAM). (*14*) Scanning ion source. (*15*) Electron energy analyzer vessel. (*16*) Detector (single or multichannel). (*17*) Specimen fracture stage. (*18*) Static, broad beam ion source. (*19*) High pressure gas reaction/catalysis cell. (*20*) Specimen transfer fork (wobble stick). (*21*) 6-specimen mobile carousel. (*22*) 10-specimen autocarousel. (*23*) Preparation vessel specimen transfer "railway". (*24*) Binocular microscope. (*25*) Alternative fast entry lock position or optional extension chamber port. (*26*) Port for monochromated electron source. (*27*) Port for specimen heating/cooling stage. Reproduced courtesy of VG Scientific Ltd, UK

71

pumps are particularly useful for rapidly evacuating large chambers. A comprehensive description of modern vacuum technology can be found in *Weissler* and *Carlson* [2.5].

Aspects of the various probes such as electron and ion guns, X-ray sources, mass analysers and electron spectrometers are dealt with in the subsequent chapters pertaining to the individual techniques. In commercially produced instruments these are normally produced by the one manufacturer. One-off research instruments often consist of a custom central chamber with a variety of accessories, often in-house constructed.

2.3 Specimen Handling

Means by which specimens of interest are introduced into vacuum can vary. Dedicated research instruments may need to be brought entirely up to atmosphere in order to change samples via unbolting flanges. The instrumental cost saving by eliminating a separate introduction chamber is impractical for analytical purposes as specimens need to be changed frequently. In any case, the subsequent baking required exposes the specimen to high pressures of residuals whilst at temperature, thereby forming a considerable contamination selvedge. The alternative fast entry systems are of two general types, one involves a specimen probe or rod onto which the sample is mounted, the rod then being pushed into the analytical region via a series of differentially pumped seals. The other, illustrated in Fig. 2.2, involves moving the specimen, mounted on a small stub, through one or two isolatable locks via rack and pinion or pulley specimen transporters. Stubs are a more convenient approach, but they can be less versatile with respect to heating and cooling of the specimen or making specific electrical connections e.g. as in attachment of a thermocouple. It is feasible to introduce up to a half-dozen stubs at once via a carousel system. With the specimen probe system electrical connections are made ex situ and the wiring travels with the specimen. The specimen translator for dedicated research instruments are usually superior at heating, cooling and can have additional degrees of mechanical freedom, such as azimuthal adjustments for angle-resolved studies. Figure 2.3 shows an internal view inside a vacuum system utilizing the stub method of sample transport.

The method of specimen mounting is largely governed by the nature of the material being examined. For most spectroscopies, good earthing of conducting specimens is essential to avoid static charging of the surface. Powders are often pressed into soft indium foil [2.6], whilst metallic specimens can be directly spot welded with suitable wire (typically nickel or stainless steel). Non-metallic conducting samples are usually silver dagged, and although quite conductive, the organic solvent in the dag can cause contamination problems. Bulk insulating samples can usually be readily glued with cyanoacrylate (super glue). Screws and clips are also employed. Indium foil is also useful for insulating powders simply as an adhesive system. Double-sided tapes have been profitably used for years, with X-ray photoelectron spectroscopy (XPS) in particular using the

Fig. 2.3. View inside a surface analysis vacuum system incorporating an autocarousel and utilizing the stub method of sample transport. Reproduced courtesy of VG Scientific Ltd, UK

tape as an energy calibrant. The tape, however, is quite gaseous under X-ray irradiation and many workers avoid its use.

Viable specimen dimensions and masses are dependent on the vacuum system and technique(s) being employed. For instance, spot electron-excited Auger analyses can be performed quite easily on particles microns across, whereas signal-to-noise requirements for XPS mean that dimensions of 5–10 mm are routine. Large specimens that cannot be cut usually pose the greatest difficulty. Some systems will accept wafers up to 50 mm in diameter, but beyond a few mm in thickness they soon become unwieldy to handle. Of particular concern is a specimen's prehistory, especially the nature of fluids with which it has been in contact, as not all materials are UHV compatible. With, say, high vapour pressure lubricants, any attempts at solvent cleaning may disturb the surface that is really of interest and the information required may be lost.

2.4 Specimen Handling: ASTM Standards

In order to avoid erroneous results it is of critical importance to handle specimens correctly. Below is a condensed/adapted version of American Society for Testing and Materials (ASTM) standards designation E 1078–85 [2.7] for specimen handling in AES and XPS, although it is equally applicable to the other surface analytical techniques. As pointed out in the standard, researchers from more traditional analytical disciplines often need to be educated in the more stringent requirements in surface analysis. The technique sensitivities are such that levels of contamination can easily occur which will lead to severe perturbations. The key concept in specimen handling is *cleanliness*. All handling of the surface in question should be minimized or preferably eliminated whenever possible. If followed carefully, the procedures outlined below will reduce the chance of undetected errors, but no procedures are foolproof and, in some instances, innovation may be required depending on the nature of the specimen.

Visual Inspection. Visual microscopic inspection prior to analysis is generally very useful. As contrast in secondary electron images may appear quite different from visual images it may be necessary to place an identifying scratch near to the region of interest in order to locate it. A visible light micrograph can also be helpful in correlating locations. Following analysis, reinspection may reveal signs of instrumentally induced artifacts, which will temper the assessment of results.

Specimen History. The environments the specimen has been subjected to will influence its handling e.g. an oil-coated mechanical component that has failed requires care more from the contamination it might impart to the spectrometer and other specimens. If the oil is removed prior to submission to the surface analyst, then the details of cleaning must be communicated. Often specimens are subjected to more conventional analytical procedures prior to surface analysis e.g. EDAX. In this case simply exposing the surface to radiation in a poor vacuum will ensure a contamination layer too thick for the surface techniques to probe. It is best to apply surface analysis prior to other techniques.

Sources of Contamination. Specimens should only be handled with *clean* tools, avoiding the surface if possible. Regular cleaning of the tools with high purity solvents prior to use is essential. The necessity for increased dexterity often means that gloves need to be employed in lieu of tools. These need to be carefully selected as they can be a serious source of contamination. Although uncomfortable and awkward the disposable polyethylene gloves are generally the cleanest. Particulate debris can be imparted to the surface by compressed gas blasts intended to remove such items. Photographic quality canned gases are convenient and generally of a good standard, but any gas stream can charge a surface making it more susceptible to picking up particulates. An ionizing nozzle on the gas stream is recommended. Contamination can also arise from within the vacuum system. Hence the need for ultrahigh vacuum as already discussed. Of course total pressure is only a guide and a partial pressure analysis is very useful due to the vast differences in sticking probability that various molecules can have. Note that nearby hot filaments can alter the gas phase chemistry and/or enhance adsorption. Radiation damage has already been considered in Sect. 1.11.1. In situ sputtering can lead to direct contamination of other specimens which may be in the chamber e.g. as in carousel systems.

Of particular concern is not only specimen contamination but also analytical chamber contamination. A contaminated specimen represents a small loss compared to a chamber which, by comparison, may not be easily cleaned. High vapour pressure elements e.g. Hg, Te, Cs, K, Na, As, I, Zn, Se, P, S etc., or alloys thereof, should be analysed with caution. Silicone compounds can also pose a problem due to contamination by surface diffusion, despite their low vapour pressures. Apparently suitable specimens may also become unsuitable as a result of radiation damage.

Specimen Storage and Transfer. Storage is a problem even in very clean laboratories. Hence the best approach is to analyse specimens as soon as possible. Glove boxes, vacuum chambers and desiccators are generally the best means of storage as they can minimize oxidation/hydrolysis effects. These can also be adapted to directly connect to entry locks to avoid any atmospheric exposure of the specimen surface. The container itself should not be a source of contamination and materials with volatile components should be kept separate to avoid cross contamination. During shipping most factors can be easily controlled except temperature, which can be quite detrimental.

Overlayer Removal. If the surface of interest lies below a coating or a layer of contamination then this outer layer will need to be removed. The most commonly employed technique is sputter profiling (see Chap. 4) through other approaches may be more suitable. Mechanical separation such as peeling is feasible if the interlayer bonding is sufficiently weak. Thick layers can also be removed by sectioning methods such as abrasive wheels, sawing, shearing, mechanical fracture, chemical or electropolishing. It needs to be remembered that the lubricants, abrasives and electrolytes all add another dimension to surface contamination, hence some in situ ion etching is invariably required. Most mounting materials used in sectioning e.g. thermoset plastics, are unsuitable for insertion into vacuum and need to be removed. Angle lapping and ball cratering have already been mentioned in the previous chapter.

In certain cases solvents and surfactant solutions can be employed to remove soluble contaminants and overlayers. Ultrasonic agitation vastly improves the effectiveness in these cases but care needs to be employed with liquid purities. Even high purity solvents can evaporate and leave behind tell-tale interference films. Although solvent soaked tissues are often used to clean tools etc., it is best to avoid tissue contact with specimen surfaces.

With refractory materials, heating can be successfully employed to remove contaminants. However, this finds most application in basic surface studies requiring initially clean and reproducible surfaces rather than in general materials analysis. For non-refractory systems the high temperatures would undesirably alter the specimen. Some overlayers may be sufficiently volatile that they can be pumped away at room temperature given sufficient time. Of course, even in UHV, if too long is required vacuum residuals end up replacing the original contamination. Degassing specimens is an extension of this problem. Normally this degassing is best performed in an auxillary chamber to prevent degrading the vacuum in the main analytical chamber.

In situ exposure techniques. Such exposure can be achieved by fracture, cleaving or scribing. Impact or tensile fracture can be achieved in specially designed chambers provided specimen geometry and materials characteristics criteria are met. Non-ideal geometries can often be accommodated by incorporation into or attachment to an ideal geometry. Ease of fracturing of metals can often be enhanced by lowering temperature or by charging intergranular regions with hydrogen. As electrical insulators may pose serious charging problems, exterior conductive coatings ought to be applied prior to in situ fracture. Scribing is useful provided the scribe mark is larger than the input probe and none of the constituents are smeared.

Specimen mounting. This has been briefly touched upon in Sect. 2.3. The most serious difficulty which might arise concerns specimens which are poor electrical conductors. In addition to the approaches pointed out in Sect. 1.11.2 to minimize charging it may also be feasible to place a conductive mask, wrap or coating about the specimen. This mask e.g. metal foil or colloidal graphite, needs to be effectively earthed and make contact as close as possible to the area to be analyzed. When sputtering is also used care must be taken not to sputter mask material onto the surface to be analyzed.

Given specimen geometries can require some innovation depending upon the nature of the analytical system. For instance, wire or fibre specimens are best clamped so as to suspend them free in space away from the main part of the sample stub. When the overhang is in the focal area of the spectrometer background, artifacts from the stub can then be avoided. Specific particles which cannot be grouped as a powder or compacted as a pellet can sometimes be floated on suitable liquids and picked up on conducting fibres.

References

2.1   W.J. Moore: *Physical Chemistry* (Longman, London 1972) p. 133
2.2   Adapted from chart devised by L. de Chernatony, GEC-AEI publication 2015–14
2.3   R.C. Weast, M.J. Astle (eds.): *CRC Handbook of Chemistry and Physics*, 63rd edn. (CRC Press, Boca Raton 1982) F-164–168
2.4   D. Menzel, J.C. Fuggle: Surf. Sci. **74**, 321 (1978)
2.5   G.L. Weissler, R.W. Carlson (eds.): *Vacuum Physics and Technology* (Academic, New York 1979)
2.6   G.E. Theriault, T.L. Barry, M.J.B. Thomas: Anal. Chem. **47**, 1492 (1975)
2.7   ASTM Standards on Surface Analysis (1986) ISBN 0-8031-0948-2

Part II

Techniques

3. Electron Microscope Techniques for Surface Characterization

P.S. Turner

With 8 Figures

Electron microscopy in its various forms has developed over the past forty years into one of the major techniques of materials science. Surface analytical techniques are more recent additions to the materials scientists' range of experimental methods for learning about materials properties. Microscopy and spectroscopy are complementary, and the use of one alone can result in an inadequate characterization of a material. In this chapter we will consider the various modes of electron microscopy and attempt to demonstrate the critical importance of applying electron optical imaging as well as surface analytical techniques in studies of surfaces.

The textbook picture of a specimen to be examined using surface analytical techniques such as XPS or SIMS is typically a perfectly flat surface of a laterally homogeneous material (Fig. 3.1a). Of course, as the chapters of this book make clear, no real sample of any significant interest is like that, and techniques such as scanning Auger microscopy can provide information about variations in composition across a surface, with resolution to < 100 nm. Nevertheless, there is a tendency to interpret surface analytical data in terms of a postulated model of the surface. Only by looking at the surface, with whatever resolution is required in order to see the relevant detail, can one be confident of the interpretation of the surface analytical data. Thus microscopy – optical, electron, scanning tunneling, etc. – must be employed in conjunction with the surface spectroscopies. For ex-

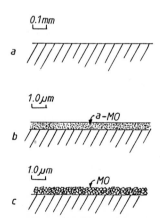

Fig. 3.1 a–c. Schematic illustration of some typical types of surface (see text for details)

ample, the two samples illustrated in Fig. 3.1b,c could give identical XPS spectra and depth profiles, whereas one consists of a continuous amorphous oxide film and the other of a fine dispersion of oxide crystallites across the surface.

The aim of this chapter is to introduce the basic concepts of electron microscopy of surfaces to readers who are not familiar with electron microscopy. We will consider the ways in which scanning and transmission electron microscopes can provide information about surfaces, and the nature of that information. Two other surface imaging systems – scanning tunneling microscopy, and scanning Auger microscopy – are covered elsewhere in this book (in Chaps. 10 and 6 respectively) and will not be considered here. An excellent introduction to more advanced aspects of the subject has been presented by *Venables* et al. [3.1]. The current state of the subject was covered in the proceedings of two recent NATO Advanced Study Institutes [3.2, 3], and at the 1989 Wickenburg Workshop [3.4].

In this short article, we cannot cover all relevant aspects of the instruments and techniques in detail. There are many excellent texts on electron microscopy, some of which are listed at the end of this chapter as sources for further reading [3.5–11]. In particular, more detailed descriptions of the major components of electron microscopes, and more comprehensive explanations of their operation and performance, should be studied by anyone proposing to learn to use electron microscopy for the characterization of surfaces. A good starting point is the short book by *Goodhew* [3.5].

In the following sections, we will consider first the nature of the information we may require from electron microscopy, then the major characteristics of electron optical systems, and the factors which determine the nature and quality of electron images. The roles of the scanning electron microscope and the transmission electron microscope in surface studies will be described and illustrated. In the final section we will look briefly at current developments in the imaging of surfaces.

3.1 What Do We Need to Know About Surface Structures?

The variety of samples which are studied using surface analysis is immense, ranging from atomically flat semiconductors grown in situ using molecular beam epitaxy, to pieces of corroded brass taps and compressed powders. For the purposes of this discussion, we will consider the problem of investigating the reactivity of a multiphase ceramic exposed to chemical attack via either vapour or liquid, and of associated separate studies of the reactivities of the major component phases in this ceramic.

We could prepare surfaces of a single phase, polycrystalline ceramic by polishing, and also by fracturing it. In order to interpret our surface analytical spectra, we will need to know some or all of:
- the grain size;
- whether there are minor phases present in the nominally single phase material;

- the orientation of grains;
- whether chemical attack results in the growth of precipitates;
- whether attack occurs preferentially at boundaries, or on certain grains;
- the topography of fractured surfaces, and whether they reveal intergranular or transgranular fracture;
- whether the sample is porous.

These questions can probably be answered using scanning microscopy on the samples used for surface analysis. A higher image resolution would be required to determine

- the width and structure of intergranular films;
- the crystal structure of minor phase or precipitates;
- whether amorphous surface layers are formed during the reactions; and so on.

If transmission microscopy is required, the preparation of samples representative of those studied by surface analysis may provide difficulties.

Similar questions will arise for the multiphase ceramic, with additional points such as

- the degree of homogeneity of the multiphase material;
- different rates of attack on different phases;
- variations in the width of intergranular films, and preferential attack at grain boundaries.

3.2 Electron Optical Imaging Systems

There are two ways in which images can be obtained using electron beams. The traditional transmission electron microscope (TEM) is a close analogue of the light optical microscope, and involves the illumination of a transparent object by a beam of electrons, and the formation of a magnified image of the electron waves emerging from the object using an objective lens and two or more projector lenses. The scanning electron microscope (SEM) uses lenses to form a demagnified image of the electron source. This fine probe of electrons is scanned across the sample, and one of a number of possible signals arising from the electron-specimen interaction is detected, amplified, and used to modulate the intensity of a TV-like image tube, which is scanned at the same rate as the electron probe. Clearly there are fundamental differences between these methods, but both use electron sources, electron lenses and electron detectors, and therefore both are limited by the performance of such devices.

The TEM and SEM techniques converge in the STEM (scanning TEM) in which a nanometre probe is scanned across a thin sample and a transmitted electron signal – often also filtered to remove inelastically scattered electrons – used to form the image. Both TEM and STEM can be operated in reflection mode, in which the electron beam is incident at a glancing angle and reflected and/or diffracted from a flat sample (REM and SREM).

In forming an image, two factors govern the information which may be obtained about the object – the resolving power of the imaging instrument (the smallest resolvable separation of two distinct points in the object) and the contrast in the image (the difference in intensity which allows us to distinguish resolved detail from a background intensity level).

The best resolution attainable with a given instrument is determined by the quality of the objective lens in the TEM or the probe-forming lens in the SEM. The wave nature of the electrons leads to a fundamental diffraction-limited resolution, but the high aberrations of magnetic electron lenses determine the actual resolution. For a given lens, the TEM resolution and the smallest probe diameter which that lens could produce are essentially identical (e.g., ~ 0.3 nm at 100 keV), but in practice a typical commercial SEM has a best resolution at least 10 times worse than a standard TEM, due to lower electron energies and larger focal lengths (e.g., 3 nm at 15 keV). In practice the useful resolution in images is often much worse, due to limitations in contrast.

The contrast of an image is limited by the random nature of the emission of electrons from the source. If we consider a particular element in the image (a picture element or pixel) with an intensity corresponding to the detection of N electrons in that pixel, then the standard deviation in the intensity will be proportional to $\sqrt{N}$. The difference ΔN between signals from nearby pixels should be at least $3\sqrt{N}$ if contrast between these pixels is to be detected in the image [3.8, 9]. In the TEM, all image points are detected simultaneously (parallel detection), and in practice it is usually easy to ensure that the noise level ($\sqrt{N}/N$) is very low. But in the SEM each pixel is detected in sequence as the probe scans across the sample (serial detection); the noise level in scanned images is a critical factor in determining the image contrast and detectable resolution.

3.2.1 Electron Sources

A critical factor in electron imaging is the current density of the electron beam at the object. The electron optical brightness, β of the system, is defined by

$$\beta = \frac{J}{\pi \alpha^2} \tag{3.1}$$

where J is the current density at a point in the illuminating beam, and α is the half-angle of the converging cone of illumination ($\pi \alpha^2$ steradian is the solid angle subtended by the illumination at that point). The significance of β is that it has a constant value below the electron gun, and it is proportional to the electron energy E. In a focussed probe of electrons we can express J in terms of the probe current I_p and the diameter d_p of the probe crossover, so

$$d_p^2 \alpha^2 \simeq \frac{4 I_p}{\pi^2 \beta} . \tag{3.2}$$

Thus for scanned imaging systems, with a given beam current, small probe sizes require high convergence angles, and vice versa.

There are three types of electron source used in contemporary EMs, differing in maximum brightness and in maximum beam current. The traditional electron gun uses a tungsten wire, heated to about 2700 K, from which electrons are emitted and accelerated to the required energy. Under optimal adjustment, a brightness $\beta = 5 \times 10^5 \, \text{A cm}^{-2} \, \text{sr}^{-1}$ is obtained at 100 keV, with a total current of up to 100 μA from an effective source diameter of about 30 μm. A more recent development is the lanthanum hexaboride (LaB_6) source, which, operating at lower temperatures, gives about ten times higher brightness, but requires better vacuum conditions. The field emission source, a pointed tungsten tip operated in UHV at room temperature, can provide brightness values as high as $10^9 \, \text{A cm}^{-2} \, \text{sr}^{-1}$ at 100 keV, but at much lower total beam currents. For details of the construction and operation of electron guns using these sources, the reader is referred to texts on electron microscopy, e.g. [3.6, 9, 10]. The standard guns for TEMs and SEMs, using W or LaB_6, are very flexible, and easily adjusted to provide for a range of beam currents at or close to the corresponding maximum brightness values.

These electron guns are almost monoenergetic, with a narrow range of electron energies around the mean, due to the thermal spread in energy of the emitted electrons. In the thermionic guns, this is about 1.5 eV, whereas for the field emission gun it is about 0.25 eV.

3.2.2 Electron Lenses

Electric and magnetic fields having cylindrical symmetry act as lenses for charged particles. Electrons emerging from a point in an object are brought to a focus in the corresponding image plane. In comparison with light optical lenses, electron lenses suffer severe, uncorrectable aberrations, particularly spherical and chromatic aberration, as a result of which a point image is blurred into a disc of diameter d_{ab} given by [3.9]

$$d_{ab}^2 = d_s^2 + d_c^2 + d_d^2 \, ,$$

with

$$d_s = 0.5 C_s \alpha^3 \, , \quad d_c = C_c \alpha \Delta E / E \quad \text{and} \quad d_d = 1.22 \lambda / \alpha^2 \, , \tag{3.3}$$

where C_s, C_c are the spherical and chromatic aberration coefficients, α is the angle between the lens axis and the electron trajectory, ΔE is the effective energy spread of electrons having mean energy E, and λ is the electron wavelength. In addition to these inherent aberrations, all electron lenses suffer from quite severe astigmatism. Manufacturers provide compensating coils to counter this; in practical electron microscopy the correction of astigmatism is a critical factor in obtaining an optimum image.

With the exception of the electrostatic lens action of the guns, all electron microscopes use electro-magnetic lenses. These consist of closed coaxial soft iron cylinders enclosing the windings, with a specially shaped gap – the pole piece – at which a strong magnetic field is constrained within the bore of the lens. The shape of the pole piece depends on the purpose of the lens. A probe-forming lens for SEM may have a conical shape, and provide a range of focal lengths from 5–50 mm. The highest resolution TEM instruments have objective lens focal lengths down to 1 mm or less, with the object immersed in the magnetic field. The values of C_s and C_c depend upon the pole piece geometry, but are of the same magnitude as the focal length. The strength of a particular lens can be varied over a considerable range by varying the current in the magnet windings.

The trajectories of electrons in magnetic fields rotate around the axis of the lens, and hence electron images rotate as the strength of a lens in varied. In today's microprocessor-controlled instruments, groups of lenses are varied in such a way as to give almost zero total rotation while image magnification is changed.

3.2.3 Detection Systems

In transmission microscopy, the image is viewed by observing a flat screen coated with a powder which fluoresces under the electron beam. Such an image may be recorded on photographic emulsion by allowing the electrons to impinge on film placed in the microscope column below the screen. Electron imaging emulsions are extremely efficient and convenient, but the fluroescent screen has limited resolution and sensitivity. For high resolution microscopy, and for studies of beam sensitive materials, transmission screens and image intensifier devices, together with sensitive TV image tubes, are becoming widely used, often in conjunction with digital frame storage systems which allow for integration of successive (noisy) TV images and hence improvements in contrast as the noise level is reduced. Subsequent digital processing of the images is also possible.

A variety of detectors are used in the scanned image systems, in order to select and amplify the required signal. We will consider the individual detectors in Sect. 3.3, but may note some general common factors. The detectors are characterized by an efficiency factor, which can be $\ll 1$. The serial nature of scanned image formation, with electronic signal detection, allows for amplification and for background subtraction before the signal is applied to the image tube. All such detection systems can add to the image noise, and the quality and performance of amplifiers can effect the speed at which scanned images may be viewed and recorded. Thus TV rate display of images, while convenient, necessarily results in lower contrast images than do slow scan rates, since there are fewer electrons per pixel in TV images. Again, the use of digital frame stores which average successive frames is becoming common in SEM.

3.3 Scanning Electron Microscopy of Surfaces

The standard SEM is the most useful electron optical system for surface imaging, and should be applied routinely to characterize all surfaces which will be, or have been analyzed by XPS, SIMS, etc. (It will usually be preferable to image after surface analysis, to avoid contamination during SEM examination.)

3.3.1 The SEM

The essential elements of an SEM are shown schematically in Fig. 3.2. The electron gun, usually fitted with W or LaB$_6$ filament, operates over the range 0.5–40 kV accelerating voltage. A condenser lens produces a demagnified image of the source, which in turn is imaged by the probe forming lens (often called the objective lens) onto the specimen. The electron path and sample chamber are evacuated. Scanning coils deflect the probe over a rectangular raster, the size of which, relative to the display screen, determines the magnification. Detectors collect the emitted electron signals, which after suitable amplification can be used to modulate the intensity of the beam of the display video screen, which is rastered in synchronism with the probe.

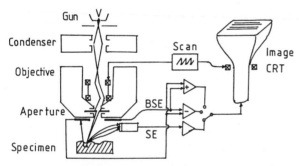

Fig. 3.2. Basic features of the SEM, showing electron gun, condenser and objective (probe-forming) lenses, scanning system, detectors and image display CRT. The diameter of the aperture and the working distance from lens to specimen determine the convergence angle of the probe. The probe at the specimen is the demagnified image of the source, broadened by lens aberrations

3.3.2 The Signals and Detectors

When an electron is incident on the surface of a thick specimen, several signals can be detected. The primary electron results in ionization of atoms along its path in the solid, which in turn can result in the ejection from the surface of secondary electrons, very close to the incident beam position. These have energies from 0–20 eV, and can be attracted to a positively charged detector with high efficiency. The secondary electron yield per primary electron is high and increases as the angle ϕ between the electron beam and the surface normal increases in proportion to $1/\cos \phi$. For example, for 10 keV electrons incident normally ($\phi = 0°$), the secondary emission coefficient is ~ 0.2 for Al, increasing to 1.0 at $\phi = 75°$ incident. The coefficient varies with incident electron energy E in proportion to $E^{-0.8}$ [3.8, 9].

The primary electrons can be deflected through large angles, and hence emerge as backscattered electrons with high energy, from a region surrounding the incident probe. Where these backscattered electrons emerge from the surface, secondary electrons are generated, so a component of the secondary signal is proportional to the backscattered yield. At sharp edges, scattered primary electrons can give rise to very high secondary signals.

By placing a large annular detector below the final lens, facing towards the sample, a proportion of the backscattered electron (BSE) signal can be collected. The scattering of primary electrons within an elemental sample of atomic number Z increases with Z, but is relatively insensitive to E over the range 5–30 keV. Thus the BSE signal from a flat, polished sample provides contrast which is composition dependent. The BSE signal is also dependent on the orientation between the beam and surface. The typical BSE annular detector is split into two semicircles; the sum of the two signals is predominantly sensitive to composition, whereas the difference signal is sensitive to topography.

The resolution of the backscattered image is typically of order 0.1–1 μm, being determined by the volume within the sample from which the detected BSE signal appears to come. This volume can be estimated using Monte Carlo simulations of the scattering process, [3.8, 9]. With increasing accelerating voltage, the volume increases; at a given voltage, the volume is smaller for heavy (strongly scattering) elements.

3.3.3 Resolution and Contrast in SEM Images

The best resolution attainable in principle is determined by the minimum probe size, which is obtained when the image of the electron source is demagnified to the greatest extent possible with the available lenses, using a final aperture α chosen to minimize the combined effects of spherical aberration ($\sim C_s \alpha^3$) and diffraction ($\sim \lambda/\alpha$). However, from the brightness relationship, (3.2), the resulting current I_p in the probe will be too low for useful imaging unless a field emission source is fitted. Thus in standard SEMs, using W or LaB$_6$ sources, it is usual to select the probe current required for the particular imaging condition of interest (i.e., required contrast level), so that the geometrical probe size d_0 and aperture angle α are related, from (3.3), by

$$d_0 = \sqrt{4 I_p / \pi^2 \beta} / \alpha \, . \tag{3.4}$$

The probe diameter d_p is then estimated by adding to d_0 the spherical, chromatic and diffraction aberration disk diameters d_s, d_c, and d_d [see (3.3)] in quadrature [3.9]:

$$d_p^2 = d_0^2 + d_d^2 + d_s^2 + d_c^2 \, . \tag{3.5}$$

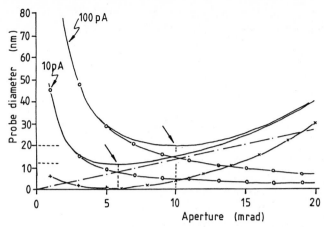

Fig. 3.3. Variation of SEM probe diameter with aperture angle, in accordance with (3.4). The contributions due to diffraction (d_d: $- + -$); spherical aberration (d_s: $- \times -$); chromatic aberration (d_c: $- \cdot -$) are shown for the case of 15 kV, 2 eV energy spread, $C_s = 7.5$ mm and $C_c = 10$ mm. Curves for the geometrical probe diameter $d_0(-o-)$ and total probe diameter d_p (——) are shown for two probe current values and gun brightness 3×10^4 A cm^{-2} sr^{-1}. The optimum probe diameters of 12 and 20 nm are obtained for apertures of 6 and 10 mrad respectively (*arrows*)

The dependence of probe diameter d_p on aperture α for a given set of operating parameters (gun voltage and brightness, objective lens focal length and aberrations) is illustrated in Fig. 3.3, for two values of the probe current. The resolution is poorer for higher beam currents, because d_0 is increased. For the higher current, the optimum resolution is achieved at a higher value for the angular aperture α. In practice α is set to one of several fixed values, determined by the diameters of the actual apertures and the lens-to-sample working distance, so the optimum operating conditions for a given probe current can only be approximated.

The resolution actually achieved in the secondary electron (SE) images can be close to the optimum, with contrast determined by the topography of the surface. For example, images of compressed pellets of a fine single phase powder (a common form of sample for routine XPS analysis) will show topographical contrast, revealing the particle size and shape, and the extent of packing of the powders (Fig. 3.4), and, at high magnification, fine details of surface topology. This information is clearly relevant to the interpretation of any depth profiles which might be recorded in XPS/SIMS/AES instruments.

If there is little or no topographical contrast (as for a polished sample), then the SE image may have contrast determined by those SE's generated by the backscattered electrons; the image will also show correspondingly poorer resolution. This effect is shown in Fig. 3.5a, for a polished surface of Synroc, the multiphase titanate ceramic for immobilization of radioactive waste [3.12]. Here most of the contrast is compositional, arising from the atomic number dependence of the SE signal derived from BS electrons. The presence of, and topography of, pores and cracks is highlighted by the very strong topographical contrast at

100μm

Fig. 3.4. SEI micrograph of a crushed single phase mineral sample, pressed onto a conducting adhesive surface, as used for some XPS characterizations

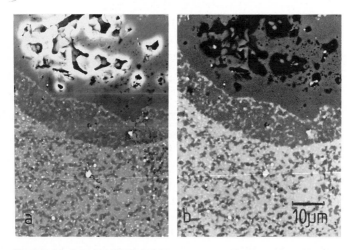

10μm

Fig. 3.5. (a) Se and (b) BSE SEM images of a polished sample of Snyroc, showing the strong edge contrast in the SE image at pores; in both images, the contrast differences between grains is due to backscattered electrons

the edges of such features. The BSE image is similar but does not show up the structure of the pores (Fig. 3.5b).

Although the secondary electron image usually has better resolution and higher contrast (relative to noise levels), the backscattered images are often more valuable. This is illustrated in Fig. 3.6, showing SE and both topological and compositional BSE images of a fracture face from the sample of Synroc. The "Compo" BSE image shows the distribution of phases; only at large cracks

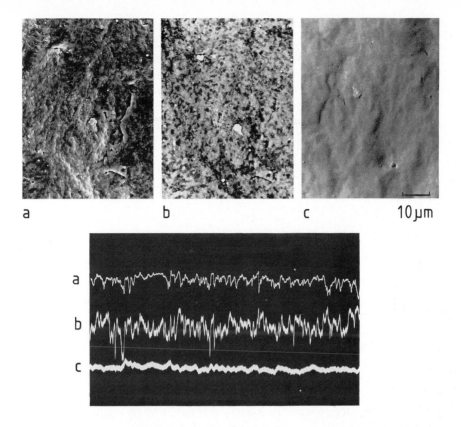

Fig. 3.6. SEM images and line traces from a fracture face of Synroc, for (**a**) SE (**b**) BSE compositional and (**c**) BSE topographical signals. The much greater noise levels in the BSE signals, are evident in the line traces; the topographical BSE image is formed by subtracting signals from the two halves of the BSE detector, thus removing the compositional contrast, but losing detail to enhanced noise levels

is there some topographical contrast. The broad features of the topology of the fracture face are revealed in the difference between the two BE signals – the compositionally sensitive signal being suppressed by subtraction. This gain in information is at the cost of increased noise level, as shown in the line traces for all three signals. The SE image shows fine details of the fracture face topography, but this can be confused by the underlying compositional signal.

3.3.4 Energy Dispersive X-Ray Spectrometry

X-rays are generated by the incident probe within a volume similar to, but rather larger than, that for the backscattered electrons. Peaks at energies characteristic of the elements within that volume can be identified and the concentrations of the elements can be calculated. Thus the bulk composition of the sample, and of the individual grains in a polycrystalline sample (provided they are larger than the X-ray excitation volume) can be determined for comparison with surface analytical data.

The dimension of the excitation volume depends on the mean atomic number of the material, the electron energy and the emitted X-ray energy. Diameters range from 0.1 μm to several μm [3.8]. Thus EDX spectra will not reveal compositional changes due to surface segregation. X-ray maps of surfaces, with spatial resolution of order 1 μm, are often valuable in revealing compositional variation across a sample, especially if scanning Auger microscopy is not available.

3.4 Transmission Electron Microscopy of Surfaces

The TEM is usually used to study the internal microstructure and crystal structure of samples which are thin enough to transmit electrons with relatively little loss of energy. This requires thicknesses in the range 20–200 nm, depending on the average atomic number of the material, using the the standard 100 keV TEM. Clearly the bulk samples used in surface analysis cannot be examined directly this way. The transmission microscope is nevertheless an important adjunct to the SEM in studies of surfaces, because of its greater resolving power (down to crystal lattice dimensions), its ability to provide surface sensitive diffraction data and images, and associated analytical capabilities through X-ray and electron energy loss spectroscopies. In addition, images and diffraction patterns may be formed in the reflection mode, off smooth surfaces or facets. The challenge is to ensure that the surfaces of samples examined in the TEM are equivalent to those studied by surface analysis. The most satisfactory approach would be to combine the surface analytical and microscopy techniques in the one instrument; some progress has been made in this direction (Sect. 3.5). However, we will consider here the more typical situation of separate instrumentation, and the consequent requirements on sample preparation to achieve our aim.

3.4.1 The Transmission Electron Microscope

The TEM is very similar to the conventional light optical microscope in terms of optical principles. The electron source is followed by two condenser lenses to provide a uniform illumination of the specimen over the area of interest, adjustable as the magnification is changed. The sample is mounted on a stage to provide suitable movement. The primary image is formed by the objective lens, which determines the resolution obtainable. The final image is projected onto a viewing screen through two or more projection lenses, and can be recorded on film placed below the screen.

In practice, these basic similarities are obscured by the very different structure of the TEM. The electron path from gun to camera must be evacuated to 10^{-5} torr or better; the magnetic electron lenses are large iron cylinders, with substantial aberrations, which restrict the angular range of the electron beam and hence result

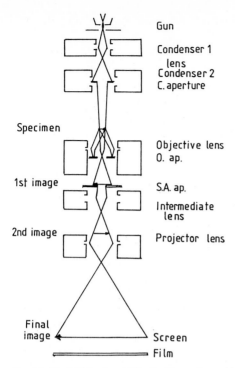

Gun

Condenser 1
lens
Condenser 2
C. aperture

Specimen

Objective lens
O. ap.

1st image

S.A. ap.

Intermediate
lens

2nd image

Projector lens

Final
image

Screen

Film

Fig. 3.7. The TEM column. The double condenser lens system forms a focussed beam at the specimen, which is immersed within the magnetic field of the objective lens. The objective aperture, placed at the back focal plane, controls the image contrast. The projector lenses magnify the image onto the screen

in a very tall instrument, with resolution far worse than the inherent diffraction limit; the strength of the lenses is changed by changing the currents in the electromagnetic coils; the very thin samples need to be tilted and translated, with very high mechanical stabilities; the image is viewed by converting the electron image into light at a fluorescent screen; and so on.

The essential features of the TEM are shown in Fig. 3.7. In addition to providing images over a magnification range from about $100\times$ to $500,000\times$, the electron diffraction pattern, formed at the back focal plane of the objective lens, can be observed on the screen and recorded, to provide information about the crystalline structure of the sample. Adjustable apertures are used to control the angular aperture of the illumination; to provide image contrast by selecting some of the diffracted electrons and preventing the rest from contributing to the image; and to select specific areas of images from which to obtain diffraction information. Magnetic deflection coils are used not only to align the electron beam, but also to control it to provide specific image modes, and to correct for astigmatism in the lenses.

3.4.2 Electron Diffraction

The wavelength of 100 keV electrons is 37 pm, which is about 50 times smaller than the typical nearest-neighbour separation in many crystals. Electrons incident on a thin sample are diffracted though angles of order 10^{-2} radian; the resulting electron diffraction pattern can be observed by imaging the back focal plane of the objective lens. Using an aperture at the level of the first image (the "selected area aperture"), diffraction data from areas down to about 300 nm can be obtained; smaller areas may be selected for diffraction analysis if the TEM has the facility to form fine illumination probes (down to 2 nm).

An amorphous film shows broad rings in its diffraction pattern, with peaks corresponding to the most probable interatomic spacings. Crystalline materials show the sharp spots corresponding to Bragg diffraction by the periodic lattice. Thus information about the degree of crystallinity, the orientation and crystal structure of individual grains, etc., can be obtained from the electron diffraction patterns. Although the dominant features of the patterns will arise from the bulk crystal, weak diffraction spots arising from surface structures can often be detected, and may be used to form dark field images of the corresponding surface detail.

3.4.3 Image Contrast and Resolution in the TEM

If all the diffracted waves were to be recombined by a perfect objective lens into a perfectly focussed image, the image would show no contrast, because the electrons are not absorbed in the thin specimen and the differences in phase are lost in the observed intensities. Defocussing the lens gives phase contrast through interference between adjacent parts of the wave (Fresnel diffraction), but in order to obtain reasonable contrast levels in TEM images, it is usually necessary to insert an aperture at the level of the back focal plane of the objective lens, to remove some fraction of the diffracted electrons from the image. If only the unscattered electrons are permitted to pass, through a small aperture centred about the optical axis, a bright field image is formed. If the incident illumination is tilted, so that one of the diffracted beams is aligned down the optic axis and passes the objective aperture, a dark-field image may be recorded. Larger apertures may be used to pass several diffracted beams; the resulting image contrast arises from interference between these beams, and shows fringes with spacings equal to those of the lattice planes of the Bragg diffracted beams included in the image. In general such images are very complex, but under certain conditions (very thin crystals, electron beam orientated down a crystal symmetry axis, a specific defocus of the objective lens) "structure images" can be recorded which relate directly to the crystal structure with a resolution corresponding to the spacing of the planes giving rise to the Bragg reflections passing the aperture.

The resolution achieved in the images is determined by a balance between the diffraction limit ($\sim \lambda/\alpha$) and the effects of spherical aberration ($\sim C_s \alpha^3$) and chromatic aberration ($\sim C_c \alpha \Delta E/E$). The simplest estimates of resolution may be made using (3.3). More accurate estimates of resolution involve consideration of the effects on the phases of the electron waves of defocus, spherical and chromatic aberration, and the objective aperture which limits the angular range of diffracted waves contributing to the image (see [3.10, 11] for detailed discussions of these topics).

3.4.4 Imaging Surface Structures in the TEM

The degree of roughness of the surfaces of perfect crystals can be revealed qualitatively using phase contrast imaging, with an objective aperture which just cuts out all the diffracted waves, so the resolution is typically around 0.3 nm. The classic examples of this mode are images of MgO cubes [3.13]; note that the crystals must be oriented to avoid any strong Bragg diffraction effects, so that there is only weak dependence of contrast on thickness (Fig. 3.8). More quantitative information can be obtained by setting the crystals to specific diffraction conditions, and recording either bright or dark field images; the resulting contrast can reveal monatomic surface steps through the thickness sensitivity of the diffracted intensity [3.14–16].

Reconstructed surfaces, having a different crystal structure from the bulk, can be imaged in dark field by selecting a surface specific diffracted beam with the objective aperture. The 7×7 superlattice structure on Si was first determined in this way [3.17].

Lattice images of crystals are not usually sensitive to minor (monatomic) changes in surface structures, since the diffracted waves will be largely deter-

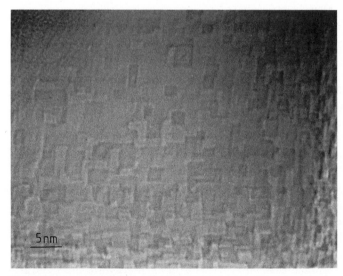

Fig. 3.8. Phase contrast image of an MgO crystal, after exposure to water vapour and electron beam irradiation; recorded at 100 keV in a TEM. Phase contrast is achieved by defocussing the image, revealing a pattern of small (1–2 nm) rectangular depressions and protrusions across the initially smooth {100} surfaces

mined by diffraction within the bulk of the crystals. The new technique of "profile imaging" [3.18] shows directly the surface structures of facets imaged at the thin edges of crystals, in atomic detail. The accurate interpretation of such images requires the matching of the experimental images with theoretical ones computed from the full theory of dynamical scattering of electrons in crystals [3.11, 19]. This method is a powerful complement to the use of scanning tunneling microscopy to determine surface structures at atomic resolution [3.2].

3.4.5 Reflection Electron Microscopy

REM images and reflection electron diffraction patterns are obtainable in the TEM or STEM by mounting the sample so that the surface of interest is almost vertical. REM images of surfaces have played a very significant role in recent years in improving knowledge of the structure of extensive "flat" surfaces [3.20]. Although severely foreshortened, these images are easily interpreted in most cases, revealing surface steps, regions with different structure, dislocations, etc. [3.3]. The classic study of the growth of the 7×7 structure on the (111) surface of silicon, using dark field REM images, established the power of this technique [3.21]. Recently, using a high resolution TEM with UHV specimen stage, the mechanisms of growth of Au on the Si-7 $\times$ 7 superlattice have been elucidated, through video recordings of the growth process in which both Si and Au surface superlattices were resolved [3.22]. In this type of research, REM is a powerful complement to LEED and STM.

3.5 Recent Developments

Scanning electron microscopes with field emission sources have recently become available, and have been shown to offer secondary image resolutions of 1 nm or less, and backscattered resolution of a few nm [3.23]. This performance will improve dramatically the applicability of SEM in surface studies, especially if the vacuum levels in the sample area can be improved further. For example, the layered structures of quantum well microelectronic devices are clearly imaged in both SE and BSE images.

A rather different approach to surface imaging has been achieved with the development of the Low-energy and the PhotoEmission EM (LEEM/PEEM) [3.24]. The sample is held at a high negative potential, so that high energy incident electrons are decelerated to a few eV before hitting the surface. The reflected/diffracted electrons are re-accelerated to form an image with resolution as high as 2 nm, and high sensitivity to surface structures and composition. LEED patterns can be displayed, and dark field images formed using selected diffraction spots. In PEEM, the sample is illuminated with an intense light source, and the emitted photoelectrons are imaged.

Important advances in surface science are likely to come from new instruments in which one or more of the surface spectroscopies are combined with imaging techniques such as REM and high resolution SEM. Such instruments must have ultra high vacuum levels in the sample chamber and facilities for treatment of the sample (e.g. heating, evaporation onto substrates without withdrawal from UHV), while retaining high resolution surface spectroscopy (AEM, SIMS), imaging and diffraction. The development of a STEM-based instrument, with the facility to obtain high resolution secondary images from both sides of a thin sample, has been reported [3.25]; this system is designed to include an Auger spectrometer.

There is an exciting future for studies involving the combination of surface analytical techniques and high resolution electron microscopy of surfaces. However, at a more routine level, the interpretation of XPS and SIMS data from a multitude of surfaces of many material will always be assisted by – and frequently requires – the additional information provided by the scanning electron microscope. Imaging and spectroscopy should be regarded as essential partners in surface science.

References

3.1   J.A. Venables, D.J. Smith, J.M. Cowley: HREM, STEM, REM SEM and STM, Surf. Sci. **181**, 235 (1987), see also J.A. Venables: Ultramicroscopy **7**, 81 (1981)
3.2   A. Howie, U. Valdre (eds.): *Surface and Interface Characterization by Electron Optical Methods*; Proc. NATO ASI, Erice 1987 (Plenum, New York 1988)
3.3   P.K. Larson, P.J. Dobson (eds.): *Reflection High-Energy Electron Diffraction and Reflection Imaging of Surfaces*, Proc. NATO ASI, Veldhoven 1987 (Plenum, New York 1988)
3.4   J.A. Venables, D.J. Smith (eds.): Proceedings of Workshop on Surfaces and Surface Reactions, Wickenberg Inn, Arizona, 1989: Ultramicroscopy **31** (1989)
3.5   P.J. Goodhew: *Electron Microscopy and Analysis* (Wykeham, London 1975); P.J. Goodhew, F.J. Humphreys: *Electron Microscopy and Analysis*, 2nd edn. (Taylor and Francis, London 1988)
3.6   P.W. Hawkes: *Electron Optics and Electron Microscopy* (Taylor and Francis, London 1972)
3.7   I.M. Watt: *The Principles and Practice of Electron Microscopy* (Cambridge University Press, Cambridge 1985)
3.8   J.I. Goldstein, D.E. Newbury, P. Echlin, D.C. Joy, C. Fiori, E. Lifshin: *Scanning Electron Microscopy and X-ray Microanalysis* (Plenum, New York 1981)
3.9   L. Reimer: *Scanning Electron Microscopy*, Springer Ser. Opt. Sci. Vol. 45 (Springer Berlin, Heidelberg 1985)
3.10 L. Reimer: *Transmission Electron Microscopy*, Springer Ser. Opt. Sci., Vol. 36 (Springer Berlin, Heidelberg 1984); second edition published in 1989, also includes reflection EM
3.11 J.M. Cowley: *Diffraction Physics*, 2nd edn. (North Holland, Amsterdam 1981)
3.12 A.E. Ringwood: *Safe Disposal of High Level Nuclear Reactor Wastes* (ANU, Canberra 1978); J.A. Cooper, D.R. Cousens, R.A. Lewis, S. Myhra, R.L. Segall, R.StC. Smart, P.S. Turner: J. Am. Ceram. Soc. **68**, 64 (1985)
3.13 A.F. Moodie, C.E. Warble: J. Cryst. Growth **10**, 26 (1971)
3.14 D. Cherns: Phil. Mag. **30**, 549 (1974)
3.15 K. Kambe, G. Lehmpfuhl: Optik **42**, 187 (1975); G. Lehmpfuhl, Y. Uchida: Ultramicroscopy **4**, 275 (1979)
3.16 M. Klaua, H. Bethge: Ultramicroscopy **11**, 125 (1983)
3.17 K. Takayanagi, Y. Tanashiro, M. Takahashi, S. Takahashi: J. Vac. Sci. Technol. **A3**, 1502 (1985)
3.18 L.D. Marks, D.J. Smith: Nature **303**, 316 (1983); D.J. Smith: Surf. Sci. **178**, 462 (1986)
3.19 L.D. Marks: Surf. Sci. **139**, 281 (1984)
3.20 For reviews of REM, see K. Yagi: J. Appl. Cryst. **20**, 147 (1987); K. Yagi: Electron microscopy of surface structure, Adv. Opt. Elec. Microsc. **11**, 57 (1989)

3.21 N. Osakabe, Y. Tanashiro, K. Yagi, G. Honjo: Japan J. Appl. Phys. **19**, L309 (1980); Surf. Sci. **109**, 353 (1981)
3.22 Y. Tanishiro, K. Takayanagi: Ultramicroscopy **31**, 20 (1989)
3.23 T. Nagatani, S. Saito: Proc. 11th Int. Conf. Elec. Microsc., Kyoto, Japan (1986) 2101; K. Ogura, M. Kersker: Proc. 46th EMSA (1988) p. 204
3.24 E. Bauer: Ultramicroscopy **17**, 51 (1985); E. Bauer, M. Mundschau, W. Swiech, W. Telieps: Ultramicroscopy **31**, 49 (1989)
3.25 G.G. Hembree, P.A. Crozier, J.S. Drucker, M. Krishnamurthy, J.A. Venables, J.M. Cowley: Ultramicroscopy **31**, 111 (1989)

4. Sputter Depth Profiling

B.V. King

With 13 Figures

The understanding and modification of surface properties of materials often requires a detailed knowledge of the spatial distribution of specific elements in both the surface plane and as a function of depth normal to the surface. This chapter will concentrate on ways of analyzing depth distributions of elements, up to a few microns deep, using ion beam sputter profiling. Sputter profiling uses the combination of a surface sensitive analytical technique, such as SIMS, LEIS, AES, XPS or SNMS [4.1], together with the continuous exposure of a new surface by ion beam sputtering. The sputtering ion beam typically is Ar^+, O_2^+ or Cs^+ with energy in the range 1–20 keV, although other inert gas or liquid metal, Ga^+ or In^+, ion sources are also used, mostly for imaging. Ar^+ is most commonly used with AES and XPS since it does not form compounds with target constituents and so does not significantly alter bulk atomic concentrations. O_2^+ and Cs^+ are favoured for SIMS since they increase secondary ion yields and reduce matrix effects. However, this is at the price of reduced sputter rates and profile distortion though compound formation and segregation. Profiling rates of up to 2 μm/h can be achieved although 0.1 μm/h is more typical.

Sputter profiling has three advantages over other methods of elemental depth profiling. Firstly, sputter profiling uses the strengths of the particular analysis technique employed (e.g. the high sensitivity of SIMS, lateral resolution of AES or the chemical identification of XPS). Secondly, sputter profiling has excellent depth resolution. Other techniques, such as RBS or NRA, which sample the whole near surface region simultaneously, have comparable depth resolutions at the surface to sputter profiling (a few nm), but the depth resolutions of RBS and NRA degrade rapidly with depth and the elemental distribution within the sample. Thirdly, sputter depth profiles are not subject to unavoidable elemental interferences from different depths.

Balanced against the above advantages are the problems of establishing depth and concentration scales. The characteristics of the various surface analytical techniques have been discussed in previous chapters. Their treatment in this chapter will be restricted to their influence on the practice of sputter profiling and the optimization of technique.

4.1 Analysis of a Sputter Depth Profile

The task for the experimentalist is to deduce the structure of the target from the measured depth profile. For example, Fig. 4.1 shows a SIMS depth profile of a 1000 Å Cr/Zr multilayer structure deposited onto a Si substrate. Analysis of this depth profile requires the establishment of accurate concentration and depth scales as well as an appreciation of whether the features in the depth profile correspond to the structures in the target or to artifacts of the sputter profilign process. For example, the Si signal rises abruptly after 400 s sputtering even though the Cr and Zr signals have not fallen. This may be due to real Si diffusion into the Cr/Zr thin film or to an artifact of the sputter profiling process. This chapter will provide a summary of possible artifacts and their effect on depth resolution.

4.1.1 Calibration of the Depth Scale

In Fig. 4.1 the elemental concentrations are presented as a function of the sputtering time. An accurate depth scale for the profile is then obtained by either (i) measuring the depth of the eroded crater using a surface profilometer or interferometry or (ii) calculating the eroded depth, z, from the sputter time. Generally both measurements are compared. In the second method, the depth sputtered during a depth profile is related to the product of the ion beam current density J (Am^{-2}), the sputtering time t (s) and the sputter yield Y (atoms ion^{-1}) of the sample by

$$z = Y J M t / 1000 e \varrho N_A n \tag{4.1}$$

where e is the electronic charge, M is the target molecular weight, ϱ is the target atomic density (kg/cm^3), N_A is Avogadro's number and n is the number of atoms

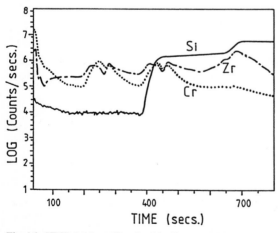

Fig. 4.1. SIMS depth profile of a thin film comprising 3 alternate layers of Cr and Zr deposited onto a Si substrate. The total film thickness is 1000 Å

in the target molecule. This method is accurate when Y does not change significantly during the analysis and so is most suitable for dilute ($< 10\%$) impurity concentrations in a uniform matrix. The Cr/Zr multilayer shown in Fig. 4.1 comprises layers of pure Cr and Zr and so the sputter yield should change markedly during analysis. Depth quantitation using (4.1) is then difficult. Care should also be taken in using this method near the surface of even elemental targets since the sputtering rate may change within a depth of the order of the ion range because of the implantation of the sputtering ions [4.2] or surface contamination. The two main areas limiting the accuracy of (4.1) are the measurement of J and unreliable data for Y even for suitable samples. The current density, J, cannot be found from simple target current integration because the emission of a considerable fraction of charged secondaries gives rise to large errors, typically a factor of 2 for keV Ar^+ on metals. For clean targets the contribution of secondary and scattered ions is small so the dominant contribution comes from secondary electrons with energies in the range 30–50 eV. These may be repelled back onto the target by biassing the target positively, +100 V, or collected in a Faraday cup, either mounted on the target holder or retractable for separate insertion into the ion beam. In these ways accuracies of about 10%, limited by incomplete suppression or unknown Faraday cup aperture sizes, are achievable [4.3].

The sputter yield, Y, is estimated either theoretically, from available compilations [4.4] or by computer simulation [4.5]. The theoretical predictions are based on linear cascade theory which gives a sputter yield Y_i for an ion i with energy E_i of the form [4.6, 7]

$$Y(E_i, \theta_i) = K_{it}\, S_n(\xi) f(\theta_i)/U_0 \tag{4.2}$$

where $\xi = E_i/E_{it}$ and E_{it} (in keV) and K_{it} [4.8] are scaling constants depending on the atomic numbers and masses of the ion and target atoms

$$E_{it} = \left((1 + M_i/M_t) Z_i Z_t \left(Z_i^{0.66} + Z_t^{0.66}\right)^{0.5}\right)/32.5 \tag{4.3}$$

$$K_{it} = \left((Z_i Z_t)^{0.833}\right)/3 \quad \text{for} \quad 1/16 < Z_t/Z_i < 5 \tag{4.4}$$

and U_0 is the surface binding energy in eV. U_0 is usually taken as the sublimation energy [4.9]. The reduced nuclear stopping cross-section, $S_n(\xi)$, has been estimated [4.10] by

$$S_n(\xi) = 0.5 \ln(1 + \xi)/\left(\xi + (\xi/383)^{0.375}\right) \tag{4.5}$$

whilst the angle of incidence dependence $f(\theta_i)$ is given by

$$f(\theta_i) = \cos^{-n}(\theta_i) \tag{4.6}$$

for $\theta_i < 80°$ with $n => 1$. For angles of incidence less than 45°, $n \simeq 1$ and $J \propto \cos\theta$ so that the erosion rate does not vary significantly with angle. Results of calculations using these formulae are shown in Fig. 4.2 for normal incidence Ar^+, O^+ and Cs^+ irradiation of Si, GaAs, Cu and Ta. The results of these formulae

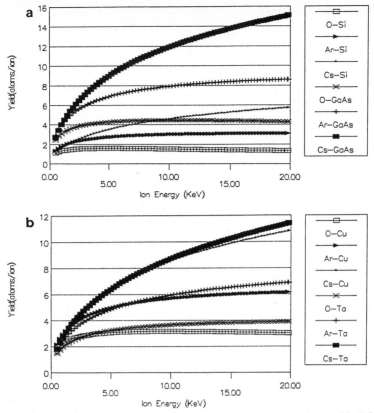

Fig. 4.2a,b. Sputter yields as calculated from (4.2–5) for normal incidence O^+, Cs^+ and Ar^+ bombardment of (**a**) Si and GaAs (**b**) Cu and Ta

are accurate to within a factor of two for typical sputtering energies and species but break down for very low ion energies (as used in SNMS), for light ions, when the energy transferred is insufficient to initiate a collision cascade or for high energy heavy ions when interactions occur between moving atoms in the cascade. They also underestimate the sputter yield of insulators when energy deposited into electronic excitation and ionization may lead to atom displacement and sputtering [4.11]. Sputter yields, energy deposition and ion ranges may also be calculated by computer simulations. The best known of these is TRIM, a Monte Carlo simulation for amorphous targets [4.5, 12].

The depth scale across an interace between two pure elements is, to a first approximation, given by the separate depth scales in the individual layers, where the interface is determined by the point at which the concentration falls to 50%. This analysis can, however, be difficult for thin layers or layers where the concentrations cannot be accurately determined. In Fig. 4.3a the Ti concentration, as measured by AES, never reaches 100% because the depth resolution of the depth profiling is comparable to the thickness of the individual Ti layers. A first depth

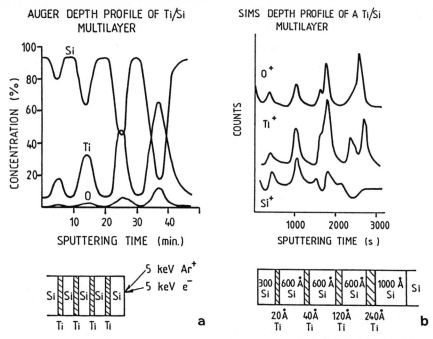

Fig. 4.3a,b. Sputter depth profiles of the Ti/Si multilayer shown using (a) AES and (b) SIMS

calibration would be found by assuming that the sample was pure Si and then estimating a total Ti thickness in each layer from the integrated Ti concentration in each layer. It is difficult to similarly treat a SIMS depth profile of the same structure (Fig. 4.3b) since unambiguous Ti and Si concentration scales cannot be assigned.

Sputter yields may also alter during depth profiling due to reactions between the ion beam and target. For example, oxides are generally formed when oxygen ion beams or oxygen flooding (inert gas bombardment at elevated oxygen partial pressures) are used during profiling, with a consequent drop in sputter yield. Figure 4.4 shows that the time to sputter through a Ti film increases by about factor of 3 when oxygen flooding is used during Ar^+ sputtering, due to the formation of TiO_2. In this case the sputter yield of TiO_2 is approximately equal to that of Ti so that the decrease in sputter rate is due the 3 fold increase in the number of atoms to be sputtered. Sputter yields for oxides [4.13] and nitrides are in most cases similar to or higher than for the corresponding metals. Exceptions are Al_2O_3 and MgO where the ratio $Y_{oxide}/Y_{metal} = 0.5$ and 0.2 respectively for 10 keV Kr^+ bombardment. The sputter yields of Si and SiO_2 are similar, ranging from about 0.5 for 1 keV Ar^+ to about 1.2 for 10 keV ar^+ bombardment. The sputter yield may also be increased if volatile compounds are formed, so called reactive ion etching [4.14]. F^+ bombarded Si has, for example, a higher sputter yield than that for Ar^+ bombardment since SiF_4 is formed and evaporates.

Sputtering of multicomponent targets presents even more difficulties in quantitation due to the uncertainty in Y and its variation with ion dose. A good review of the sputtering of multicomponent materials has been given by *Betz* and *Wehner* [4.15]. Briefly, the sputter yield, Y, of a metal alloy $A_x B_{1-x}$ or multilayer of the same average composition is assumed to be related to the elemental yields Y_A and Y_B (with $Y_B > Y_A$) by

$$xY_A + (1 - x)Y_B < Y < Y_B . \tag{4.7}$$

There are exceptions to the above rule. For example, 6 keV Xe^+ bombardment of 60% Ag-Au alloy gives a sputter yield of 17.2 [4.16] which is higher than the component sputter yields of 16 for Ag and 13.8 for Au. For more detail on many of the above aspects, the reader is directed to an extensive review of quantitative sputtering [4.17].

4.1.2 Calibration of the Concentration Scale

The signal measured in any one of the surface techniques used with sputter profiling is related to target elemental concentrations by factors which have been discussed in other chapters. In summary the counts, n_i, measured in time t for species i are given as

$$n_i = At\gamma c_i \chi_i \tag{4.8}$$

where A is the analyzed area (cm^{-2}), γ is the instrument transmission, c_i is the species concentration (cm^{-3}) and χ_i is the cross-section for production of the species i above the surface

$$\chi_i = JY\chi_{\text{ion}}/N \qquad \text{for SIMS} \tag{4.9}$$
$$\chi_i = dI\chi_{\text{el}} \qquad \text{for AES or XPS} \tag{4.10}$$

where J is the sputter ion current density (cm^{-2}), Y is the sputter yield, N is the atomic density (cm^{-3}), χ_{ion} is the cross-section for secondary ion production, d is the Auger or XPS information depth, I is the primary electron or photon flux (s^{-1}) and χ_{el} is the cross-section for Auger electron or photoelectron production. In practice SIMS is more sensitive than AES or XPS because of the high secondary electron background in the latter techniques.

In sputter depth profiling, quantification is often difficult since the sensitivity factors, Y, χ_{ion}, d and χ_{el}, generally change with sputter time. For example, in Fig. 4.3, depth profiling using AES shows four peaks in the Ti^+ concentration whereas a SIMS profile shows 5 Ti peaks which do not correspond to dips in the Si^+ signal. The two peaked structure from the deepest Ti layer is characteristic of SIMS profiles of concentrated layers (see, for example, the Cr peaks in Fig. 4.1) and is due to the dependence of secondary ion yields on the matrix. In the above case, the Ti rich layer has been found to contain $TiSi_2$ by Auger lineshape analysis. The double peaked structure then results from the difference in yield of

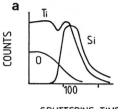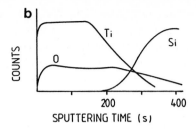

Fig. 4.4a,b. SIMS sputter depth profiles of a Ti layer deposited onto a Si substrate. The ion beam was 5 keV Ar$^+$ at an incidence angle of 45°. The analyses were performed at oxygen partial pressures of **(a)** 10^{-8} Pa and **(b)** 2×10^{-6} Pa

Ti$^+$ sputtered from a TiSi$_2$ matrix or from a predominately Si matrix. The presence of peaks in Si$^+$ signal at the position of the shallower Ti layers is probably due to oxygen enhancement of the Si$^+$ yield. Oxygen ion bombardment or oxygen flooding during inert gas bombardment would reduce these SIMS matrix effects by maintaining a relatively constant oxygen concentration throughout the depth profile. For example, in Fig. 4.4, the Ti$^+$ profile shows a dip then peak at the Ti-Si interface when profiled by Ar$^+$. When oxygen flooding is used the Ti$^+$ is shown to vary smoothly across the deposited film as expected. The Si$^+$ profile also corresponds better to the known distribution when flooding is used.

Quantification of the concentration is not only related to the matrix effects and interferences referred to above and in other chapters but is also determined by factors related to ion sputtering. The most important are ion beam induced compositional changes, sample contamination and preferential sputtering.

For AES and XPS, ion beam induced loss or gain of elements at the surface or in the bulk of the target can markedly affect bulk quantification. For example, ion irradiation causes loss of O from Ta$_2$O$_5$ to form a mean composition of Ta$_2$ [4.18]. Similar losses have been seen for HfO$_2$,Nb$_2$O$_5$,TiO$_2$ but not for Al$_2$O$_3$ [4.19].

Contamination of the sample may arise from i) impurities in the ion beam, ii) adsorption from the residual gas in the analysis chamber and iii) redeposition of previously sputtered material onto the target during profiling. Implantation into the target of impurities in the ion beam may cause erroneous depth profiles. These impurities may be ions which pass magnetic mass separation in the ion beam line, e.g. ArH$^+$ during depth profiling using Ar ion beams, or neutrals formed in the beam line and transported with the ion beam. To minimize significant sample contamination from adsorption of molecules from the chamber vacuum, the incident ion flux J (ions cm^{-2} s^{-1}), must be greater than ($10\gamma C/Y$) where C is the arrival rate of the gas molecules, given by

$$C = 2.6 \times 10^{24} \, P/MT \qquad \text{(particles cm}^{-2}\text{ s}^{-1}\text{)} \qquad (4.11)$$

and P, M and T are the pressure (in Pa), molecular mass of the gas and temperature respectively and γ and Y are the sticking probability and resputtering yield of the adsorbate [4.20]. A flux of 6×10^{-6} A cm^{-2} s^{-1} is then needed to avoid target contamination for a chamber pressure of 10^{-7} Pa with unity sticking probability. This flux corresponds to an average erosion rate of $1\,\text{Å}\,\text{s}^{-1}$. Redeposition of sputtered atoms into the analysis crater occurs due to gas scattering or, more commonly, resputtering from nearby surfaces. The high chamber pressures $(10^{-3}$–10^{-4} Pa) associated with oxygen flooding for SIMS as well as the use of non-differentially pumped ion guns in AES and XPS enhances these memory effects with ions or atoms being scattered as they leave the surface and subsequently redeposited into the crater. Such memory effects also occur if surfaces of ion lenses, for example, are within about 1 cm of the sample so that sputtered material is collected on the surface and redeposited into the crater. The latter effect is particularly severe for quantitative trace element analysis in semiconductors and can only be overcome by moving the analyzer further from the target or having a method of replacing contaminated surfaces.

Preferential sputtering is defined as the difference between the composition of the flux of the sputtered particles and the composition of the outermost layers of the sample. Consider a homogenous binary alloy sample with bulk compositions, c_A^b and c_B^b for the constituents A and B. The equilbrium sputtered fluxes of A and B from such a target, which are proportional to $Y_A c_A^s$ and $Y_B c_B^s$ respectively, must have proportions of A and B identical to the bulk. Then

$$c_A^s / c_B^s = \left(c_A^b / c_B^b\right) Y_A / Y_B \qquad (4.12)$$

at equilibrium. As a result the surface composition of A would change from the bulk value to the value in (4.12) as sputtering proceeded. The result of preferential sputtering is that transients occur in measured profiles of surfaces and interfaces. In Fig. 4.5 $Y_{AG} > Y_{Ta}$ so that Ag is depleted at the surface with sputtering as Ta is enriched. These transients may however also be caused by other processes – preferential recoil implantation of light elements or radiation enhanced surface

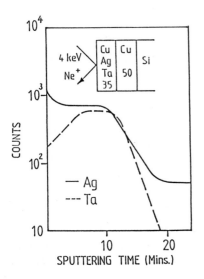

Fig. 4.5. LEIS sputter depth profile using $4\,\text{keV}$ Ne^+ at $45°$ incidence of a Cu/Ta/Ag thin film deposited onto a Si substrate. There is an initial transient in the signal from Ag and Ta due to preferential sputtering. The width of the transient is about twice the projected range of the sputtering ion beam

segregation. Preferential recoil implantation is seen for alloys with large mass differences between the components, i.e. $m_A/m_B > 5$. The depth over which this effect is evident if of the order of twice R_p. Surface segregation will be further discussed in a later section.

The above artifacts can be mostly overcome by the use of standards for calibration of the concentration and depth scales. For SIMS depth profiles implanted standards are used. If the implant is laterally homogeneous, both in implant dose and depth distribution [4.21], agreement to within 10% can be obtained for analysis of B in Si by different SIMS instruments. Thin film standards for calibration of depth scales in AES have also been developed, e.g. by National Physical Laboratory [4.22] and National Bureau of Standards [4.23, 24].

4.2 The Depth Resolution of Sputter Profiling

Optimization of the depth resolution is of great importance in accurate profiling so that, in particular, artifacts can be identified. The depth resolution of sputter profiling is determined by a combination of instrumental effects, effects related to the analysis technique (information depth), ion sputtering surface effects (preferential sputtering, surface topography) and bulk effects (ion mixing and enhanced diffusion and segregation). It should be noted that sputter profiling is a high ion fluence process. Other techniques, such as static SIMS, rely on excellent sensitivity to obtain measurements with very low ion doses. In that case only fractions of a monolayer of the sample surface are sputtered so monolayer depth resolution can be attained. If the dose is increased, to measure subsurface concentrations, the depth resolution is degraded due to the same processes as for conventional depth profiling.

4.2.1 Specification of the Depth Resolution

The depth resolution is generally described in terms of the measured width of an interface between two dissimilar layers. Figure 4.6a compares the ideal profile, which is rectangular in shape, with the measured profile which is generally represented as an error function. The interface width, Δz, is usually defined as the interval where the intensity drops from 84% to 16% of the maximum signal. This is equivalent to two standard deviations (2σ) of the error curve. If the layers are not thick then Δz will be of the same order of magnitude as the layer thickness, d (Fig. 4.6b). In the case illustrated, the variation in concentration, ΔC, is about half the peak height. This corresponds to Δz equal to 80% of the layer thickness. As Δz is increased ΔC will decrease [4.25] until effectively all resolution is lost at $\Delta z = 1.5d$. If the layer thickness is further reduced we have a marker geometry (Fig. 4.6c). Calculated [4.26] and measured [4.27] sputter profiles indicate that the measured profile of B is not gaussian but is asymmetric about the original interface with an extended tail at larger depths. The peak position is moved

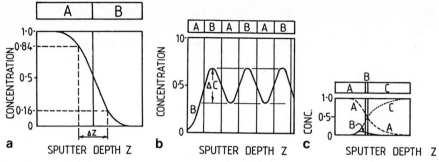

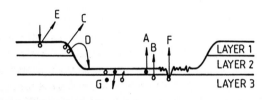

Fig. 4.6. Schematic representation of the depth profiles of elements A, B and C for (a) an AB bilayer (b) an AB multilayer and (c) a B maker layer sandwiched between A and C

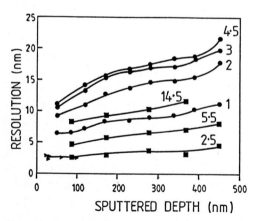

Fig. 4.7. Summary of the mechanisms which can lead to a loss of depth resolution in sputter profiling. (A) ejection of particles from the surface of a smooth crater; (B) subsurface ejection due to a larger escape depth; (C) analysis of material sputtered from the crater walls; (D) redeposition from crater walls into the analyzed crater; (E) analysis of material sputtered by neutrals in the ion beam; (F) ejection of particles from a rough crater; (G) atomic mixing within the target

Fig. 4.8. The depth resolution Δz as a function of sputtered depth z for (●) 1, 2, 3 and 4.5 keV Ar$^+$ irradiated NiCr mulitlayers [4.29] (■) 2.5, 5.5 and 14.5 keV Cs$^+$ irradiated GaAs [4.30] (▶) SiO$_2$/Si [4.31]

106

towards the surface with respect to the original marker position. In this case, the Δz cannot be strictly used with equations to specify the resolution, although it is a useful approximation. Rather, beyond the marker depth, the concentration decreases exponentially with distance

$$C \propto e^{-x/\delta} . \tag{4.13}$$

Optimization of the depth resolution requires a knowledge of all the factors which influence this quantity, illustrated in Fig. 4.7. They include instrumental effects (C, D, E), redistribution of material onto the surface (B) and within the bulk (G), and, finally, the formation and analysis of nonuniform surfaces (F).

Assuming that the individual contributions Δz_i are mutually independent and each results in broadening of a delta function initial distribution to a gaussian then the total depth resolution is given by $\Delta z^2 = \sum(\Delta z_i^2)$. Compilations of many Δz values as a function of sputtered depth, z, have been presented [4.28] for argon ion sputtering using AES and SIMS. The depth resolution, Δz, ranges from less than 1 nm for $z = 2$ nm to about 50 nm at $z = 5$ microns. The trend of the data indicates that $\Delta z \propto z^{0.5}$ for a wide range of z. However, differences in Δz for different matrices indicate changes in the relative importance of different factors affecting depth resolution. Data for Ni/Cr, GaAs/GaAlAs and Si/SiO$_2$ are presented in Fig. 4.8. For 3 keV Ar$^+$ irradiation of Ni/Cr multilayers, for example, Δz is much larger than the ion projected range, R_p, of 26 Å. In contrast, Δz for 14.5 keV Cs$^+$ irradiation of GaAs and 1 keV Ar$^+$ irradiation of Si are approximately equal to their respective values of R_p, 80 Å and 20 Å.

4.2.2 Instrumental Factors Determining the Depth Resolution

Depth resolution can be lost due to instrumental problems including those due to (i) the ion beam, (ii) the instrument geometry and stability and (iii) sample contamination.

Ion beam effects can be further subdivided into crater edge effects, electron stimulated desorption during AES and surface charging of insulating targets. By far the most important of these factors in determining the depth resolution is the flatness of the ion beam crater over the analysis area. Normally the ion beam is rastered and the signal electronically gated to accept only the signal from the central 4–20% of the crater so as to avoid analyzing the crater edges. Checkerboard gating is also used in more recent SIMS instruments. In this technique, the signal from the central crater region is divided into 8 × 8 or more pixels. The depth profiles from the individual pixels can be excluded from the total profile if they correspond to contaminated regions within the crater [4.32]. In conjunction with the minichip technique, where the ion beam is rastered over the entire small sample (typically 0.5 × 0.5 mm^2) to avoid crater edges, dynamic ranges of over 7 decades can be achieved [4.21]. In AES and XPS, the analysis area is determined by the size of electron or X-ray beams. For AES, since the analyzed area determined by the incident electron beam is typically only a few microns in diameter, the crater depth is generally uniform over the analyzed area so crater edge effects are small. They can, however, be significant in XPS profiling since the analyzed area is typically much larger. In general the depth resolution due to crater formation is gaussian with an interface broadening typically given by $\Delta z = az$ where a is 0.2–5% for SIMS data and 1% for AES. A crater edge effect is then most likely responsible for the rise in the Si signal seen in Fig. 4.1. The ion beam used for this profile was found to have two lobes which sputtered at twice the rate of the central beam. Si^+ sputtered from these side craters was detected even though electronic gating should have selected only the central 5% of the sputtered ions.

Non-uniform craters may also result from the impact of neutrals which are formed along the beam line and hit the sample as an unfocussed beam which is not affected by the rastering of the final deflection plates. Formation of neutrals may occur by ion scattering at apertures. However, the largest contribution in HV systems is found to come from the interaction of the primary ion beam with the residual gas [4.33]. Neutral beam contamination is particularly important in SIMS profiling since the neutrals can produce a signal and degrade the dynamic range which is a feature of SIMS depth profiles. Secondary ions liberated from outside the raster scanned area by the energetic neutrals will be recorded if they originate from within the field of view of the analysis optics [4.34]. The region around the crater is also likely to contain higher oxygen concentrations than inside the crater so the ionization probability and hence SIMS yield would increase. *Wittmaack* [4.35] has estimated that to achieve a dynamic range of 10^6 in depth profiling the pressure, P, of residual gas that is contained in the length, L, of beamline which can be seen by the target should correspond to an upper limit for PL of 4×10^{-6} cm Pa. This implies that a pressure lower than about 10^{-7} Pa is required for profiling with a high dynamic range in most analysis systems. Figure 4.9 shows an amalgamation of different experimental results for the influence of instrumental effects on SIMS depth profiles of B implanted into Si.

The other beam effects, electron stimulated desorption and beam induced charging, are most important for insulating samples. The removal of sample atoms by the electron beam during AES profiling can cause a crater to form in the middle of the ion beam crater. For a 1% resolution in Δz due to electron stimulated desorption effects, the ion beam diameter should be greater than 50 μm for an electron beam current of 0.25 μA [4.37] when stable oxides are profiled. For many organic compounds the effect is increased whereas for metals it is reduced. Ion beam induced sample charging causes dramatic changes in the energies of secondary ions and so in apparent yield of secondary or scattered ions and electrons. It can also cause field assisted migration of mobile species like Na within the sample. Surface charging is usually overcome by flooding the target with low energy electrons.

One of the other instrumental contributors to a loss in depth resolution, sample contamination has been previously discussed. The final factor, instrument geometry and stability, is small for properly designed instruments. For example,

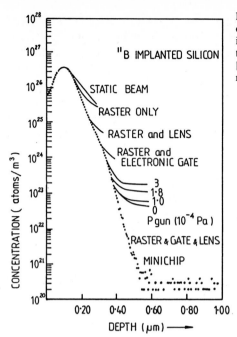

Fig. 4.9. Combination of various depth profiles of B implanted into Si showing the decrease in the background signal caused by (i) ion rastering, electronic gating and extraction lenses [4.36] (ii) residual gas pressure [4.33] and (ii) minichip sample preparation [4.32]

only when the misalignment between the ion and electron beams in an AES instrument is greater than 0.8 times the variance of a static gaussian ion beam does the loss in depth resolution become significant in comparison to other factors [4.38].

4.2.3 Surface Effects Determining the Depth Resolution

These include the development of surface topography, preferential sputtering and the escape depth of detected particles. Of these, the development of surface topography is the most important in determining the overall depth resolution especially for polycrystalline metals after long sputter times.

Initially surface roughness occurs due to the statistical process of removing atoms by sputtering. This effect stabilizes rapidly to give a resolution of about 1 nm. For higher ion doses roughening increases even in almost contaminant free conditions. For example, conical forests a few microns high develop on the (11, 3, 1) Cu surface on extended bombardment [4.39]. For this surface it is thought that local variations in Y, caused by subsurface defects and dislocations initiated by the ion cascade, are responsible for the initiation of roughness. The roughness increases due to this processes and the variation in Y along different ejection directions [4.40]. For polycrystalline metals the resolution Δz has been empirically found to be

$$\Delta z = az^{0.5} \tag{4.14}$$

where $a = 0.86 \pm 0.22$ and Δz and z are in nm [4.25]. Δz has also been found to be relatively insensitive to the initial surface roughness for initially rough surfaces [4.41]. Although there is a large variety of surface structures cones are generally aligned with the ion direction so that sample rotation during profiling lessens the loss of resolution due to roughening. The use of reactive gas ion beams can also improve resolution by forming amorphous compounds on the surface which exhibit no preferential ejection directions. Use of liquid metal ion beams can also lead to lower roughening [4.42]. However it is difficult to make general statements about the extent of experimental surface roughening, so it is preferable to measure the roughness after profiling.

The ultimate depth resolution of sputter profiling is determined by the information depth of the particular surface analysis techniques used. This contribution to the depth resolution, typically up to a few monolayers, is however usually small in comparison to other sources even at the start of a depth profile. For LEIS, the signal comes from the top monolayer if inert gas ions are scattered, or a few monolayers for electropositive ions like Li^+. The majority of secondary ions detected in SIMS escape from the top 1–2 ML [4.43]. Typically the attenuation lengths of Auger electrons are 5–50 Å. They have been given more precisely [4.44] as

$$\lambda = A_i E^{-2} + B_i E^{0.5} \quad nm \tag{4.15}$$

where $A_i = 538a$, $2170a$ and $4900/\varrho$ and $B_i = 0.41a^{1.5}$, $0.72a^{1.5}$ and $110/\varrho$ for elements, inorganic compounds and organic compounds respectively. The parameter a is the average monolayer thickness (in nm) defined by

$$a^3 = 10^{24} A/\varrho n N_A \tag{4.16}$$

where ϱ is the bulk density ($kg\,cm^{-3}$), A is the atomic or molecular weight and n is the number of atoms in the target molecule.

4.2.4 Bulk Effects Affecting the Depth Resolution

These are processes driven by the energy deposited into the sample by the ion beam as well as the implanted ions and include atomic mixing and radiation enhanced thermal diffusion and segregation.

All the processes which affect depth resolution except ion beam mixing may be circumvented by choice of target and ion species or by correct experimental procedure. However, the signal observed during sputter profiling is always affected by the radiation damage caused by the sputtering particle beam, since the target atom will be successively displaced, typically a few Å at a time, by many incident ion cascades before it is sputtered. The concentration, $C(x, z)$ of a species at a depth x below the surface after sputtering to a depth z is then different to the initial distribution of the species $C(x, 0)$. The measured depth profile, assumed proportional to the surface concentration, $C(0, z)$, is then a convolution

$$C(0, z) = \int C(x, 0)g(z, x)dx \qquad (4.17)$$

of the original depth distribution and a function $g(z, x)$ which depends on the amount of ion mixing. The effect of ion mixing is to smooth the profile where concentrations change rapidly and to bring the peaks toward the surface with respect to their original positions. The redistribution of atoms during sputter profiling leads to a broadening of a delta function impurity distribution into a gaussian impurity distribution at a certain depth x (if sputtering and matrix relocation is ignored) with variance [4.45]

$$\sigma^2 = 2D(x)t = 0.608\phi F_d R_c^2 / N E_d \qquad (4.18)$$

where $F_d x(eV/\text{Å})$ is the energy deposited into collisional processes at depth x, N is the atomic density (atoms/Å^3), (R_c^2/E_c) is a factor which only depends on the target (Å^2/eV), ϕ is the ion fluence (ions/Å^2) and $D(x)$ is the diffusion constant (Å^2/s) over an irradiation time t. The diffusion formalism allows a parameter $(Dt/\phi F_d)$ to be used as a measure of mixing efficiency – the rate of mixing per unit of deposited energy. Experimental values for $(Dt/\phi F_d)$ typically are in the range 1–200 Å^5/eV. Low values are found for targets of high cohesive energy in which the cascade energy density is low [4.46]. The lower limit also corresponds to results from analytic theories of ballistic mixing [4.45, 47]. The depth resolution Δz due to ion mixing may be found from $(Dt/\phi F_d)$ by solution of a diffusion equation together with associated boundary conditions [4.48]. It is found that Δz depends on the degree of preferential sputtering as well as $(Dt/\phi F_d)$ but that the depth resolution is not a sensitive function of $(Dt/\phi F_d)$. For example, for values of $(Dt/\phi F_d)$ in the range 10–200 Å^5/eV, λ varies from 50–80 Å for 4 keV Ne$^+$ irradiation of Cu at 45° incidence [4.49]. Calculations from ion mixing theory indicate that $\Delta z = 1.8\lambda$, so that Δz varies from 90–130 Å. The projected range of 4 keV Ne$^+$ in Cu is 45 Å which is less than the experimental values of Δz. For impurities in Si (Fig. 4.10) λ is found to be of the same order of magnitude as R_p. For the intermixing of Ge and Si, *Kirschner* and *Etzhorn* [4.53] also estimate Δz to be about 1.5 R_p. As a guide then, if ion mixing is dominant, Δz is about 1–2 R_p. Values of R_p calculated from TRIM are shown for common metals and semiconductors in Fig. 4.11.

In the absence of chemical or sputter effects incorporation of ions into the target will cause the apparent magnitude of mixing to decrease [4.26]. This should be the case for self-implantation. Since the interaction between the inert gas and the target atoms is small, it is reasonable to disregard the influence of inert gas implantation on measured depth profiles. Chemical reactions between different target elements and between implanted ions and target atoms will however alter depth profiles by either aiding or negating diffusional mixing. In general impurities with the high heats of reaction will preferentially segregate. *Hues* and *Williams* [4.51] in a study of the effect of an O_2 jet on ion mixing, found that impurities, like Ca, with higher oxygen affinities than Si, segregated to the surface whilst Ag with lower oxygen affinity than Si were displaced further into the bulk.

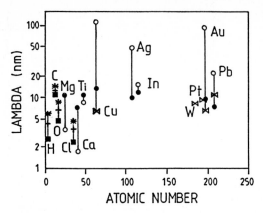

Fig. 4.10. Values of λ found from sputtering markers of the elements shown. The marker layers were either deposited onto Si substrates or sandwiched between deposited Si layers. For most impurities λ is in the range 2–10 nm (■, +, *) 4, 8, 12 keV Cs⁺, 2° [4.49]; (o, •) 8 keV Ar⁺, 38° with and without oxygen ambient [4.50];(⋈) 5 keV Ne⁺, 53° [4.51]

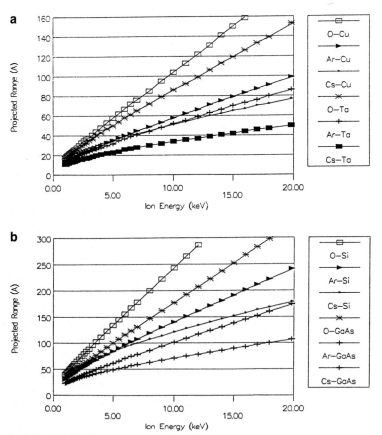

Fig. 4.11. Calculated projected ranges of O^+, Ar^+ and Cs^+ in Cu, Ta (**a**) and in Si and GaAs (**b**) as a function of ion energy

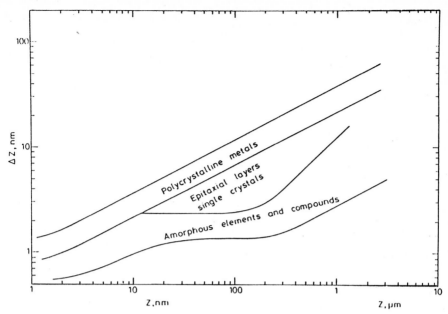

Fig. 4.12. Typical values of Δz for different targets as a function of the depth, z, eroded by keV Ar$^+$ [4.52]

4.2.5 How Can the Depth Resolution Be Minimized?

There are two possibilities:

i) reduce the ion range;

ii) reduce topographical effects.

Strategies for optimizing the depth resolution depend on which factors predominately determine the resolution. Figure 4.12 shows typical values of Δz for keV Ar sputtering of the target types shown. For amorphous materials, where topographical effects are minimized, Δz is determined primarily by atomic mixing but also by information depth, and for small values of z, the information depth can be minimized by the use of LEIS or SIMS. Δz for atomic mixing is primarily determined by the ion projected range as discussed above. In this case, the depth resolution Δz decreases with decreasing ion energy and increasing ion mass as a result of the decrease in the projected range. For Ge/Si interfaces [4.53] Δz decreases from 10 nm for 5 keV He$^+$ to 3 nm for 0.5 keV Xe$^+$. The deposited energy density, which is higher for heavier ions of the same energy, is of secondary importance, at least over the keV energy range typically used in sputtering. Thus SNMS, which uses low energy ion bombardment, has inherently good depth resolution. Lower values of R_p and thus depth resolution can also be obtained by using glancing incidence ($\theta < 70°$) ion irradiation. This has been seen both in 8 keV Cs$^+$ [4.54] and in 1 keV Ar$^+$ (Fig. 4.13) profiling.

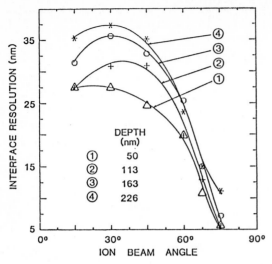

Fig. 4.13. Variation of Δz with ion beam angle to the target normal after sputtering to the indicated depths using 4 keV Ar$^+$ [4.29]

For metal films and after longer sputter times, topographical effects become important. The depth resolution is then minimized for smooth surfaces, which are best achieved by the use of active gases as the source of ions, especially for metal targets. Oblique angles of ion incidence (Fig. 4.13) can also reduce surface roughening since Fig. 4.12 indicates that roughening should dominate the resolution for polycrystalline metal films. If the resolution function $g(z, x)$ in (4.17) is known then the original depth distribution $C(x, 0)$ can, in principle, be found from the measured depth profile $C(0, z)$. When the initial impurity distribution is constant in the near surface regions where $g(z, x)$ is appreciable then (4.17) can be written as a convolution integral

$$C(0, z) = \int C(x, 0)g(z - x)dx \ . \tag{4.19}$$

The resolution function depends in general on all the contributions to the depth resolution i.e. mixing, escape depth, topography etc. If the resolution function can be approximated by a gaussian then the above equation can be solved by Laplace transforms [4.55], Fourier transforms or iterative procedures [4.25]. The procedure is simplified if $C(0, z)$ has the shape of an error function, i.e. $C(0, z) = \mathrm{erf}(z/\Delta z)$. In that case $C(x, 0) = \mathrm{erf}(z/\Delta z_0)$ where $\Delta z_0^2 = \Delta z^2 - \Delta z_g^2$ and Δz_g is appropriate parameter in the gaussian resolution function. To obtain the original profile the broadening, Δz, of the measured profile is then corrected using the value Δz_g from the resolution function.

4.3 Conclusion

The theoretical developments associated with the study of effects due to the ion beam probe and also effects due to ion irradiation in general suggest that a reasonable understanding of the processes is emerging. Ion irradiation with the deliberate aim of inducing mixing and recoil implantation is likely to be a very significant, practical form of surface modification technology. As a consequence, there is a need to develop the available models in greater detail, for subsequent use in systems of practical importance.

References

4.1   H. Oechsner: In *Thin Film and Depth Profile Analysis*, ed. by H. Oechsner, Topics Curr. Phys. Vol. 37 (Springer, Berlin, Heidelberg 1984) p. 63
4.2   J. Kirschner, H.W. Etzhorn: Appl. Surf. Sci. **3**, 251 (1979)
4.3   M.P. Seah: J. Vac. Sci. Technol. A **3**, 1330 (1979)
4.4   H.H. Andersen, H.L. Bay: In *Sputtering by Particle Bombardment I*, ed. by R. Behrisch, Topics Appl. Phys. Vol. 47 (Springer, Berlin, Heidelberg 1981) p. 145
4.5   J.P. Biersack, W. Eckstein: Appl. Phys. A **34**, 73 (1984)
4.6   P. Sigmund: Phys. Rev. **184**, 383 (1969)
4.7   P. Sigmund: Phys. Rev. **187**, 768 (1969)
4.8   P.C. Zalm: J. Appl. Phys. **54**, 2660 (1983)
4.9   K.A. Gschneidner, Jr.: Solid State Phys. **16**, 344 (1964)
4.10 W.D. Wilson, L.G. Haggmark, J.P. Biersack: Phys. Rev. B **15**, 2458 (1977)
4.11 J. Schou: Nucl. Instr. Meth. B **27**, 188 (1987)
4.12 J.F. Ziegler, J.P. Biersack, U. Littmark: In *The Stopping and Range of Ions in Solids* (Pergamon, Oxford 1985)
4.13 R. Kelly, N.Q. Lam: Radiat. Eff. **19**, 39 (1973)
4.14 J.W. Coburn, H.F. Winters: J. Vac. Sci. Technol. **16**, 391 (1979)
4.15 G. Betz, G.K. Wehner: *Sputtering by Particle Bombardment II*, ed. by R. Behrisch, Topics Appl. Phys. Vol. 52 (Springer, Berlin, Heidelberg 1983) p. 11
4.16 M. Szymonski, R.S. Bhattacharya, H. Overeijnder, A.E. de Vries: J. Phys. D **11**, 75 (1978)
4.17 P. Zalm: Surf. Interface Anal. **11**, 1 (1988)
4.18 E. Taglauer: Appl. Surf. Sci. **13**, 1980 (1982)
4.19 S. Hofmann, J.M. Sanz: J. Trace Microprobe Tech. **1**, 213 (1982)
4.20 J.F. O'Hanlon: *A Users Guide to Vacuum Technology* (Wiley, New York 1980) p. 125
4.21 S.M. Hues, R.J. Colton: Surf. Int. Anal. **14**, 101 (1989)
4.22 M.P. Seah, C.P. Hunt, M.T. Anthony: Surf. Int. Anal. **6**, 92 (1984)
4.23 J. Fine, B. Navinsek: J. Vac. Sci. Technol. A **3**, 1408 (1985)
4.24 D.E. Newbury, D. Simons: In *Secondary Ion Mass Spectrometry SIMS IV*, ed. by A. Benninghoven, J. Okano, R. Shimizu, H.W. Werner, Springer Ser. Chem. Phys. Vol. 36 (Springer, Berlin, Heidelberg 1984) p. 101
4.25 S. Hofmann, J.M. Sanz: In *Thin Films and Depth Profile Analysis*, ed. by H. Oechsner, Topics Curr. Phys. Vol. 37 (Springer, Berlin, Heidelberg 1984) p. 141
4.26 U. Littmark, W.O. Hofer: Nucl. Instr. Meth. **168**, 329 (1980)
4.27 B.V. King, D.G. Tonn, I.S.T. Tsong, J.A. Leavitt: Mater. Res. Soc. Symp. Proc. **27**, 103 (1984)
4.28 M.P. Seah, C.P. Hunt: Surf. Int. Anal. **5**, 33 (1983)
4.29 J. Fine, P.A. Lindfors, M.E. Gorman, R.L. Gerlach, B. Navinsek, D.F. Mitchell, G.P. Chambers: J. Vac. Sci. Tech. A **3**, 1413 (1985)
4.30 M. Gauneau, R. Chaplain, A. Regreny, M. Salvi, C. Guillemot, R. Azoulay, N. Duhamel: Surf. Int. Anal. **11**, 545 (1988)
4.31 R. Helms, N.M. Johnson, S.A. Schwarz, W.E. Spicer: J. Appl. Phys. **50**, 7007 (1979)
4.32 R.v. Criegern, I. Weitzel, J. Fottner: In *Secondary Ion Mass Spectrometry SIMS IV*, ed. by A. Benninghoven, J. Okano, R. Shimizu, Springer Ser. Chem. Phys. Vol. 37 (Springer, Berlin, Heidelberg 1984) p. 308

4.33   K. Wittmaack, J.B. Clegg: Appl. Phys. Lett. **37**, 283 (1980)
4.34   P. Williams, C.A. Evans, Jr.: Int. J. Mass. Spectrom. Ion Phys. **22**, 327 (1976)
4.35   K. Wittmaack: Radiat. Effects **63**, 205 (1982)
4.36   C.W. Magee, W.L. Harrington, R.E. Honig: Rev. Sci. Instr. **49**, 477 (1978)
4.37   C.G. Pantano, T.E. Madey: Appl. Surf. Sci. **7**, 115 (1981)
4.38   S. Duncan, R. Smith, D.E. Sykes, J.M. Walls: Surf. Int. Anal. **5**, 71 (1979)
4.39   G. Carter, B. Navinsek, J.L. Whitton: In *Sputtering by Particle Bombardment I*, ed. by R. Behrisch, Topics Appl. Phys. Vol.47 (Springer, Berlin, Heidelberg 1981)
4.40   H.E. Roosendaal: In *Sputtering by Particle Bombardment I*, ed. by R. Behrisch, Topics Appl. Phys. Vol.47 (Springer, Berlin, Heidelberg 1981)
4.41   A. Zalar, S. Hofmann: Surf. Int. Anal. **2**, 183 (1980)
4.42   F.G. Rudenauer: In *Secondary Ion Mass Spectrometry SIMS IV*, ed. by A. Benninghoven, J. Okano, R. Shimizu, H.W. Werner, Springer Ser. Chem. Phys. Vol.36 (Springer, Berlin, Heidelberg 1984) p.133
4.43   G. Falcone, P. Sigmund: Appl. Phys. **25**, 307 (1981)
4.44   M.P. Seah, W.A. Dench: Surf. Interface Anal. **1**, 2 (1979)
4.45   P. Sigmund, A. Gras-Marti: Nucl. Instr. Meth. **182/3**, 5 (1981)
4.46   R.S. Averback: Nucl. Instr. Meth. B **15**, 675 (1986)
4.47   U. Littmark: Nucl. Instr. Meth. B **7/8**, 684 (1985)
4.48   B.V. King, I.S.T. Tsong: J. Vac. Sci. Technol. A **2**, 1443 (1984)
4.49   B.V. King, S.G. Puranik, M.A. Sobhan, R.J. MacDonald: Nucl. Instr. Meth. B **39**, 153 (1989)
4.50   K. Wittmaack: Nucl. Instr. Meth. **209/210**, 191 (1983)
4.51   S.M. Hues, P. Williams: Nucl. Instr. Meth. B **15**, 206 (1986)
4.52   M.P. Seah: Vacuum **34**, 463 (1984)
4.53   B.V. King, S.G. Puranik, R.J. MacDonald: Nucl. Instr. Meth. B **33**, 657 (1988)
4.54   J. Kirschner, H.-W. Etzkorn: In *Thin Film and Depth Profile Analysis*, ed. by H. Oechsner, Topics Curr. Phys. Vol.37 (Springer, Berlin, Heidelberg 1984) p.103
4.55   K. Wittmaack: J. Vac. Sci. Technol. A **3**, 1350 (1985)
4.56   B.V. King, I.S.T. Tsong: Nucl. Instr. Meth. B **7/8**, 793 (1985)

5. SIMS – Secondary Ion Mass Spectrometry

R.J. MacDonald and B.V. King

With 22 Figures

Secondary ion mass spectroscopy (SIMS) is an ion beam analysis technique useful for characterizing the top few micrometres of samples. Primary ions of energy 0.5–20 keV, commonly O^-, Cs^+, Xe^+, Ar^+, Ga^+ and O_2^+, are used to erode the sample surface and the secondary elemental or cluster ions formed from the target atoms by the impact are extracted from the surface by an electric field and then energy and mass analyzed. The ions are then detected by a Faraday cup or electron multiplier and the resulting secondary ion distribution displayed as a function of mass, surface location or depth into the sample (Fig. 5.1).

SIMS as a technique for surface analysis has been available for thirty years or more. The pioneering work of researchers such as Honig, Liebl and Slodzian, to name but a few, suggested the potential of SIMS as an analytical tool, particularly in the area of surface composition analysis. In the last decade or so, recognition of the importance of the surface in many aspects of materials science and the growing importance of thin film processes and solid state electronics have all lead to a dramatic increase in the sophistication of surface analysis instrumentation, including SIMS. SIMS is probably the most sensitive technique

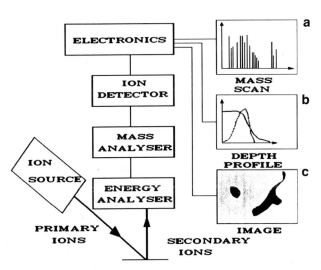

Fig. 5.1a–c. Schematic representation of the main components of a SIMS experiment. The signal due to the detected secondary ions may be presented as a function of (a) the ion mass, (b) the duration of ion bombardment or (c) the location of the surface

currently available for compositional analysis. For example, detection limits are 10^{13}–10^{18} atoms cm^{-3} for impurities in semiconductors. The technique is relatively simple in concept, is readily adapted to an "add-on" configuration and provides a large amount of information which at first sight is straightforward to interpret. It is however frustratingly difficult to relate the strength of the SIMS signal to the composition of the surface – SIMS has not yet been provided with a theory allowing absolute or relative quantification. Most quantitative analysis using SIMS will involve the use of well-characterized standards and even then the accuracy of compositional analysis done by comparison with a standard will be far less than the sensitivity of the technique for trace analysis.

We will attempt in this chapter to give a balanced view of SIMS. We will firstly summarize the general practice of SIMS (Sect. 5.1) and then discuss our understanding of the way in which the secondary ion is created and the factors which affect the magnitude of the yield of secondary ions. We will summarize the properties of secondary ions and indicate the way in which those properties influence the equipment designed to utilize secondary ions for compositional analysis (Sect. 5.2). We will consider some of the methods proposed to quantify the SIMS signal in terms of surface composition (Sect. 5.3). Finally we will consider the problems associated with SIMS analysis and provide some examples which will illustrate both the usefulness of SIMS as a tool and the complexity of interpretation of the information available (Sect. 5.4).

This review will not be exhaustive. For a fuller treatment the reader is referred to recent books dedicated to SIMS [5.1, 2], SIMS depth profiling [5.3] and static SIMS [5.4]. There are also proceedings of biennial conferences devoted to SIMS which contain papers on both the theory and practice of SIMS [5.5].

5.1 The Practice of SIMS

5.1.1 Overview

Secondary ions are ejected from a surface subject to bombardment by a primary ion by a process known as sputtering. The complex nature of the interaction between an incident energetic ion or neutral and a solid surface is shown schematically in Fig. 5.2. The incident ions, with an energy in the range 1–50 keV, generate collision cascades in the region of the surface of the solid. Within the cascade a small part of the momentum may be redirected towards the surface, providing sufficient energy for the ejection of atoms and molecules from the outermost surface layers. A small proportion of the ejected atoms or molecules may be in a charged state and remain in that state until detected in a mass spectrometer. The ejected ionized particles may be singly or multiply ionized, they may be atomic or molecular in structure and they may be positive or negative in charge state. These are the ions which provide the signal which is the basis of SIMS.

SIMS information may be presented in three ways. Firstly, secondary ion intensities may be collected as the mass of the secondary ions which are accepted

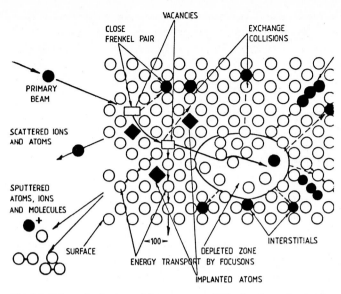

Fig. 5.2. Schematic diagram of the processes which take place after ion impact onto a surface. SIMS is concerned with analysis of the sputtered ions. The impact of one primary ion can cause many target atom displacements which affect the surface seen by subsequent ion impacts

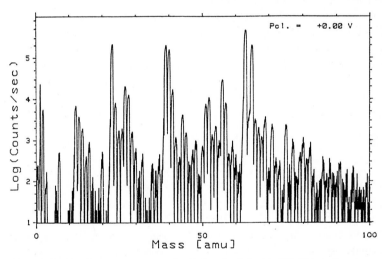

Fig. 5.3. Positive ion mass spectrum for Cu bombarded with 12 keV Ar$^+$. Prominent peaks are seen for Na, K, Ca, and the two isotopes of Cu

by the spectrometer is scanned. The resulting mass spectrum (Fig. 5.3) is typically obtained for low primary ion doses and is representative of the top monolayer of the surface (static SIMS). Secondly, the secondary ions from selected elements may be monitored as a function of sputter time (Fig. 5.4) as the primary ion beam erodes up to a few microns into the sample (dynamic SIMS). The secondary ions

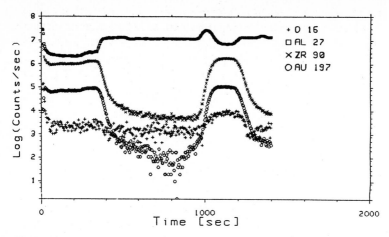

Fig. 5.4. Depth profile of O, Al, Zr and Au as a function of primary ion bombardment time for a thin film sample comprising two layers of mixed Zr, Au and Al separated by a layer of "pure" Al. Note that all signals are high initially, due to enhancement of the ion yields by an increased O concentration at the surface

detected in SIMS generally come from the top monolayer of the surface. The primary ion beam is then used for two purposes in dynamic SIMS: to eject secondary ions from the surface and to sputter the target, continuously exposing a new surface for SIMS analysis. In that way a profile of the target elemental concentrations is found as a function of depth into the target. The practice of depth profiling is considered further in another chapter. Thirdly, the secondary ion signal may be presented as a function of position on the surface in a similar way to scanning electron microscopy (Fig. 5.22).

5.1.2 Advantages and Disadvantages of SIMS

The major advantages of the SIMS technique are:

i) excellent sensitivity (< 1 ppm) and a wide dynamic range are found for most elements;

ii) the ability to distinguish different isotopes of the same element;

iii) the ability to obtain information on chemical bonding at the surface by monitoring molecular ions.

The primary disadvantage is the variation of detection sensitivities over orders of magnitude for different elements in one substrate and for the same element in different substrates. This may be overcome to a certain extent, as will be discussed later, by the use of reactive primary ions (O_2^+ or Cs^+) which enhance the yields of positive and negative ions respectively. In spite of this sensitivity variation, semi-quantitative analysis may be performed using standards in a static SIMS mode, i.e. with low primary ion doses. However during dynamic SIMS measurements additional problems may arise because the chemical nature of the surface, and thus secondary ion yields can be altered by primary ion implantation or changes in the target surface concentration, either from surface segregation of elements from the target bulk and from the vacuum system, or from large changes with depth in elemental concentrations in the original target. For example, in Fig. 5.4 the aluminium signal peaks after sputtering for 1000 s, due to the change in the elemental concentrations in the second impurity layer.

Other disadvantages of SIMS analysis include

i) the requirement in some SIMS analyzers that only one mass peak may be measured at any one time. Multi-element analysis then must be done sequentially by cycling the mass analyzer through the different peaks to be measured. This procedure is time consuming, especially if static conditions are required;

ii) the alteration of the target by radiation-induced processes during dynamic SIMS measurements. These include ion-beam-induced intermixing of target elements, radiation-induced thermal segregation, surface topography changes and redeposition of other material into the analyzed region. These aspects are dealt with more fully in Chap. 4 on depth profiling.

5.1.3 The Yield of Secondary Ions

The yield of secondary ions of a given element, ejected from a solid surface, will depend on a number of parameters. We could write

$$Y_A \, dE \, d\Theta = J_i S_A N_A \sigma_A P_A T \, dE \, d\Theta \,, \qquad (5.1)$$

where the symbols have the following meanings

$Y_A \, dE \, d\Theta$: yield of secondary ions of element A detected by the spectrometer
J_i: incident ion beam density (ions cm^{-2})
S_A: sputter yield of atoms of element A in the matrix comprising the surface layer of the target
N_A: surface density of atoms of element A (atoms cm^{-2})
σ_A: cross-section for ionization of an atom of element A as it is ejected from the surface
P_A: probability that the ion of element A, once formed, survives in that ionized state to be detected
T: transmission of the spectrometer system for ions of element A
dE: acceptance width in energy of the spectrometer system
$d\Theta$: acceptance angle of the spectrometer system.

We can consider each of these factors in turn.

a) The Incident Ion Beam Density. This is a function of the ion source and the experimental measurement to be made. Static SIMS measurements typically use Ar or Xe beams to avoid surface chemical modification by implantation. Atom beams are generally preferred to ion beams since they minimize surface charging which distorts secondary ion emisison by changing the accelerating fields used for secondary ion extraction. Static SIMS requires a very low primary beam density, perhaps 10^9 particles $cm^{-2} s^{-1}$ deposited over a few mm^2. The total dose in any experiment is normally kept below $10^{13} cm^{-2}$ to maintain static conditions. For static SIMS measurements of organic targets, where the object is to determine the mass of the organic molecules making up the target, low ion doses are required so that the organic molecules are not fragmented by prior ion bombardment before they are ejected from the surface and analyzed. If each ion impact affects an area of $10 nm^2$ then only 10^{12} impacts cm^{-2} will cause 10% of the top monolayer to be damaged. For a particle beam current of $1 nA cm^{-2}$, this damage threshold would be reached after 160 s.

In dynamic SIMS, or depth profiling mode, beam current densities of 0.1–$10 mA cm^{-2}$ are typical although the exact value used depends on the time required to profile the complete structure. For example, to profile Si to a depth of $1 \mu m$ in 20 min requires a current density of about $1 mA cm^{-2}$. Typically O_2^+, Cs^+, or Ar^+ beams are used with beam diameters of 1–$10 \mu m$. The best depth resolution however is obtained when the detected ions all come from the same depth i.e. when the eroded crater is perfectly flat. Typically this is achieved by scanning the primary ion beam over areas of about $250 \mu m \times 250 \mu m$ on the sample. The requirement of rapid depth profiling places constraints on the size of the ion beam used. For example, if the ion beam used in the above Si depth profile was scanned in this way, a beam current of $0.6 \mu A$ would be required to achieve the desired current density of $1 mA cm^{-2}$. Such a beam current could only be achieved for Ar^+, O_2^+ or Cs^+ ions by using a beam of at least $50 \mu m$ diameter.

Imaging SIMS may be performed using an ion microscope or imaging microprobe. In microscope mode, the spatial distribution of ions emitted from the surface is retained through the secondary ion transport optics and measured with about $1 \mu m$ resolution by a position sensitive detector or on a fluorescent screen. In microprobe mode, a well focussed primary ion beam is rastered over the surface, as in the dynamic SIMS mode discussed above, and the secondary ion emission from the small bombarded area measured as a function of the beam position. This mode is inherently inefficient compared to the microscope mode since only a small part of the surface is being detected at any one time. The liquid metal ion sources, often used for imaging SIMS, produce current densities of 1–$10 A cm^{-2}$ so that, even in microprobe mode, rapid ion images may be produced.

In most circumstances encountered, the density of the incident ion beam is such that the collision cascades initiated are not overlapping so that the sputtered secondary ion signal is linear with incident ion beam density. If, however, primary beams of molecular ions like O_2^+ are used, the molecule dissociates upon impact. The energetic atoms produced then initiate collision cascades in the target which can overlap in space and time. This will lead to nonlinear relationships between the incident ion beam density and the yield of secondary ions.

b) The Sputter Yield of Elements. The sputter yield of an element A, S_A, is related to the erosion rate, U, of its ion bombarded surface by

$$U = \frac{J_i S_A}{N_A} ,$$
(5.2)

where J_i is the ion current density and N_A is the elemental atomic density. The sputter yield of a target depends on the ion energy and species as well as the target atomic number, surface crystallinity and topography. However, theoretical sputter yields calculated for smooth amorphous targets generally give good agreement with values found from experimental depth profiles of single element targets. Therefore theoretical sputter yields can be profitably used to calculate erosion rates in dynamic SIMS measurements. This then allows the SIMS depth profile which is given in terms of sputter time (Fig. 5.4) to be recast in terms of sputtered depth.

A theory of the sputtering of atoms from the smooth surface of an amorphous solid consisting of a single atomic species has been well developed by *Sigmund* [5.6]. In this case, the sputter yield of atoms, S, from a target bombarded by ions of mass M_1 and energy E is proportional to the nuclear stopping power, $F_d(E)$ (i.e. the energy spent in elastic collisions between atoms.)

$$S = 0.042 \frac{\alpha(M_2/M_1)F_d(E)}{E_b} ,$$
(5.3)

where α is a function only of the ratio of the incident ion mass, M_1, to target atom mass, M_2 and E_b is the surface binding energy of a target atom to the surface. E_b is typically taken equal to the heat of sublimation of the target element. Energy spent in electronic excitation in most solids is not available for sputtering, though electronic excitation in some insulating materials may lead to atom ejection [5.7] and an increase in S by a factor of 10 or more. There do exist compilations of sputtering yield data for amorphous single element solids and for some compounds and alloys, e.g. [5.8, 9]. Empirical relationships also exist as discussed in Chap. 4. There are also a variety of computer codes, recently reviewed by *Andersen* [5.10], of which TRIM is probably the best known example [5.11], which can calculate many parameters associated with the ion-surface interaction and the collision cascade, including the sputtering yield. Their accuracy in reproducing experimental values for the sputtering yields of elemental targets is good, but there exists little experimental data for comparison in the case of multi-element matrices. In general, however, we can predict the sputtering yield of a given element in almost any target matrix to better accuracy than we can account for some other parameters in (5.1). In general, S increases with primary ion energy and mass to a maximum of 0.1–10 at a primary ion energy of about 10 keV or more. S also increases with increasing ion incidence angle, θ, up to about 70° to the surface normal. Over this range S is approximately proportional to $\cos^{-1} \theta$ for inert gas bombardment. However, this is not necessarily the case for active ion bombardment since for O_2^+ bombardment of Si use of a normal incidence beam leads to the formation of SiO_2 and a sputter yield of only 0.27 atoms/ion. For 60° incidence the yield increases by a factor of 10.

In general, the sputtering yield of an alloy may be approximated as the sputtering yield of the pure element A, multiplied by the fractional composition of A in the matrix. Thus, for an alloy target $A_a B_{(1-a)}$ a first approximation to the total sputter yield is

$$S = aS_A + (1-a)S_B . \qquad (5.4)$$

More details on sputter yields are included in Chap. 4.

c) The Surface Density of Atoms. We expect the ion yield of a given species of ion to be related to the surface density of atoms of that type. However the surface density of a given element may not correspond to the bulk composition and so may not be well known. The alteration may or may not be caused by the primary ion beam. For example, the primary ion irradiation process may modify the surface concentrations by ion implantation or ion-induced segregation, diffusion or mixing. In all cases the composition of the top monolayer will differ from the subsurface composition. This will lead to a time-dependent secondary ion yield in a dynamic SIMS experiment which complicates the determination of the bulk elemental composition from the measured SIMS signal of a particular element.

The surface density of atoms of a given element will be particularly affected by the process of preferential sputtering. If, in a sample of two or more elements, the probabilities of sputtering atoms of different elements are different, there will be a modification of the surface layer concentrations as a result of sputtering. Atoms with higher probabilities of sputtering will be removed preferentially and hence the surface concentration of the atoms with lower probability of sputtering will increase with time. This process of modification of the surface layer concentration by preferential sputtering will continue until the sputtered particle flux has the bulk concentration and a concentration gradient with depth into the target will be established. Processes of preferential sputtering have been reviewed by *Betz* and *Wehner* [5.12] and are discussed further in Chap. 4.

d) The Cross-Section for Ionization of a Sputtered Atom and the Probability of Survival of an Ionized Atom. The cross-section for ionization of a sputtered atom is often expressed as a probability that an atom will be ejected from the surface in an ionized state. There exists a variety of models for the cross-section or the probability function, so we will use the two terms as loosely interchangeable. There are several problems to be solved.

i) In the case of ejection of a secondary ion from the surface of a clean metal or semiconductor, are we dealing with:

a) a two-stage process in which the secondary ion is formed in the collision which causes its ejection and is then subject to a charge-exchange process which might neutralize the ionized atom along its exit trajectory or,

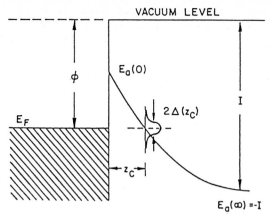

VACUUM LEVEL

$E_a(0)$

ϕ

$2\Delta(z_c)$

E_F

I

z_c

$E_a(\infty) = -I$

Fig. 5.5. Schematic energy diagram of an atom leaving a metal surface. E_F is the Fermi level and ϕ is the work function. Initially the atomic level E_i is broad and may lie above E_F before the atom is sputtered. The variation of the image potential causes E_a to lower with separation until $E_a = E_F$ at the crossing point. Electrons in the metal can tunnel out to fill the atomic level once $E_a < E_F$ beyond the crossing point

b) a single-stage process in which the secondary ion is the result of an undifferentiated charge-exchange process along the outgoing trajectory corresponding to ejection?

ii) In the presence of an adsorbate layer of active gas involving chemisorption to the target surface, the secondary ion yield of positive ions of the target and sometimes of the negative ions, is enhanced by up to two orders of magnitude over that observed from a clean surface. Is a new mechanism of ionization introduced or is the process of ionization and/or of subsequent neutralization observed in emission from a clean surface simply modified in some way to increase the cross-section for ionization or decrease the probability of neutralization?

There is no clear evidence to distinguish between the one-stage or two-stage process of secondary ion emission from clean surfaces. There are parallels to be drawn with models of charge exchange during the scattering of low energy inert gas ions [5.13], but these are also inconclusive. There are some facts we can state with certainty as a result of experimentation, e.g. the secondary ion yield is a function of the work function of the surface. *Ming Yu* and *Lang* have recently reviewed this evidence in a very precise way [5.14] for both negative and positive ions. The model which is currently the more favoured [5.14] considers a single-stage process which can be represented by the energy level diagram shown in Fig. 5.5. This represents an energy level on an atom leaving a solid as a function of distance from the surface of the solid. For positive ion formation this level would correspond in depth below the vacuum level to the ionization energy of a given state of the isolated atom, while for formation of a negative ion it would correspond to the electron affinity. The depth of the level below the vacuum level would be a function of the distance from the surface, because of the

125

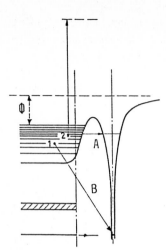

Fig. 5.6. Possible charge exchange processes involving an atom or ion moving away from a surface. (*A*) Resonant charge exchange (*B*) one type of Auger process

existence of the image potential and possible polarization effects. The transition of electrons between the metal and the ion/atom is only possible under a resonant condition involving levels of equal energy with respect to the vacuum level. The model thus neglects the possibility of Auger exchange processes which were very much features of *Hagstrum*'s earlier model of charge exchange to low energy ions incident on a surface [5.15] and are believed to be strong contributors to neutralization of low energy ions scattered from a surface.

According to Fig. 5.5, the sputtered atom will be ejected in an ionized state, thus giving rise to an image potential, and the probability of neutralization along the outgoing trajectory will be associated with resonance of the ionization level and occupied states of the electron energy distributions in the solid, i.e. with states below the Fermi energy. The maximum chance of exchange will then occur when resonance occurs between the ionization level and states near the Fermi level. The ionization level is broadened as a result of the finite lifetime of the level and, since the wavefunction of the electron in the metal decays exponentially with distance z from the surface, we can approximate the level width by:

$$\Delta(z) = \Delta_0 \, e^{-\gamma z} \, , \tag{5.5}$$

where γ is a characteristic length.

This model can be quantified, and for the case in which a crossing of the ionization level E_a with the Fermi level E_F occurs at a distance z_c from the surface,

$$P^\pm = \exp\left[-2\Delta(z_c)/\hbar\gamma V_\perp\right] \tag{5.6}$$

where $\Delta(z_c)$ is the width of the ionization level at the distance z_c at which the level E_a crosses E_F. $V_\perp$ is the velocity component of the ejected atom or ion normal to the surface. This dependence of P^+ on the perpendicular component of the ion exit velocity has been observed experimentally [5.16].

126

The basic difference between this one-stage process and the two-stage process is that in the two-stage process the collision between the two atoms at the surface would be assigned an ionization probability and then the neutralization processes which can occur on the outgoing trajectory would be considered. These neutralization processes are shown in Fig. 5.6, and include the possibility of Auger exchange. Analysis of the neutralization events suggest that the probability for an atom in an ionized state at the surface surviving in that ionized state to be detected by a spectrometer would be

$$P^+ = e^{-V_c/V_\perp} ,$$ (5.7)

where V_c is a characteristic constant and $V_\perp$ is as defined above. Thus experimentally there is little to distinguish these models. In either case, it is difficult to calculate the parameters involved in the expressions for the probability.

The situation when an active gas is present, either in the form of the incident beam (e.g. O^-, O_2^+) or as a background gas in the chamber, is more complex. Secondary ion yields increase by one to two orders of magnitude, and cannot be explained by changes in the work function. Earlier models proposed that the electron exchange event was blocked as a result of the modification of the surface by adsorption. This led to shifts in the Fermi level and crossing distances with the ionization level which were either very large or such that crossing did not occur at all. The enhancement of the ion yield has, however, been shown to be linearly dependent on the surface coverage of adsorbate, down to coverages which are a small fraction of a monolayer, where the modification of the surface electronic energy levels would be minimal. The enhancement is then realized on the adsorbate site. To explain this, a bond-breaking model has been proposed [5.17]. The oxygen-surface atom bond is broken in the collision process resulting in ejection of the surface atom. The possibilities may be represented on a potential energy diagram as shown schematically in Fig. 5.7. The path to be followed in this diagram will depend on the energy loss in the collisions, but it may give rise to the ejection of ionized particles. The logical suggestion is that, for a surface coverage which is small so that there is essentially M^+O^- bonding, the number of M^+ ions ought to equal the number of O^- ions.

It is obvious from the above that the general picture of secondary ion formation is known, but that an accurate analytical model allowing calculation of a cross-section for ionization in (5.1) is not available. This remains the major barrier to quantification of SIMS.

e) **Transmission of the Mass Spectrometer** This is an instrument function which can in principle be modelled by computer simulation or measured experimentally. Typically the transmission of a quadrupole analyzer is 0.1% or less whilst a magnetic sector instrument has a transmission of 1–10% and a time-of-flight instrument 10% or more. This will be discussed in a later instrumental section.

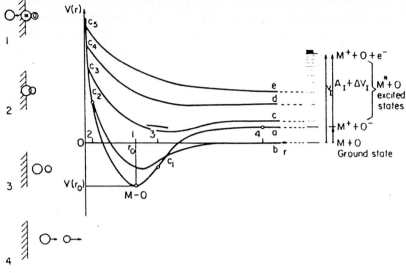

Fig. 5.7. Schematic diagram of the M–O system during its dissociation with increasing distance, r, from the surface. In this model of ionization of sputtered atoms, energy is transferred from the collision cascade to a quasi-molecule comprising a substrate metal atom with an adsorbed oxygen atom (inset *1*). As the metal and oxygen atoms move together (inset *2*) the system moves to the left up the repulsive side of the potential energy curve. If the energy imparted in the collision is great enough the system may cross onto curves c, d or e at points c_3, c_4 or c_5 respectively, and excited metal atoms or $M^+ + O$ will be sputtered. For the more probable low energy collisions, the system will return to the right along curves a or b and so metal and oxygen ions or atoms will be sputtered (inset *4*)

f) The Energy and Angular Distributions of Secondary Ions. Equation (5.1) contains terms in dE and $d\theta$, related to the energy distribution and the angular distribution of the sputtered ions. The energy distribution will also enter the equation because the cross-section for ionization and the probability of neutralization are both likely to be energy dependent. The transmission of the instrument may also be energy dependent.

The angular distribution is important in that maximum sensitivity for measurement will occur at the maximum in the angular distribution of the secondary ion yield. The acceptance angle of the mass spectrometer should clearly be related to the halfwidth of the sputtered ion angular distribution.

The Energy Spectrum of Sputtered Ions. The energy spectrum of sputtered atoms is fairly well understood. *Thompson* [5.18] showed that the energy spectrum of atoms sputtered from a random target is given by

$$N(E)dE \propto \frac{E}{(E + E_b)^3} dE , \qquad (5.8)$$

where $N(E)dE$ is the yield of ions in the energy range dE at E and E_b is the binding energy of the atom to the surface. *Sigmund* [5.6] has derived the same result for the same conditions of the Thompson model.

This spectrum will peak at an energy of a few eV and will fall off as $1/E^2$ at higher energies, independent of the primary ion energy. It will be related to the ion yield detected, through the cross-section for ionization and the probability of subsequent neutralization. These cross-sections have the form of an exponential of $(-V_c/V_\perp)$ so the energy spectrum of the secondary ions will be of the form

$$N(E)dE \propto \frac{E}{(E + E_b)^3} e^{-V_c/V_\perp} . \tag{5.9}$$

The neutralization parameter will favour slower ions, hence it is likely that the energy spectrum will have its peak shifted to higher energies relative to the energy spectrum of sputtered neutrals, and a tail which shows a dependence close to $1/E^2$.

Angular Distribution of Secondary Ions. The angular distribution of secondary ions due to sputtering has not been studied in detail: In the case of sputtered neutrals, under circumstances in which the irradiation is normal to the surface, the angular distribution is close to a cosine. Taking into account the effect of the normal component of the secondary ion velocity, one would expect the removal of the ion by neutralization to occur preferentially along trajectories close to the surface. Thus the distribution is likely to be of the form $\cos^n \theta$, where θ is the angle from the surface normal.

In the case of non-normal incident ions, the angular distribution of sputtered atoms is a little more complex. It is approximately a cosine distribution about the specular direction. Secondary ion distributions due to sputtering by non-normal incident ions will be more complex, due to the neutralization or ionization cross-section dependence on the perpendicular component of the exit velocity. As the exit trajectory approaches the surface, the dependence on $V_\perp$ will lead to strong preferential neutralization of the seondary ions. There has been little reported in the literature on such angular distributions.

Angular distributions will also be significantly modified if the analyzing equipment uses strong extraction fields to collect as many secondary ions as possible. The extraction field will not influence the angular distribution to any great extent close to the ejection point where the atomic exchange processes between secondary ion and surface occur. Further from the surface, however, the trajectories of all ejected ions will have been significantly changed by the extraction field.

5.2 Construction of a Secondary Ion Mass Spectrometer

The main elements of a SIMS analyzer are (i) a primary ion source (ii) optics to transfer these ions to the target surface and to collect secondary ions from the surface and direct them to (iii) a mass spectrometer.

All SIMS machines must have a source of ions available to initiate the sputtering process. In general, the ions will have energy in the range 1–20 keV, but

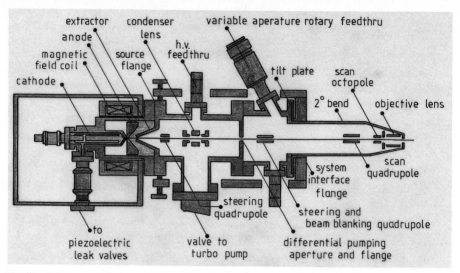

Fig. 5.8. Schematic diagram of a PHI duoplasmatron source and primary ion optics

other characteristics of the incident beam will depend on the type of application. Many add-on SIMS analyzers will use a reasonably simple ion beam, with a beam diameter of order 0.1–1 mm and with the facility to raster the beam across larger areas. Dedicated SIMS analyzers on the other hand often have sophisticated ion beam systems, particularly if the analyzer is likely to be used in microscope mode.

All ions guns comprise an ion source and a method of transferring the ions to the sample. This is normally done by floating the ion source at the accelerating voltage and leaving the target at ground potential. In some instruments, e.g. the Cameca IMS-3F, the target may also be floated at a high potential. There are four common types of ion sources used, namely electron bombardment ion sources with and without plasma formation (duoplasmatron, hollow cathode), surface ionization sources and field ionization sources (liquid metal ion sources).

In duoplasmatron sources, ions are extracted in a strong external electrostatic field from a high density plasma which is magnetically confined between an anode and intermediate electrode. These sources are bright (10^6–10^7 A m^{-2} Sr^{-1}) with an energy spread of about 10 eV, thus providing current densities of about 10 mA cm^{-2} in beams of diameter 1–50 μm. Both inert and active gas beams can be produced by this source. Figure 5.8 shows a schematic diagram of the PHI duoplasmatron source and ion transport optics.

In a surface ionization source, alkali metals such as caesium are varporized from a porous metal surface which has a high work function. When the temperature of the surface is high so that the caesium surface concentration is low, then the caesium is almost all desorbed as ions. The brightness of the source is about 10^6 A m^{-2} Sr^{-1} but the source has a very low energy spread (< 1 eV) and so current densities and spot sizes comparable to duoplasmatrons can be achieved.

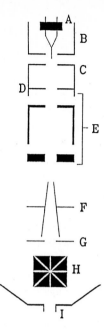

Fig. 5.9. Schematic diagram of the FEI liquid metal ion source. A–filament, B–suppressor, C–extractor, D–beam defining aperture, E–lens, F–blanking plates, G–blanking aperture, H–octapole, I–differential pumping aperture

In liquid metal ion sources a low melting point metal (Ga, In, Cs) is supplied as a liquid over a metal tip which is subject to a high electric field. Electrons tunnel from surface atoms to the tip, leaving the metal ions to be extracted. These sources have high brightness ($10^{10}\,\mathrm{A\,m^{-2}\,Sr^{-1}}$) and allow nA currents to be delivered into spots of diameter less than 50 nm. Figure 5.9 shows a schematic diagram of the layout of the FEI single lens liquid metal ion source.

Duoplasmatron sources are used mainly for the production of O_2^+ and Ar^+ for depth profiling and mass scans. Surface ionization sources allow depth profiling using Cs^+ with the attendant advantages of increased negative ion yields and decreased matrix effects in positive ion depth profiling. Liquid metal ion sources are exclusively used for imaging SIMS since currents are normally too small for rapid depth profiling.

Primary ion optics, for transferring ions onto the sample consists of usually two electrostatic lenses for beam focussing, xy deflection for beam positioning, rastering and neutral rejection, a stigmator for focus correction independently in x and y directions when small spot sizes are used, and a mass filter to remove beam contamination, i.e. multiply charged ions, clusters or impurity ions. The most commonly used mass filters are the Wien filter and the magnetic sector. In a Wien filter crossed magnetic and electric fields produce equal and opposite forces on ions of a certain M/Z ratio which then pass through the filter undeflected. Ions of other M/Z ratios do not pass through the filter. The electric fields may be shaped by shims to produce additional focussing. In a magnetic sector, ions are bent in a uniform magnetic field. Ions of energy eV_0 and a specific M/Z ratio are bent in the horizontal plane with a radius of curvature

$$R = \sqrt{\frac{2eV_0}{M}} ,$$ (5.10)

which, if it matches the radius of curvature of the instrument, allows the ions to be transmitted through the magnet. Focussing in the vertical plane can be achieved by rotating the magnetic boundaries with respect to the particle trajectory.

The dynamic range and depth resolution achievable in depth profiling is dependent on the crater flatness and the background signal reaching the detector. Flat craters are achieved by scanning the ion beam over typically five beam diameters. Electronic gating of the detected ions then ensures that only the ions emitted from the centre of the crater are counted. For this procedure to work correctly, neutrals arriving with the ion beam must be eliminated since they will generate secondary ions independent of any raster gating. A bend of at least $1°$ followed by apertures must be included in the section of the ion optical path to reject neutrals from the beam striking the sample.

SIMS analysis of insulating samples, common in static SIMS, creates problems due to surface charging. The injection of positive ions and ejection of secondary electrons (with an efficiency of almost unity at SIMS energies) can cause surface potentials to vary with time to a few hundred volts which can completely stop secondary negative ion emission and alter the energy spectrum of emitted positive ions making optimization of ion optics difficult. Charging may be reduced by using thin samples, low ion current densities, electron flooding or fast atom bombardment. Electrons with energies of a few eV to a few hundred eV can be generated from just a filament or a simple gun. If the electron current is matched to the ion current, then the surface charging can be reduced. This process is self-stabilizing since a positive surface potential will reduce the loss of secondary electrons which will in turn decrease the surface potential.

In the application of SIMS to analysis, the information on surface composition is contained in a range of secondary ions of variable mass, distributed over a broad angle and energy range. In order for the equipment to have a reasonable mass resolution, the energy range of particles accepted into the mass analysis section of the spectrometer must be narrow. This energy range is given by ΔE which is the spread in the ion energy E seen by the mass spectrometer. The energy E is equal to $qV + E_0$, where q and E_0 are the charge and energy, respectively, of the secondary ion and V the extraction potential. A spread in E then results from the spread in the initial energy of the secondary ions and from any drift in V. This energy resolution must be smaller if a larger mass resolution $M/\Delta M$ is desired. A SIMS mass spectrometer must therefore consist of an energy analyzer preceding the mass analyzer. The secondary ions may be extracted into this energy analyzer by a high ($< 1\,\text{kV}\,\text{mm}^{-1}$), or low ($< 10\,\text{V}\,\text{mm}^{-1}$) electric field. Higher transmission results from the use of a high extraction field since more ions are collected into the analyzer and the relative energy spread of the secondary ions ($\Delta E/qV$) is less. Typical electrostatic energy analyzers have high transmission over a relatively narrow energy spread, typically

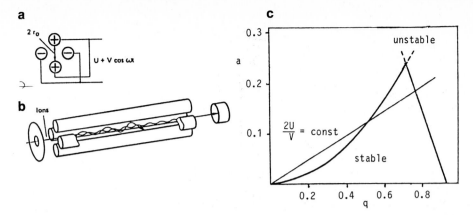

Fig. 5.10a–c. Diagram of a rf quadrupole. (a) A combination of a dc voltage U and ac voltage V is applied to opposite pairs of rods. (b) The parameters a and q are given by $(4eU/m\omega^2 r_0^2)$ and $(2eV/m\omega^2 r_0^2)$ respectively. The ratio a/q is independent of m/e, the ion's mass to charge ratio. Therefore the orbits of all ions will lie on the straight line going through the origin. However, only ions lying within a narrow region of that line near $q = 0.7$ have stable trajectories. The width of this stable region may be changed by altering U/V. The value of m/e corresponding to this stable region is adjusted by altering U and V whilst keeping their ratio constant. (c) The ion will then travel through the rods and be detected. Ions with different mass-to-charge ratios will have diverging trajectories and strike rods or aperture

1–2%. To match this transmission to an energy spread ΔE of about 100 eV requires an extraction potential of 5–10 kV. If lower potentials are used, only a fraction of the total secondary ion spectrum can be collected and analyzed.

There are three main types of mass spectrometer: rf quadrupole, magnetic sector and time of flight. In an rf quadrupole analyzer, low energy (< 50 eV) secondary ions travel down through the centre of four circular rods (Fig. 5.10) which are electrically connected in opposite pairs. A combination of dc and rf (~ 1 MHz) voltages is applied to one set of rods and an equal but opposite combination is applied to the other pair. Only ions of a certain M/Z ratio, where the ion charge q is given by $q = Ze$ with e the electronic charge, have a stable trajectory and are transmitted through the quadrupole. Ions of different M/Z ratios move in unstable trajectories and strike the rods or apertures. The mass of the transmitted ion is selected by increasing the rf and dc voltages whilst keeping their ratio constant. Only one M/Z ratio may be transmitted at any one time. Mass analysis of many elements must therefore be performed sequentially. The mass resolution, $M/\Delta M$ of the quadrupole is normally about 1000 over a restricted mass range. This resolution can be increased by using larger diameter rods at the expense of a lower obtainable mass range. Transmissions of the order of 1% and mass analyzes up to $M/Z \sim 2000$ are typical for rf quadrupoles. They are useful for adding SIMS analysis onto other analysis systems or where space is critical. A further discussion of rf quadrupole systems is given in [5.19].

Magnetic sector spectrometers send ions through a perpendicular magnetic field. The radius of curvature of the ions then depends on $(M/Z)^{1/2}$ as found previously. The mass resolution of the magnetic sector is proportional to the

133

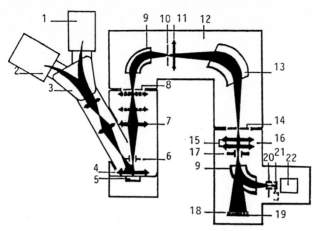

Fig. 5.11. Schematic diagram of the ion optics of a Cameca IMS-4F ion microscope. The electrostatic sector selects the ion energies passed to the double focussing magnetic sector. The ions may then be displayed on a fluorescent screen in microscope mode or collected in a Faraday cup or electron multiplier for quantitative static or dynamic analysis. The elements of the analyzer are *(1,2)* – ion sources, *(3)* – primary beam mass filter, *(4)* – immersion lens, *(5)* – specimen, *(6,17,20)* – deflectors, *(7,11,15)* – lenses, *(8,10,14)* – slits, *(9)* – electrostatic sector, *(12)* – spectrometer, *(13)* – electromagnet, *(16)* – projection display and detection, *(18)* – channel plate, *(19)* – fluorescent screen, *(21)* – Faraday cup, *(22)* – electron multiplier

radius of curvature of the path but is sensitive to angular and energy spread in the incoming secondary ions. The loss in resolution may be compensated by energy analysis in an electrostatic sector. The transmission of the instruments is high ($> 10\%$) and the mass resolution is high ($M/\Delta M > 10^4$) over a large mass range. The high mass resolution is important for many types of analysis. For example, a typical problem in semiconductors is to analyze P in Si. The mass of P^{31} is 30.9859 amu whereas the mass of $Si^{30}H$ is 30.9736 amu. If the analyzer has a mass resolution $M/\Delta M$ less than 4000, the two peaks will overlap and unambiguous determination of low concentrations of P in Si will be made impossible.

Figure 5.11 shows a schematic diagram of a common magnetic sector instrument, the CAMECA IMS-3F system. This is a SIMS microprobe which has been available for a number of years and has progressed through several generations of different analyzers. It is a microscope system capable of producing images of about 1 μm spatial resolution. It has a high mass resolution (up to 10^4 in spectrometer mode) which is achieved by the use of a high extraction potential (about 5 kV), secondary ion energy and mass selection in 90° electrostatic and magnetic sectors. The combination of transfer lenses and sectors forms a stigmatic image of up to a 400 μm × 400 μm area of the target onto the fluorescent screen. Alternatively, the secondary ions can be directed into a electron multiplier for static and dynamic SIMS analysis.

Time of flight instruments rely on the measurement of the drift time of secondary ions along a flight tube. If the energy of the ions is known, eV_0, by

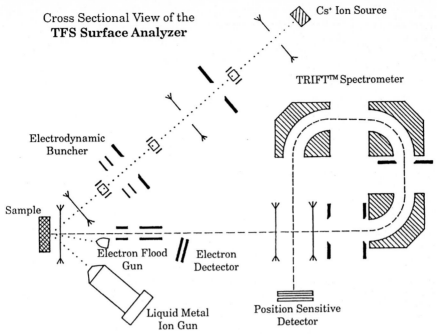

Cs⁺ Ion Source

TRIFT™ Spectrometer

Electrodynamic
Buncher

Sample

Electron Flood
Gun

Electron
Dectector

Liquid Metal
Ion Gun

Position Sensitive
Detector

Fig. 5.12. Schematic diagram of a direct imaging time-of-flight SIMS analyzer. The 270° analyzer filters the energy of the secondary ions, allowing a mass resolution of more than 3000. The analyzer can be used in microscope mode, with a lateral resolution of 1–2 μm or in microprobe mode (using a liquid metal ion source) with a 0.1 μm resolution

accelerating them through a high potential difference, then the mass to charge ratio of the ion is related to the time taken to drift a distance d by

$$M/z = 2eV_0 t^2/d^2 . \tag{5.11}$$

Typical flight times are 10 μs–1 ms depending on the ion mass and the flight path. The secondary ions must, however, all start at the same time. This requires that the formation of the secondary ions must occur in short pulses, typically 10–50 ns in duration. The time between pulses is set by the maximum flight time plus the data processing time. The primary ion beam is usually pulsed by rapid deflection past slits in the primary ion column. Alternatively, secondary ion formation above the surface from sputtered neutrals could be pulsed, for example, by laser ionization. The transmission of time of flight instruments is very high (0.5–1) but is dependent on the energy spread of the secondary ions. This may be circumvented by incorporating an energy-compensation section or an energy filtering stage in the flight tube so that higher energy ions travel further than low energy ions of the same mass to charge ratio and ions of different energy arrive at the detector simultaneously [5.20]. For example, a 270° spherical sector energy filter has been used in the TFS surface analyzer of Charles Evans & Associates (Fig. 5.12).

A detector system will then follow the mass filter, with the type of detector depending on the likely signal strength, i.e. it will consist of an ion counting system for low count rates or a low current amplifier if the count rate is above about 10^7 counts per second.

5.3 Special Experimental Parameters in SIMS Analysis

5.3.1 Signal Enhancement by Surface Adsorption

It was indicated in Sect. 5.1.3d that the magnitude of the SIMS signal for a given ion was significantly affected by the presence of active gas adsorbates on the surface of the target. This is because the cross-section for ionization is very dependent on the chemical state of the surface. The signal enhancement factor due to the presence of an active gas layer can be more than 100 times that of the ion yield from a clean surface. For example, the yield of positive Al^+ secondary ions under 8 keV Ar^+ bombardment [5.21] increases by up to three orders of magnitude when the target surface is saturated with oxygen (Fig. 5.13). This change is not uniform even between cluster and multiply charged ions

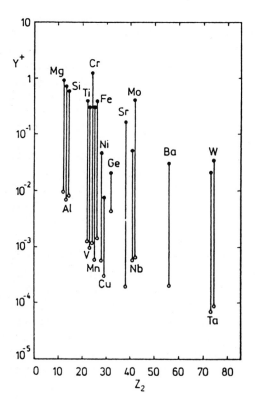

Fig. 5.13. Secondary ion yields for the indicated elements with (*filled circles*) or without (*open circles*) oxygen

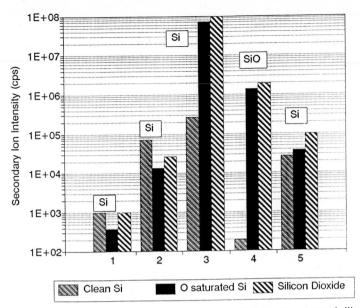

Fig. 5.14. Ion yields of silicon secondary ions for clean Si, oxygen-saturated silicon or silicon dioxide surfaces

from the same element since, as the oxygen coverage of a silicon surface is changed from zero to a saturation coverage, the yield of Si^+ increases by 2.3 orders of magnitude, but SiO^{2+}, Si^{2+} and Si_2^+ remain the same or decrease [5.22] (Fig. 5.14). Negative ion yields can be similarly increased by introducing an electropositive element like Cs^+ onto the surface, typically by Cs^+ bombardment [5.23] (Fig. 5.15). However, it is still difficult to quantify concentrations from SIMS yields enhanced by reactive elements because

i) the enhanced yield is still a widely varying number from element to element as shown in Fig. 5.15. Systematic studies of the secondary ion yield from a wide range of elements have not been undertaken, in part because of the sensitivity of such yields to experimental parameters such as the cleanliness of the surface, the background gas concentration, the transmission characteristics of the SIMS analyzer, etc.

ii) when an active gas is used to enhance the ion yield and hence the sensitivity, the measured ion yield is a function of the active gas concentration at the surface. This concentration of active gas may be established by irradiating the surface in a background environment of the active gas or alternatively, the incident ion beam may consist of ions of the active gas itself. Thus some SIMS machines will involve bombardment of a surface with an inert gas ion beam such as Ar^+ or Ne^+, in a low pressure ($\sim 10^{-6}$ torr) environment of O for example. Others may use ion guns producing beams of active elements such as O or Cs in an ultra high vacuum environment. In both cases the enhancement of secondary ion

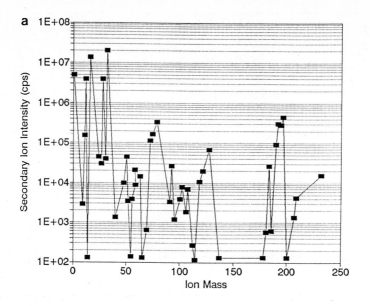

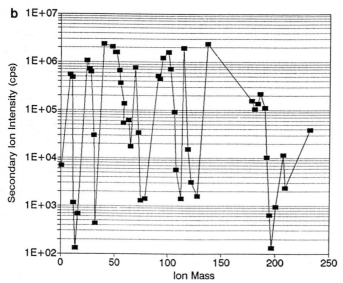

Fig. 5.15. (a) Positive ion yields for the indicated targets bombarded with 13.5 keV O⁻ at normal incidence. (b) Negative ion yields for the indicated targets bombarded with 16.5 keV Cs⁺ at normal incidence

yield will depend on the arrival rate of active gas, the adsorption probability, the incident ion beam, current density and the sputter rate of active gas atoms or molecules and the consequent coverage of the surface. Transient effects in the secondary ion signal will occur, with the halflife of the transient dependent on the time taken to establish an equilibrium surface concentration of active gas and the variation of enhancement factor with surface concentration of adsorbate. The surface concentration and enhancement factor depends in many cases on the bulk composition of the target. For example, for oxygen bombardment of metal alloys the ionization probability again varies considerably with composition. Generally, if element A forms a stronger oxide bond than B, the presence of A in an alloy enhances the ionization of B, whereas the presence of B decreases the ionization of A.

Ion enhancement effects are not restricted to deliberate changes in the surface brought about by adsorption or bombardment. Changing matrix element compositions during, for example, depth profiling, may also alter ion yields. It is possible to circumvent some of these uncertainties to gain quantitative information about concentrations by the use of standards. If dilute concentrations are to be measured, the easiest method is to implant a known small dose of the impurity into an identical matrix. In this case, the ionization probabilities of the impurity in the two samples should be the same and so quantification of the unknown concentration then follows from equation (5.1).

5.3.2 Improvement of Mass Discrimination Using the Energy of the Secondary Ions

Many SIMS experiments do not require mass resolution equivalent to isotope separation, but it is often necessary to detect the signals due to atomic and molecular ions. This can be done on the basis of the energy spectrum of the secondary ions. The energy spectrum of molecular ions, e.g. M_n^+ or $M_nO_x^+$, maximize yield at a lower energy than that of the atomic species M^+. In addition, the energy spectrum is substantially narrower than that of the atomic ion. This is shown in Fig. 5.16. In cases where overlap of an elemental peak of interest by a molecular ion peak is a possible problem, the case can often be resolved by changing the pass energy of the energy analyzer preceding the mass analysis

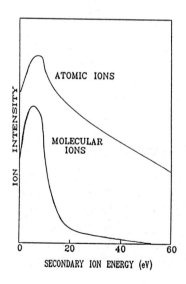

ATOMIC IONS

MOLECULAR IONS

ION INTENSITY

0   20   40   60
SECONDARY ION ENERGY (eV)

Fig. 5.16. Schematic diagram of the energy spectrum of either atomic or molecular secondary ions. Typically the energy spectrum of the molecular ions gets sharper as the number of atoms in the molecule increases

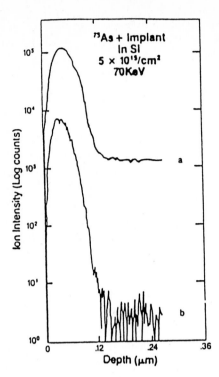

Fig. 5.17. Biassing the secondary ion optics to only accept high energy secondaries excludes molecular ions which interfere with and limit the dynamic range of dopant profiling. The lower curve was obtained by offsetting the secondary ion optics by 40 V to eliminate a molecular ion

section of the instrument. The molecular ion contribution to any signal will be preferentially removed as the pass energy increases. Such a procedure is particularly useful in depth profiling dopants in semiconductors where molecular ion interferences cause constant backgrounds which limit the dynamic range available. For example, in Fig. 5.17, the detection limit for $^{75}As^+$ implanted in Si is much lower when a 40 V offset is put on the secondary ion optics so as to only accept high energy secondaries [5.24]. This is because the molecular ion $^{29}Si^{30}Si^{16}O^+$ also has mass of 75 amu and, since it comes from the bulk Si being sputtered by O_2^+, has constant intensity with depth. The energy spectrum of the molecular ion is much narrower than for the elemental ion and so the molecular ion is discriminated against when high energy secondaries are selected.

5.4 Some Examples of Applications of SIMS Analysis

5.4.1 The Use of Static SIMS

a) Trace Element Identification in MBE Grown Cadmium Telluride. SIMS is most easily used to identify differences in trace element concentrations between two targets. In this case, absolute concentration determinations are not required, thus avoiding SIMS ion yield uncertainties. In the following example, an epitaxial

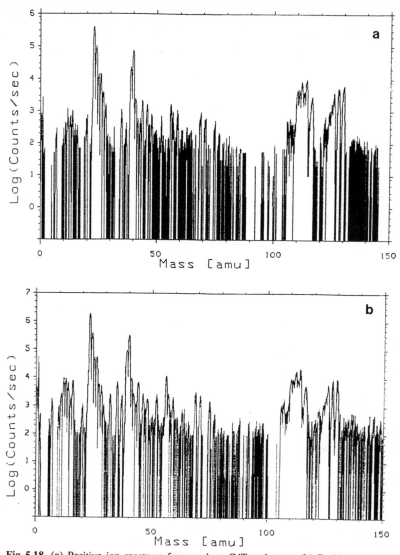

Fig. 5.18. (a) Positive ion spectrum from a clean CdTe substrate. (b) Positive ion spectrum for a MBE–grown CdTe thin film

CdTe film had been grown on a CdTe(111) substrate by molecular beam epitaxy. The film was a single crystal, but measurement of optical properties indicated the presence of impurities in the film. Figure 5.18a, b shows SIMS mass spectra of a CdTe substrate and the CdTe film, respectively, which were taken on a Riber MIQ 156 RF quadrupole-based analyzer. The level of oxygen and hydrocarbons (masses 12–16) in the CdTe substrate is much lower than in the CdTe film. There is also increased Fe (mass 56) in the deposited film. These are probably the impurities responsible for the altered optical properties.

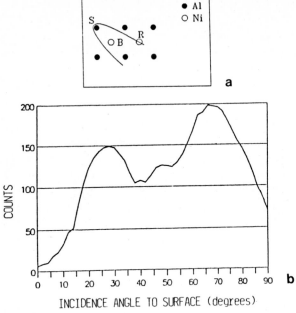

Fig. 5.19. (a) A schematic diagram of the atomic positions in the top three layers of a Ni₃Al(001) surface for one [010] plane. The other [010] plane is comprized solely of Ni atoms in the same positions as shown in this diagram. (**b**) The yield of Al⁺ as a function of the angle of the 5 keV Ar⁺ beam to the surface of Ni₃Al in the ⟨100⟩ azimuth. The two peaks in the scan correspond to the primary ions being focussed by atom S in the top monolayer onto atoms R and B in the second monolayer

b) Analysis of the Surface Structure of Ni₃Al Using Angular Resolved SIMS.

Static SIMS may also be used to determine the atomic geometry of crystalline surfaces. This is because the yield of atoms and ions ejected from an ion-bombarded single crystal surface depends strongly on the angle of incidence of the primary ion beam. In particular, if a parallel ion beam is directed towards a surface atom (e.g. S in Fig. 5.19) [5.25] the beam will be scattered from the atom with the result that there is a cone-shaped region behind the atom where the ions do not penetrate. The primary ion beam will be focussed into a region just at the edge of this "shadow cone". If the edge of the shadow cone intersects another surface or near surface atom (R in Fig. 5.19), the ion which has been focussed onto R will undergo a violent collision with R and transfer momentum to the surrounding atoms. As a result target atoms may escape from the surface as secondary ions. This process may alternatively be interpreted in terms of (5.3) in that the intersection of the shadow cone with R causes an increase in $F_d(E)$ at the surface and thus an increase in S_A.

The shape of the shadow cone may be found from empirical relations [5.26]. If the ion incident angle ϑ corresponding to a peak in the SIMS yield is known from experiment, then the interatomic spacings may be found from simple trigonometry. For example, Fig. 5.19 shows the Al⁺ yield from Ni₃Al(001) at a scattering

142

angle of 45° as the angle of the ion beam to the sample surface is changed from 0° to 90° in the ⟨100⟩ azimuth. There are large peaks at incidence angles of 25° and 68°. These peaks correspond to the incident ions being focussed from top-layer Ni atoms onto second-layer Ni atoms and top-layer Ni and Al atoms onto second-layer Ni atoms, respectively, in an unreconstructed surface. More details on the use of this technique for surface structure analysis can be found in the work of *Chang* and *Winograd* [5.27].

c) Quantitative Composition Analysis Using Static SIMS – The Andersen-Hinthorne Model of Secondary Ion Formation.

It has been indicated above that it is difficult to obtain quantitative analysis of a sample from a basic theory of secondary ion formation. One reasonably successful attempt at quantitative analysis was that due to *Andersen* and *Hinthorne* [5.28]. They suggested that the ion-bombarded region of a solid surface from which the secondary ion originated behaved as a plasma in local thermodynamic equilibrium. The ion yield could then be expressed in terms of the Saha-Eggert equation

$$\frac{N^+ N_e}{N_0} = \frac{2}{h^2} \frac{M^+ M_e}{M_0} kT^{3/2} \frac{B^+ B_e}{B_0} \exp\left(\frac{I - \phi}{kJ}\right) , \tag{5.12}$$

where the symbols have the following meanings

N_m^+, N_e, N_0: number densities of singly charged ions, free electrons and neutral atoms in the plasma.

M^+, M_e, M_0: masses of ion, electron and atom

B^+, B_e, B_0: partition functions of the ion, the free electron and the neutral atom

I: ionization potential of the atom

T: plasma temperature

$\Delta\phi$: potential depression due to screening in the plasma.

Application of the Saha-Eggert equation required knowledge of the concentrations of two elements in the matrix under test. These concentrations were used to calculate T, the plasma temperature and N_e, the free electron density. Concentrations of other components of the matrix could then be calculated.

This method gave very reasonable compositional analysis results, particularly when applied to glasses and minerals (Fig. 5.20). The theoretical basis of the concept of a plasma in local thermodynamic equilibrium could be checked spectroscopically. This was done in a variety of ways and it was demonstrated that the ion bombarded surface behaved spectroscopically as a plasma in LTE. The temperature T and free electron density N_e derived spectroscopically, were consistent with the values obtained from the known concentrations using the secondary ion yields. A method of analysis was demonstrated using combined photon and secondary ion emission [5.29].

The LTE plasma model was eventually shown to be a fortuitous description of the emission process. Physically, the necessary equilibrium between exci-

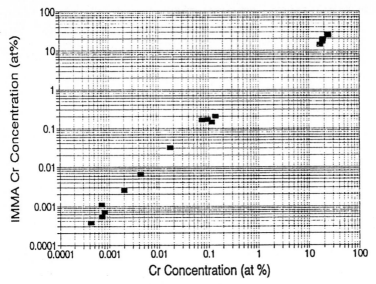

Fig. 5.20. The concentration of Cr as found from ion microprobe analysis using (5.12) compared to that found from conventional analysis

tations and de-excitation or ionization and neutralization could not exist. The "plasma temperature" measured spectroscopically was found to be a function of the excitation energy of the levels concerned. The reason for the apparent success of the Saha-Eggert equation arises from the observation of the constancy of the pre-exponential terms over the effective plasma temperatures measured, so the equation (5.12) reduced to the form

$$N^+/N_0 = Ae^{I/E_c} \qquad (5.13)$$

where the left hand side is the probability of ionization, A is a constant and E_c is a characteristic energy term. While the physical basis of the method is obviously questionable if interpreted too rigorously, the applicability of the method to some analysis situations as a fitting procedure is not questioned.

5.4.2 The Use of Dynamic SIMS Isotopic Ratios in Plasma Deposited Cu Thin Films

Many examples exist of the use of SIMS in depth profiling, especially in semi-conductors where the sensitivity of SIMS to trace element analysis is well suited to dopant analysis. Examples of SIMS depth profiles are given in Chap. 4. Depth profiling may also be required where mass spectra cannot be acquired with low enough ion doses to maintain static conditions. For example, the ratio of Cu^{63} to Cu^{65} was required to be measured for Cu thin films which were made by erosion of a Cu target in a glow discharge. To achieve the 0.1% precision needed

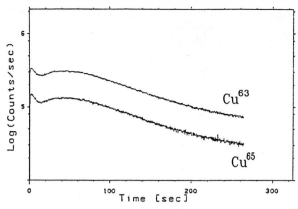

Fig. 5.21. Depth profile of Cu^{63} and Cu^{65} for a Cu thin film profiled with $12\,keV\,Ar^+$

in the measurement required a $12\,keV\,Ar^+$ dose which caused some sputtering of the target. The oxygen concentration in the target material decreased away from the surface. In the rf quadrupole system used, sequential analysis of Cu^{63} and Cu^{65} would then have led to more oxygen-induced enhancement of the Cu^{63} signal than the Cu^{65} signal. The measured ratio would thus be higher than the real value. This problem was circumvented by rapidly switching between the two peaks in a depth profiling mode (Fig. 5.21). The decreases in the signals of both peaks with sputter time can be seen. However, the ratio of the two isotopes is constant over this depth range. The ratio of the integrated counts under each curve then gives a correct measure of the isotopic ratio.

5.4.3 The Use of Imaging SIMS
– Analysis of Adhesive Bond Failure

The use of structural adhesives is becoming more important for bonding components in aircraft. The strength and durability of an adhesive bond depends on the interface between the polymer adhesive and the metal. The failure mechanism of an epoxy resin bond between two aluminium plates has been identified [5.30] with the help of imaging SIMS.

In this study, a SIMS instrument incorporating a Kratos $10\,keV$ Ga ion gun, Balzers rf quadrupole and Kratos energy filter was used to collect SIMS maps of fractured surfaces of aluminium plates which had undergone a Boeing wedge test. Prior to the test, 2024 aluminium alloy plates were grit blasted and treated with an acqueous silane solution. The inset of Fig. 5.22 shows the structural formula of the silane. The plates were then loaded in a wedge configuration until failure. Specimens which were loaded slowly over several weeks in a humid environment show no evidence of Si in the fractured surface. This indicates that the failure occurred in the weakened aluminium oxide film. Samples loaded more quickly show regions of surface Al, surface Si and mixtures of the two, as well as regions lacking in both Al and Si (Fig. 5.22). Regions containing both Al and

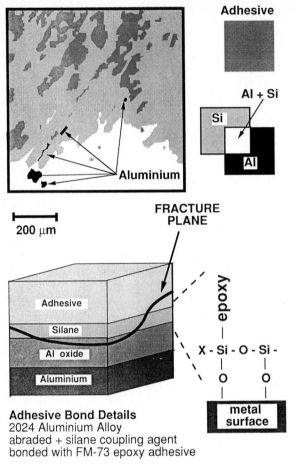

Adhesive

Al + Si

Si

Al

Aluminium

200 μm

FRACTURE PLANE

Adhesive

Silane

Al oxide

Aluminium

epoxy

X - Si - O - Si -
| |
O   O
| |

metal surface

Adhesive Bond Details
2024 Aluminium Alloy
abraded + silane coupling agent
bonded with FM-73 epoxy adhesive

Fig. 5.22. SIMS composition map of an adhesive bond fracture on 2024 aluminium treated with silane, showing the locus of fracture based on the distribution of Si and Al. The dark grey areas in the map contain no Al or Si and so correspond to adhesive at the surface. The white areas contain both Al and Si so correspond to the surface fracturing at the silane-oxide interface

Si are due to the plate cracking at the silane–aluminium oxide interface. The presence of regions lacking both Si and Al points to failure within the adhesive. Surface regions containing Si or Al alone point to failure in the silane film or aluminium substrate respectively. SIMS imaging then indicates that the fracture wanders across the interface between the metal and adhesive.

References

5.1   A. Benninghoven, F.G. Ruedenauer, H.W. Werner: *Secondary Ion Mass Spectrometry* (Wiley, New York 1987)

5.2   J.V. Vickerman, A. Brown, N. Reed (eds.): *Secondary Ion Mass Spectrometry* (Oxford University Press, Oxford 1989)

5.3   R.A. Wilson, F.A. Stevie, C.W. Magee: *Secondary Ion Mass Spectrometry* (Wiley, New York 1989)

5.4   D. Briggs, A. Brown, J.C. Vickerman: *Handbook of Static Secondary Ion Mass Spectrometry (SIMS)* (Wiley, Chichester 1989)

5.5   A. Benninghoven, C.A. Evans, K.D. McKeegan, H.A. Storms, H.W. Werner (eds.): The most recent proceedings is "SIMS VII" (Wiley, Chichester 1990)

5.6   P. Sigmund: Phys. Rev. **184**, 383 (1969)

5.7   J. Schou: Nucl. Instr. Meth. B **27**, 188 (1987)

5.8   N. Matsunami, Y. Yamamura, Y. Itikawa, N. Itoh, Y. Kazamuta, S. Miyagawa, K. Morita, R. Shimizu, H. Tawara: Atomic Data and Nuclear Data Tables **31** (1984)

5.9   H.H. Andersen, H.L. Bay: In *Sputtering by Particle Bombardment I*, ed. by R. Behrisch (Springer, Berlin, Heidelberg 1981) p. 145

5.10 H.H. Andersen: Nucl. Instr. Meth. B **18**, 321 (1987)

5.11 J.P. Biersack, L.G. Haggmark: Nucl. Instr. Meth. **174**, 250 (1980)

5.12 G. Betz, G.K. Wehner: In *Sputtering by Particle Bombardment II*, ed. by R. Behrisch (Springer, Berlin, Heidelberg 1983) p. 11

5.13 D.J. O'Connor, R.J. MacDonald: In *Ion Beams for Materials Analysis*, ed. by J.R. Bird, J.S. Williams (Academic, Sydney 1989) p. 373

5.14 M.L. Yu, N.D. Lang: Nucl. Instr. Meth. B **14**, 403 (1986)

5.15 H.D. Hagstrum: In *Inelastic Ion Surface Interactions*, ed. by N.H. Tolk, J.C. Tully, W. Heiland, C.W. White (Academic, New York 1977)

5.16 R.J. MacDonald, E. Taglauer, W. Heiland: Appl. Surf. Sci. **5**, 197 (1980)

5.17 I.S.T. Tsong: In *Inelastic Particle-Surface Collisions*, ed. by E. Taglauer, W. Heiland (Springer, Berlin, Heidelberg 1981) p. 258

5.18 M.W. Thompson: Phil. Mag. **18**, 337 (1968)

5.19 P.H. Dawson: In *Quadrupole Mass Spectrometry* (Elsevier, Amsterdam 1976) p. 9

5.20 W.P. Poschenrieder: Int. J. Mass. Spectrom. Ion Phys. **9**, 357 (1972)

5.21 A. Benninghoven: Surf. Sci. **53**, 596 (1975)

5.22 J.L. Maul, K. Wittmaack: Surf. Sci. **47**, 358 (1975)

5.23 H.A. Storms, K.F. Brown, J.D. Stein: Anal. Chem. **49**, 2023 (1977)

5.24 PHI Applications NOte 8507, Perkin Elmer Physical Electronics Division, Eden Prairie, USA

5.25 S. Johnson, B.V. King: Proc. Sixth Australian Conference on Nuclear Techniques of Analysis (Australian Institute of Nuclear Science and Engineering, Lucas Heights, Australia 1989) p. 18

5.26 O.S. Oen: Surf. Sci. **131**, L407 (1983)

5.27 C.-C. Chang, N. Winograd: Phys. Rev. B **39**, 3467 (1989

5.28 C.A. Andersen, J.R. Hinthorne: Anal. Chem. **45**, 1421 (1973)

5.29 R.J. MacDonald, R.F. Garrett: Surf. Sci. **78**, 371 (1978)

5.30 D.R. Arnott: Proc. Australian Aeronautical Conference, Melbourne (1989)

6. Auger Spectroscopy and Scanning Auger Microscopy

R. Browning

With 8 Figures

Auger electron spectroscopy (AES) is a widely used technique for the analysis of surfaces, and it has application in many fields of materials science. AES is highly surface specific, and for most materials it is only sensitive to the top few monolayers. Minute amounts of material at the surface can be analyzed. Under the right conditions 10^{-3} of a monolayer [6.1], and in an Auger microscope as few as 10^4 atoms, with a spatial resolution of 20 nm can be detected [6.2]. The qualitative interpretation of AES spectra is straightforward for many materials, and all elements except for hydrogen and helium are detectable. Any radiation with sufficient energy to ionize an atomic core level will stimulate Auger electron emission, but, by convention, AES is normally understood to use an electron beam.

The combination of surface sensitivity, high spatial resolution, and often unambigous interpretation, are the factors which make AES uniquely suited to many types of materials analysis problems. In many cases, AES is a source of information that could be provided in no other way. Examples can be found in metallurgy, in the analysis of thin films, in the processing of semiconductor related materials, in corrosion and passivation studies, in catalysis, in the analysis of ceramic materials, and in tribology. AES is applicable to all fields where the surface properties of a material (this includes internal surfaces such as grain boundaries) strongly influence the overall properties of the material.

In this chapter we will introduce the basis of Auger electron spectroscopy from a solid surface, and we then show how this can be used as a powerful materials analysis technique by citing examples from a variety of fields. The development of AES, both in the understanding of the Auger signal from the surface, and in the instrumental techniques, is still ongoing. Therefore in the last part of the chapter, we will discuss not only what are the present advantages and limitations of the technique, but also the eventual possibilities.

6.1 Auger Electron Spectroscopy

Auger electron spectroscopy (AES) is based on the Auger effect independently discovered by *Meitner* [6.3] and *Auger* [6.4] in 1923. They found that if an atom is ionized by the removal of an electron from an atomic core level, a second electron can be ejected with an energy characteristic of the atom. The second electron is

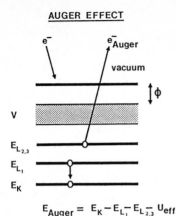

AUGER EFFECT

$$E_{Auger} = E_K - E_{L_1} - E_{L_{2,3}} U_{eff}$$

Fig. 6.1. Schematic energy level diagram illustrating the origin of the Auger effect. Incident radiation, such as primary electron beam, causes ionization of a core level. The hole in the core level is then filled with an electron from a higher shell. The energy released is carried away by the emission of a second electron

the Auger electron, and its energy is independent of the energy of the ionizing source. This allows the identification of the element from the measurement of the emitted electron energies.

The origin and nature of the Auger process can be understood from a diagram of the electron energy levels in Fig. 6.1. Ionizing radiation ejects an electron from an atom in the solid, leaving a hole in one of the atomic core levels. This core level is quickly filled by an electron from a higher level shell, and energy is released. This energy can leave the atom in the form of an X-ray, or there can be a competing process where another electron gains the energy, and is ejected from the atom. This second ejected electron is the Auger electron, and its energy depends on the energy of the atomic level involved in its production, not on the energy of the initial ionizing radiation. The energy of the Auger electrons can therefore be used for elemental analysis. Only H and He do not give rise to Auger electrons as two electron shells are needed. Although Li only has one electron per atom in the L shell it can share its valency electrons in a solid.

The energy of the Auger electron emitted from a solid surface is largely determined by the binding energies of the atomic energy levels in the participating atom. There are also contributions to the Auger electron binding energy from the screened Coulomb interaction of the final state holes, and relaxation of the surrounding electrons. This relaxation involves both the electrons in the atom itself [6.5], and in the surrounding material [6.6].

If the Auger electron energy is written in terms of experimental photoelectron binding energies referenced to the Fermi level, the energy can be written:

$$E_{Auger} = E_x - E_y - E_z - U_{eff} \qquad (6.1)$$

where E_x, E_y, E_z are the binding energies of the three participating electrons. The term U_{eff} is a term that expresses the extra energy needed to remove an electron from a doubly ionized atom, and the dynamic relaxation of the electrons during the two electron emission process. A review of these effects is given by *Weissmann* and *Muller* [6.7].

In the routine analysis of materials it is usually not necessary to understand the origin of the Auger transitions in detail, as the major Auger energies are tabulated and presented in standard spectra in reference handbooks [6.8, 9]. These reference spectra are very useful for the initial interpretation of unknown Auger spectra. However, they are limited in that only elemental spectra, or a very small number of spectra from simple compounds are given. Chemical effects can change both the shape, relative intensity, and energy of Auger transitions [6.7, 10].

The nomenclature used to describe an Auger transition is derived from X-ray terms. The inner electron shell being the K, and higher shells being L, M, N and V to denote valence states. For example the strong carbon KLL Auger feature around 270 eV is formed by an initial ionization of the inner K shell $1s$ electrons. The valence electrons of the carbon atom are in the two L shells. One valence electron fills the K shell hole and the other is ejected as the Auger electron. The carbon spectrum is very sensitive to the chemical environment, and the density of states that the valence electrons occupy [6.7, 11]. For carbon there are differences between the different structural forms such as graphite and diamond [6.12, 13]. The Auger transition nomenclature can be made more rigorous by including with the X-ray terms, the partial terms for the angular momentum. For s, p and d the terms 1, 2, 3 are used, eg KL_1L_2.

6.2 Electron Spectra from a Solid Surface

The energy distribution of electrons from a sample under excitation by a primary electron beam is shown in Fig. 6.2. At the high energy side there are the reflected (elastically backscattered) primary electrons. Just to the low energy side of this peak are the primary electrons that have suffered small energy losses, including losses to both surface and bulk plasmons. There are also loss processes that entail transitions from bound atomic core states to unfilled band states. Below the loss electrons is a long tail of high energy electrons that have lost energy in multiple collisions. Superimposed on this distribution of rediffused primary electrons, are features due to Auger electrons. Below each Auger line is a tail of Auger electrons from deeper in the solid that have lost energy before being emitted. Finally, at the lowest energies are the so-called "true" secondary electrons which are ejected from the valence and conduction bands. The energies of the Auger electrons used for surface spectroscopy are in the range 20–2000 eV. Electrons in this energy can travel only a few monolayers in a solid before losing energy. Therefore Auger spectroscopy is very surface sensitive [6.14].

The background slope is largely determined by the Auger electron flux as this is the major internal source of electrons apart from the backscattered primary electrons. The Auger electrons lose energy as they pass through the sample, and the lower energy electrons in turn lose energy, leading to a cascade of secondary electrons. From diffusion arguments this cascade of electrons with energies lower than the Auger energies can be described by an exponential function, and the i regions between the Auger transitions can be described [6.15]:

151

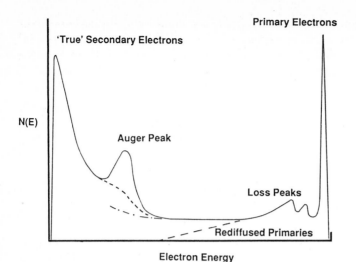

Fig. 6.2. Schematic of the secondary electron distribution from a solid surface. At high energy are the reflected primary electrons. Just below the elastic peak in energy, are primary electrons which have suffered discrete losses to plasmons. At low energy there is a long tail of rediffused primary electrons that have suffered many losses, and superimposed on this background are Auger electron features. The Auger electrons initiate further contributions to the background. At the lowest energy are the true secondary electrons from the conduction and valence bands

$$N(E) = A_i E^{-m_i} \tag{6.2}$$

where the A_i are constants of proportionality found from experiment, and the exponents m_i are close to 1. The actual values of m_i are determined by the balance between the inelastic and elastic scattering in the material [6.16]. The exponential dependence has been found to be reasonable for many materials when the component of background due to the losses from the primary electrons can be ignored ($E < E_{\text{primary}}/2$). The background below an isolated Auger peak due to higher energy transitions can therefore be removed by fitting (6.2) to the spectrum on the high energy side of the peak. In experimental data the exponents m_i can be most readily observed in log-log plots of $N(E)$ against E. In Fig. 6.3 such a plot of the $N(E)$ distribution from a 4 nm carbon film on Ag is shown. It can be seen that the regions above and below the carbon peak can be fitted linearly. The change in slope is due to the cascade caused by carbon electrons losing energy in escaping from the subsurface and bulk. The region under the Auger peak also includes loss electrons, and this contribution, $S(E)$, can therefore be approximated by a loss curve from each part of the Auger peak of the form of (6.2) [6.17]:

$$S(E) = A_2 E^{-m_2} \int_{E_T}^{E} N_A(E) dE \tag{6.3}$$

where $N_A(E)$ is the Auger current at energy E, and E_T is the threshold above which N_A is zero. E_T is approximately 290 eV in Fig. 6.3. This can be solved

152

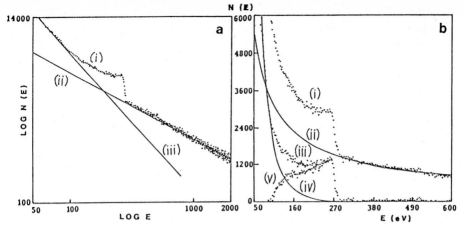

Fig. 6.3a,b. Results from a 4 nm C film on an Ag substrate. Taken with 20 keV primary energy. (a) log $N(E)$ vs log(E) plot. The two straight lines are of the form (6.2). (b) $N(E)$ vs E plot showing (*i*) raw data; (*ii*) high energy cascade; (*ii*) after removal of (*iii*); (*iv*) low energy cascade; and (*v*) final C KVV Auger line which was integrated. After [6.21]

iteratively by the simplifying assumption that the total cascade can be considered to be the sum of the two cascades. Removing both the contribution of (6.3) and the low energy linearized cascade gives the background stripped Auger signal, shown in Fig. 6.3. An alternative method is to successively remove a loss function derived from the experimental electron energy loss spectrum at the Auger energy [6.18]. At present, the form of the $N(E)$ is not fully understood, but future trends in background analysis, and synthesis may play an increasingly important part in the understanding of Auger quantification [6.19].

6.3 Instrumentation

The basic AES instrument consists of an electron gun to irradiate the sample surface, and an electron energy analyzer to collect, and analyze the secondary electron distribution. The two most popular analyzer designs for AES currently in use, are the cylindrical mirror analyzer (CMA), and the concentric hemispherical analyzer (CHA). A third type of analyzer, the four grid low energy electron diffraction (LEED) analyzer, which was initially used extensively for Auger studies [6.20], is still used for experiments on single crystal surfaces, where a monitor of both surface structure, and chemistry are required. However, because the CMA, and CHA have higher efficiencies, they are favoured for materials analysis. These analyzers are illustrated schematically in Fig. 6.4.

The CMA shown in Fig. 6.4a is the analyzer most often used for routine AES. It consists of two concentric cylinders, the inner cylinder at ground, and the outer cylinder at a negative voltage. Electrons entering the analyzer at 42° from the

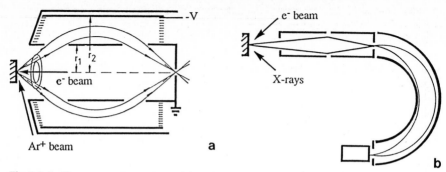

Fig. 6.4a,b. Electron spectrometers used for AES. (a) CMA cylindrical mirror analyzer. (b) CHA, concentric hemispherical analyzer

axis, and having the correct energy, will be focused through the analyzer. The pass energy is scanned by ramping the outer cylinder voltage. An area defining aperture is placed at the exit focus, and the transmitted electrons are detected by an electron multiplier. The multiplier is used either for single electron pulse counting, or current amplification. CMAs normally have an electron gun inside the inner cylinder so that the instrument is very compact. The CMA has a well defined focal point, and the distance of the sample surface from the end of the analyzer is important for accurate electron energy measurement. The transmission and energy resolution of the analyzer are also influenced by the sample position, and the size of the area on the sample excited by the primary electron beam. The ultimate resolving power of the analyzer is fixed by the angle of acceptance around the input angle, but a variable exit aperture can be used to increase the signal at the expense of the energy resolution.

The CMA has a fixed resolving power. That is, $\varrho = E/\Delta E$ is constant, where ΔE is the width of the energy window transmitted, E is the transmitted energy and ϱ is the resolving power. For a CMA, ϱ is typically 200 : 1. When the outer cylinder voltage is ramped to collect a spectrum from the emitted secondary electron distribution $N(E)$, then the detected signal is proportional to $E\,N(E)$.

A second type of spectrometer used in AES is the concentric hemispherical analyzer (CHA), shown in Fig. 6.4b. This analyzer usually consists of an input lens, and two concentric hemispheres. Electrons from the sample are focused by the input lens onto a slit at the entrance of the hemispheres. A potential difference between the hemispheres is arranged to give a $1/r^2$ field, and electrons with the correct energy are focused into a slit at the exit of the hemispheres. The resolving power of the hemispheres is determined by the slit size, the mid hemisphere radius, and the entrance angle. However, the resolving power of the total system can be changed by changing the input lens retardation. This gives the CHA great flexibility, and high energy resolution. The resolution is also not so strongly influenced by the source size, and alignment, as in the CMA.

Auger spectra are normally displayed in one of two ways. The direct spectra, $E\,N(E)$, can be plotted as shown in Fig. 6.5a, but historically the derivative spectra, $d/dE[E\,N(E)]$ shown in Fig. 6.5b has been preferred. This is because

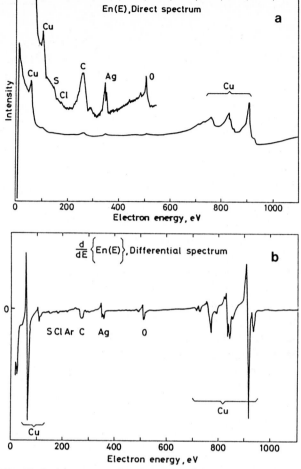

Fig. 6.5a,b. Auger spectra from contaminated copper. (a) The direct spectrum $E N(E)$. The surface contaminants of the copper show up as very small peaks on a large secondary background. (b) The differential spectrum obtained by sinusoidal energy modulation. The Auger peaks now appear as swings around zero, and the background slope is suppressed. After [6.26]

the electron current is amplified by an electron multiplier, and the signal is at high voltage with respect to earth. To overcome this problem the spectrometer energy is modulated at several kilohertz. The modulated signal can be detected at earth potential by using a filter capacitor. The modulated signal is detected using a lock-in amplifier as the differential of the $E N(E)$ signal. This has the further advantage that the large, slowly varying, inelastic background under the Auger peaks is suppressed.

6.4 Quantification of Auger Spectra

The Auger spectrum from a surface is a very complex signal, and depends strongly on many factors, besides the elemental concentrations. To quantify a spectrum, or even to compare spectra, a close understanding of these contributions to the Auger signals is necessary. Unfortunately, at present, there are many unresolved issues in the quantification process, although these can be reduced to two principle areas: the accurate measurement of the secondary electron distribution, and its interpretation.

All instruments used for AES are to some extent unique, and have output signals that do not strictly represent the true secondary electron distribution with energy. One major factor in quantification, therefore, is the detailed understanding of the properties of the specific analyzer being used in the measurement. A corollary of this is, that unless the signal response of a spectrometer system can be accurately characterized, then some difficulty will be found in determining the meaning of the signal. Although CMA based systems are presently far more popular for AES anlysis, accurately characterized CHA analyzer systems appear to be more suitable for general quantitative problems [6.21, 22]. For quantitative work, CHA spectrometers have the advantages that the source distribution, and positioning are not highly critical, and that the spectrometers can be operated in a constant pass energy mode. This mode allows the multiplier efficiencies to be keep constant with spectrum energy. CHA spectrometers can be used to measure the direct $N(E)$ spectrum, and it may be, that ultimately, the other quantification issues related to signal interpretation, can only be resolved from the measurement of the true secondary electron distribution, not a function that includes measurement artifacts such as the differential spectrum. For a review of these measurement artifacts see *Seah* [6.23].

In practice Auger spectra are quantified with different levels of sophistication. The level of accuracy depends very much on the both the materials system, and the instrument. There is no general method of quantifying Auger spectra. Very different approaches can be used in the different circumstances of dilute alloys, multicomponent systems, or surface films [6.24]. There are several reviews of the subject in the literature [6.23, 25, 26] and here we will only follow one basic approach which uses sensitivity factors derived from pure elemental standards.

Using bulk elemental standards, the atomic concentration, C_i, of an element i, can be calculated from:

$$C_i = \frac{I_i/I_i^\infty}{\sum_j F_{ij} I_j / I_i^\infty} , \qquad (6.4)$$

where j runs from 1 to the number of elements present at the surface, I_i is the Auger signal of element i measured from the unknown, I_i^∞ is the corresponding signal from the pure element, and F_{ij} is the matrix factor. The matrix factor includes changes in backscattering, attenuation length, and atomic density, between the pure elemental standard, and the matrix. The use of matrix factors is reviewed by Seah [6.26]. The relative elemental intensities can be obtained from compilations [6.8, 27], but it is preferable to use elemental standards measured on the same instrument as the unknown. For semiquantitative analysis (20–30% error) it is not critical how the Auger peak intensity is measured, as long as this is done consistently within an experiment. Conventionally, the peak-to-peak height difference of the negative and positive excursions in the differential spectra are used. But if there are overlapping peaks or chemical shifts between surface phases, just the one excursion can be used. A simple measure of the Auger signals from $E N(E)$ spectra and $N(E)$ can be made by choosing a point on the background just above the peaks, and subtracting this signal from the intensity at the peak maximum. On a background with a steep slope, the slope can be approximately corrected for by a linear extrapolation of the background for $E N(E)$ spectra and an E^{-1} linear extrapolation for $N(E)$ spectra.

To take Auger quantification further than the analysis above, many factors must be taken into account. These factors include the interaction of the primary electron beam with the target, the crossection of the Auger transition, the interaction of the outgoing Auger electrons with the target, the properties and response of the electron spectrometer system, and the variation in the spectra due to changes in secondary electron distributions, and overlapping Auger transitions. Unfortunately, each of the above mentioned factors influencing the yield can also be complex. For example, the interaction of the primary electron beam with the target depends on the angle of incidence, and is therefore dependent on the microtopography of the target [6.28], the backscattering of primary electrons from the target subsurface [6.29], and the diffraction of the primary electrons into the target [6.30, 31]. The yield is also dependent on the energy of the primary electrons, and on the spectrometer alignment. Thus, understanding the possible contributions to the yield is vital to performing a meaningful analysis.

6.5 Materials Analysis Using AES

6.5.1 Thin Film Analysis Using Depth Profiling

Auger spectroscopy can be used in conjunction with ion sputtering in order to obtain depth profiles of elemental distribution in thin films. This is a widely used technique in the semiconductor industry [6.32–37], and it has also been applied in many other fields for the analysis of thin films and surface distributions (e.g. solid state diffusion [6.38, 39] oxidation layers [6.40], and passivation of the surface [6.41]). Depth profiling uses an ion beam to sputter the specimen surface layer by layer. The surface sensitivity of AES means that only the top few monolayers that have been exposed by the ion beam are analyzed, and as the ion beam erodes the sample, the intensity of several Auger transitions are followed. This gives an intensity profile as a function of time. With a knowledge of the sputtering rate, and the elemental sensitivities, this can be converted into concentrations with depth.

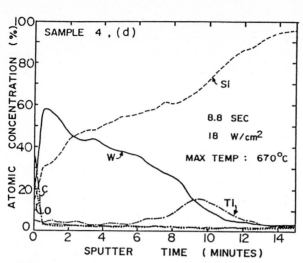

Fig. 6.6. Auger depth profile from a study of the formation of tungsten silicide on Si. A Ti film has been interposed between the W surface film, and the Si substrate. From the study it was found that, on annealing, the formation of WSi_2 was enhanced by the Ti film, the surface roughness was decreased, and the adhesion improved. From [6.37]

A typical depth profile is shown in Fig. 6.6. This data is from a study of the influence of a Ti barrier film on the reaction between a W overlayer, and a Si substrate. The depth distribution of the reactants shows that the Ti film persists on annealing, and enhances the formation of tungsten silicide. This technique is extremely valuable in analyzing films between 10 nm and 1 μm. Normally an Ar ion beam with an energy between 0.5–5 keV is used for sputtering. The energy is chosen either to increase the sputtering rate, by choosing a high energy, or to reduce the effects of ion beam mixing on the depth resolution by using a lower energy. The depth resolution of the profiles from most materials, which is limited by ion beam mixing of the surface layers, is 5 nm at best.

Ion beam bombardment is not a neutral process. Although inert gases are used to minimize the chemical interaction between sputtering ions and the target, alterations to the composition of the sample can still occur. For example, if the sputtering yield of two or more elements in the same material differ significantly, then the element that sputters most rapidly becomes depleted in the near surface region, until its sputtering rate matches the rate of the slowest sputtering element [6.42]. Therefore, as material is sputtered away, a depletion front is formed until the sputtering of all elements reaches a steady state. In some circumstances this can be corrected for [6.43]. However in general, unless the system being sputtered has been very well characterized, some caution should always be used in reporting depth profiles quantitatively. This can be more complex, because it may not only be the elemental sputtering rates that determine the composition of the sputtered surface [6.42], but other factors such as the chemistry of the surface, or even ion induced segregation [6.44].

Depth profiles can also be made using chemical wedges and ball cratering. These techniques can be used to obviate the problems of slow sputtering through oxides, and the loss of depth resolution because of the uneven topography formed after prolonged ion sputtering [6.45–47].

6.5.2 Scanning Auger Microscopy

Scanning Auger microscopy (SAM) is a powerful tool for materials microanalysis. The spatial resolution of presently available commercial instruments is in the range 50—200 nm, and 20 nm has been demonstrated on an experimental microscope [6.2]. This compares well with the 1–2 μm resolution of an X-ray microprobe, which is the workhorse of materials microanalysis. Figure 6.7 shows a comparision of optical, backscattered, X-ray wavelength dispersive, and Auger microscope images of sintered SiC from the same study [6.48]. It is only in the Auger microprobe that a highly dispersed second phase component, boron nitride, has been detected. The Auger microprobe compares well with the X-ray

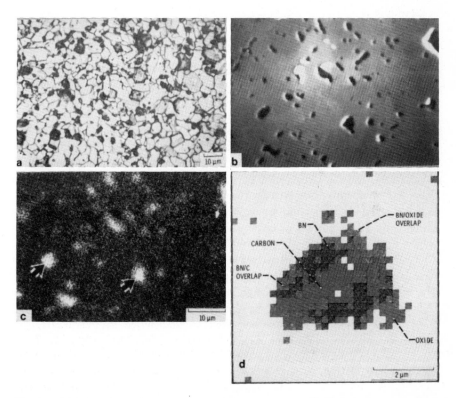

Fig. 6.7a–d. Micrographs of SiC sintered in a nitrogen atmosphere. (a) Optical micrograph of polished and etched microstructure. (b) 15 keV backscatter electron image. (c) Carbon X-ray map by wavelength dispersive analysis. (d) Monochrome reproduction of chemical phase image, derived from Auger imaging. The phase image is of one particle at the triple point between SiC grains. Note the different scales between this and the X-ray image in (c)

microprobe, both in sensitivity to light elements, and in its spatial resolution. Limitations are that the quantification methods have not reached the same level of confidence the X-ray microprobe has attained on many systems. Also, because of the surface sensitivity, coating insulating specimens is not possible. Geological applications of SAM are possible [6.49]. AES has been used to show that surface Fe enrichment by the Solar wind is a likely cause of the darkening of Lunar rocks [6.50,51], and the ability to analyze thin surface layers makes accretion studies possible in less than geological times [6.52–54]. On semiconductors [6.55], superconducting oxides [6.56,57], and engineering ceramics [6.58,59], Auger microscopy is used for both multiphase microanalysis, and for surface studies in corrosion and fracture.

Because of its surface sensitivity, scanning Auger microscopy has become an important tool in segregation, and embrittlement studies, where the chemistry of internal surfaces can dominate the properties of a material. SAM can be used both to study the fracture surfaces of a material, to show how grain boundary segregation influences fracture, and in surface segregation experiments, to show how tempering treatments and additives are likely to influence grain boundary segregation. If the fracture is done within the vacuum chamber, the surface of the fracture can be analyzed directly. An outstanding advantage of SAM for this work is that the intragranular fracture can be separated from the transgranular fracture using the SEM capabilties. Sub-monolayer amounts of segregants to the fracture surface can be analyzed using the Auger microprobe, and the influence of temper and additives can be determined. A wide range of materials has been studied using this technique, for example, steels [6.60,61], high temperature alloys [6.62], magnetic alloys [6.63], ceramics [6.58,59,64], and superconductors [6.57]. AES has been used to confirm the long suspected competitive grain boundary segregation in steel of P and C, as embrittling and de-brittling agents respectively, and the effects of Cr, Mn and V as elements that remove carbon from this competitive segregation [6.60,61]. These studies very nicely show, that the effects of annealing temperature on intergranular fracture, are related to the competitive process of enhanced diffusion rate, and the increase in both solubility, and interaction of segregants with temperature [6.60].

Grain boundary segregation can also be simulated by monitoring vacuum (free) surface segregation while heating within the scanning Auger microscope. The effects of concentration change, annealing time, and temperature, can be followed. Also the distribution on different metallic phases can be observed [6.61,65–68]. This technique has been used to simulate the "rare earth effect" in the oxide scale adhesion of the high temperature superalloys [6.69–71], and to follow intial oxidation, and scale formation of high temperature alloys [6.72,73]. The superalloys, such as NiCrAl, are used in turbine blades, and at high temperatures form a protective alumina scale. Without additions of a rare earth, the oxide scale completely spalls off during thermal cycling. The spalling reduces the life of the turbine blade. The effect of adding Zr and Y dopants to the alloy is dramatic. Figure 6.8 shows the segregation of these dopants, and the suppression of S segregation using AES on the vacuum NiCrAl surface [6.70]. The

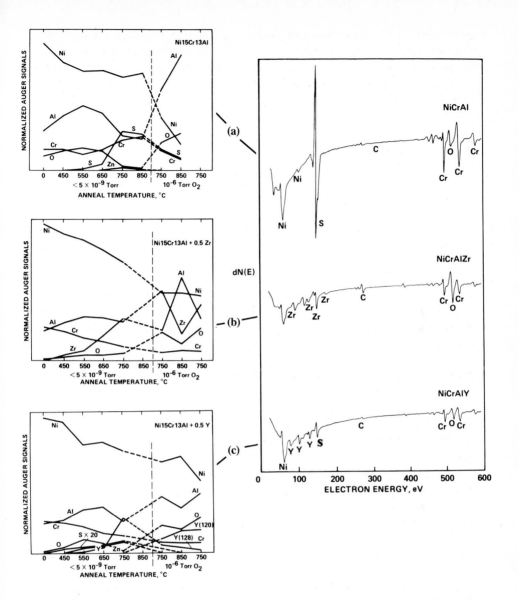

Fig. 6.8a–c. Competitive segregation of S, Zr, and Y with temperature and oxidation conditions to the vacuum surface of Ni15Cr13Al alloy monitor using AES. Left hand side of anneal profiles show UHV anneal, the right hand side an oxidizing anneal in 10^{-6} Torr O_2. AES spectra correspond to 705°C anneal in UHV. (a) Auger signal intensity from undoped NiCrAl. (b) Auger signal intensity from 0.5% Zr doped NiCrAl. (c) Auger signal intensity from 0.5% Y doped NiCrAl. After [6.70]

assumption in this work is that the segregation to the free surface, simulates the segregation to the oxide/metal interface, and thus conclusions about the scale adhesion can be drawn. Although there is still some discussion about the validity of the conclusions [6.71], the experiments provide a novel look at these complex solid state processes.

6.6 Future Trends

AES is still a developing field, both in techniques and in applications. For materials science the most important trends are towards quantification of spectra, and depth profiles, and the increasing spatial resolution of Auger microscopy. The quantification issues are complex and international efforts are being made to tackle them. The Versailles Project on Advanced Materials and Standards (VAMAS), has a surface analysis working party, and recently a standard reference spectrum $N(E)$ for Cu has been established [6.74]. This reference standard will facilite the characterization of spectrometer transmission functions. Better understanding of the $N(E)$ spectrum, and possibilities in the computer simulation and synthesis of spectra, also open up new approaches to quantification.

Another important trend are the advances in Auger microscopy. On complex heterogeneous samples, multichannel image collection and processing techniques, promise to give a very detailed, and quantitative, view of materials microstructure [6.75]. An instrument capable of Auger imaging, backscatter imaging, SEM, and EDX imaging simultaneously has recently been built [6.76]. The spatial resolution of SAM has not reached its limit, and sub 20 nm resolution is possible. This would be over 50 times the spatial resolution of an X-ray microprobe. These developments are sure to stimulate further applications of the technique in materials science.

References

6.1   L.A. Larson: MRS Symposia, Materials Characterization, Vol. 69 (1986) p. 129
6.2   H. Todokoro, Y. Sakitani, S. Fukuhara, Y. Okajima: J. Electron Microsc. 30, 107 (1981)
6.3   L. Meitner: Z. Phys. 17, 54 (1923)
6.4   P. Auger: Comptes Rendus 177, 169 (1923)
6.5   D.A. Shirley: Phys. Rev. A 7, 1520 (1973)
6.6   R. Hoogewijs, L. Fiermans, J. Vennik: Surf. Sci. 69, 273 (1977)
6.7   R. Weissmann, K. Muller: Surf. Sci. Rep. 105, 251 (1981)
6.8   L.E. Davis, N.C. MacDonald, P.W. Palmberg, G.E. Riach, R.E. Weber: *Handbook of Auger Electron Spectroscopy* (Physical Electronics, Eden Prairie, MN)
6.9   T. Sekine et al.: *Handbook of Auger Electron Spectroscopy* (JEOL Ltd. Tokyo, Japan)
6.10 P. Morgen, B. Jorgensen: Surf. Sci. 208, 306 (1989)
6.11 P.J. Lurie, J.M. Wilson: Surf. Sci. 65, 476 (1977)
6.12 T.W. Hass, J.T. Grant, G.J. Dooley III: J. Appl. Phys. 43, 1853 (1972)
6.13 Y. Mizokawa, T. Miyasato, S. Nakamura, K.M. Geib, C.W. Wilmsen: Surf. Sci. 182, 431 (1987)

6.14　M.P. Seah, W.A. Dench: Surf. Int. Anal. **1**, 2 (1979)
6.15　E.N. Sickafus: Phys. Rev. B **16**, 1436 (1977)
6.16　J.A.D. Matthew, M. Prutton, M.M. El Gomati, D.C. Peacock: Surf. Int. Anal. **11**, 173 (1988)
6.17　E.N. Sickafus: Surf. Sci. **100**, 529 (1980)
6.18　C. Burrell, N.R. Armstrong: Appl. Surf. Sci. **17**, 53 (1983)
6.19　S. Ichimura, D. Ze-Jun, R. Shimizu: Surf. Int. Anal. **13**, 149 (1988)
6.20　C.C. Chang: Surf. Sci. **25**, 53 (1971)
6.21　M. Prutton, M.M. El Gomati: Surf. Int. Anal. **9**, 99 (1986)
6.22　C.G.H. Walker, D.C. Peacock, M. Prutton, M.M. El Gomati: Surf. Int. Anal. **11**, 266 (1988)
6.23　M.P. Seah: Surf. Int. Anal. **9**, 85 (1986)
6.24　P. Lejcek: Surf. Sci. **202**, 493 (1988)
6.25　J.T. Grant: Surf. Int. Anal. **14**, 271 (1989)
6.26　M.P. Seah: SEM 83, SEM Inc. AMF O'Hare, Chicago (1983) p. 521–536
6.27　T. Sekine, A. Mogami, M. Kudoh, K. Hirata: Vacuum **34**, 631 (1984)
6.28　P.H. Holloway: J. Electron Spectrosc. Relat. Phenom. **7**, 215 (1975)
6.29　S. Ichimura, R. Shimizu: Surf. Sci. **112**, 386 (1981)
6.30　H.E. Bishop, B. Chornik, C. LeGressus, A. LeMoel: Surf. Int. Anal. **6**, 116 (1984)
6.31　F.E. Doern, L. Kover, N.S. McIntyre: Surf. Int. Anal. **6**, 282 (1984)
6.32　L.P. Erickson, B.F. Phillips: J. Vac. Sci. Technol. B **1**, 158 (1983)
6.33　H.H. Busta, C.H. Tang: J. Electrochem. Soc. **133**, 1195 (1986)
6.34　F. Marchetti, M. Dapor, S. Girardi, F. Giacomozzi, A. Cavalleri: Mater Sci. Eng. A **115**, 217 (1989)
6.35　R. Pantel, F.A. D'Avitaya: Thin Solid Films **140**, 177 (1986)
6.36　S.J. Pearton, A.B. Emerson, U.K. Chakrabarti, E. Lane, K.S. Jones, K.T. Short, A.E. White, T.R. Fullowan: J. Appl. Phys. **66**, 3839 (1989)
6.37　C.S. Wei, M. Setton, J. Van der Spiegel, J. Santiago: J. Appl. Phys. **61**, 1429 (1987)
6.38　W. Palmer: Appl. Phys. A **42**, 219 (1987)
6.39　B.M. Clemens: J. Non Cryst. Solids **61–62**, 817 (1984)
6.40　T.T. Huang, B. Peterson, D.A. Shores, E. Pender: Corrosion Science **24**, 167 (1984)
6.41　S. Mathieu: La Revue de Metallurgie, January 1989, p. 73
6.42　P.H. Holloway: Surf. Sci. **66**, 479 (1977)
6.43　T. Sekine, A. Mogami, J.D. Geller, SEM, SEM Inc. AMF O'Hare (1981) p. 245
6.44　R. Li, L. Tu, Y. Sun: Surf. Sci. **163**, 67 (1985)
6.45　J.M. Walls, I.K. Brown, D.D. Hall: Appl. Surf. Sci. **15**, 93 (1983)
6.46　I. Lhermitte-Sebire, M. LaHaye, R. Colmet, R. Naslain: Thin Solid Films **138**, 209 (1986)
6.47　J. F. Bresse: Surf. Sci. **168**, 810 (1986)
6.48　R. Browning, J.L. Smialek, N.S. Jacobson: Advanced Ceramic Materials **2**, 773 (1987)
6.49　M.H. Hochella, M.F. Turner, D.W. Harris: SEM II (1986) p. 337
6.50　T. Gold, E. Bilson, R.L. Baron: Proc. 5th Lunar Sci. Conf. (1975) p. 2413
6.51　T. Gold, E. Bilson, R.L. Baron: Proc. 6th Lunar Sci. Conf. (1975) p. 3285
6.52　J.W. Morse, A. Mucci, L.M. Walter, M.S. Kaminsky: Science **205**, 904 (1979)
6.53　A. Mucci, J.W. Morse, M.S. Kaminsky: Am. J. Sci. **285**, 289 (1985)
6.54　A. Mucci, J.W. Morse, M.S. Kaminsky: Am. J. Sci. **285**, 306 (1985)
6.55　J.F. Jongste, F.E. Prins, G.C.A.M. Janssen: Matt. Lett. **8**, 273 (1989)
6.56　M.G. Ransey, F.P. Netzer: Mat. Sci. Eng. B **2**, 269 (1989)
6.57　A. Roshko, Y.M. Chiang: J. Appl. Phys. **66**, 3710 (1989)
6.58　H.H. Madden, W.O. Wallace: Surf. Sci. **172**, 641 (1986)
6.59　R. Sherman: J. Am. Ceram. Soc. **68**, C7 (1985)
6.60　H.J. Grabke: ISIJ **29**, 529 (1989)
6.61　M. Militzer, J. Wieting: Acta Metall. **37**, 2585 (1989)
6.62　M. Takeyama, C.T. Liu: Acta Metall. **37**, 2681 (1989)
6.63　W.E. Wallace, S.G. Sankar, J.M. Elbiki, S.F. Cheng: Mat. Sci. Eng. B **3**, 351 (1989)
6.64　R. Hamminger, G. Grathwohl, F. Thummler: J. Mater. Sci. **18**, 3154 (1983)
6.65　D.C. Peacock: Appl. Surf. Sci. **26**, 306 (1986)
6.66　D.C. Peacock: Appl. Surf. Sci. **27**, 58 (1986)
6.67　P. Humbert, A. Mosser: Surf. Sci. **126**, 708 (1983)
6.68　J. Du Plessis, P.E. Vijoen, F. Bezuidenhout: Surf. Sci. **138**, 26 (1984)
6.69　J.G. Smeggil, A.W. Funkenbush, N.S. Bornstein: Metall. Trans. **17A**, 923 (1986)
6.70　J.L. Smialek, R. Browning: Proc. Symp. High Temp. Material Chemistry-III, Electrochem. Soc. (1986) p. 258

6.71   K.L. Luthra, C.L. Briant: Mater. Sci. Forum **43**, 299 (1989)
6.72   K. Sato, Y. Inoue: ISIJ **29**, 246 (1989)
6.73   M. Tomita, T. Tanabe, S. Imoto: Surf. Sci. **209**, 173 (1989)
6.74   G.C. Smith, M.P. Seah: Surf. Int. Anal. **12**, 105 (1988)
6.75   R. Browning: MRS Bull. **12**, 75 (1987)
6.76   M. Prutton, C.G.H. Walker, J.C. Greenwood, P.G. Kenny, J.C. Dee, I.R. Barkshire, R.H. Roberts: A third-generation Auger microscope using Parallel Multispectral Data Acquisition and Analysis, Sur. Int. Analy. **17**, 71–84 (1991)

7. X-Ray Photoelectron Spectroscopy

M.H. Kibel

With 14 Figures

The detection and energy analysis of photoelectrons produced by radiation whose energy exceeds their binding energies is the subject of an extensively-used technique known as photoelectron (PE) spectroscopy. This technique can be conveniently divided into two broad areas, the first employing ultraviolet radiation, hence called ultraviolet photoelectron spectroscopy (UPS), and the second using X-rays, termed X-ray photoelectron spectroscopy (XPS). The latter spectroscopy is the subject of this present chapter, while UPS is discussed in Chap. 14.

The history of the development of XPS is a fascinating topic by itself, but involved discussion is not appropriate here. For a detailed account, the reader is referred to the excellent work of Jenkin et. al. [7.1–4]. XPS is also known as electron spectroscopy for chemical analysis (ESCA), a name coined by *Siegbahn* in order to emphasize the presence of both photo- and Auger electron peaks in the XP spectrum [7.5, 6].

7.1 Basic Principles

7.1.1 Theory

Presented schematically in Fig. 7.1 are the related processes that are involved in the ejection of a photo- or Auger electron. XPS involves the removal of a single core electron, while AES is a two-electron process subsequent to the removal of the core electron, with the Auger electron ejected following reorganization within the atom. Auger electrons are produced in XPS along with photoelectrons, and these can complicate the interpretation of the subsequent spectra as will be discussed later in this chapter.

The photoemission process is shown on an energy level diagram in Fig. 7.2. The sample is irradiated with X-rays of known energy, $h\nu$, and electrons of binding energy (BE) E_b are ejected, where $E_b < h\nu$. These electrons have a kinetic energy (KE) E_k which can be measured in the spectrometer, and is given by

$$E_k = h\nu - E_b - \Phi_{sp} , \qquad (7.1)$$

where Φ_{sp} is the spectrometer work function, and is the combination of the sample work function, Φ_s, and the work function induced by the analyzer. Since

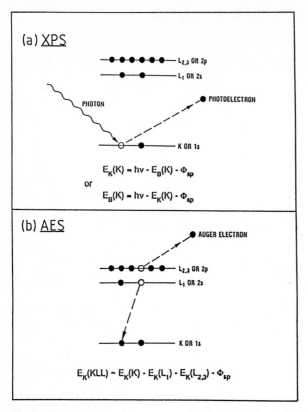

Fig. 7.1. Representation of the processes involved in XPS and AES. XPS involves the removal of a single core electron while AES is a two-electron process following the removal of the core electron

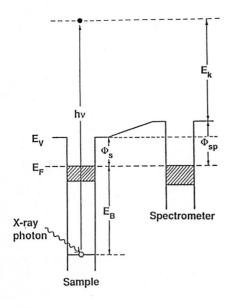

Fig. 7.2. Schematic diagram of the photoemision process. The sample is irradiated with X-rays of known energy, $h\nu$, and electrons of binding energy E_b are ejected. These electrons have a kinetic energy, E_k, which can be measured in the spectrometer, and is given by (7.1). Φ_s is the work function of the sample, and Φ_{sp} is the spectrometer work function

we can compensate for the work function term electronically, it can be eliminated, leaving

$$E_k = h\nu - E_b \tag{7.2}$$

or

$$E_b = h\nu - E_k \ . \tag{7.3}$$

Thus by measuring the KE of the photoelectrons, (7.3) can be used to translate this energy into the BE of the electrons.

7.1.2 Typical Spectrum

An XP spectrum is generated by plotting the measured photoelectron intensity as a function of BE, as shown in Fig. 7.3, which is the Mg $K\alpha$ ($h\nu = 1253.6\,\text{eV}$) XP spectrum of a HgCdTe film [7.7]. The resulting series of lines are superimposed on a background caused by the Bremsstrahlung radiation inherent in non-monochromatic X-ray sources. The BEs of these lines are characteristic for each element, and are a direct representation of the atomic orbital energies. Published tables of photoelectron BEs for all elements can be used to assist in the assignment of peaks in XP spectra [7.8, 9]. Note that the carbon $1s$ orbital is

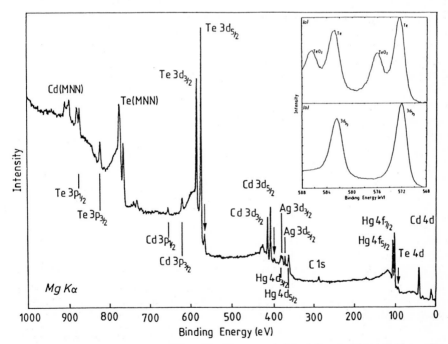

Fig. 7.3a,b. Mg $K\alpha$ XP spectrum of $Hg_{0.6}Cd_{0.4}Te$ grown via Metal Organic Chemical Vapour Deposition (MOCVD). Inset: (a) Te $3d$ region of full spectrum before cleaning, showing the presence of oxide. (b) Te $3d$ region after ion etching, which has removed the oxide

represented by a single line in the spectrum, while the various p, d and f orbitals produce two lines. These doublets are due to spin-orbit splitting, and their occurrence will be explained in Sect. 7.3.1. An expansion, of the Te $4d$ region is presented in Fig. 7.3a (inset), which shows that there are actually four components apparent for Te. The higher intensity peaks represent elemental Te in the zero oxidation state, with the other two peaks representing TeO_2, i.e. Te in an oxidation state of +4. This is a typical example of a chemical shift in the BE, which will be the topic of Sect. 7.3.2. From Fig. 7.3b the effect of a light argon-ion bombardment can be seen, with the complete removal of the TeO_2, showing how XPS can be employed to monitor surface cleanliness.

Returning to Fig. 7.3, note also the presence of Cd and Te MNN Auger lines. While these groups of peaks can sometimes be considered a hinderence to spectral interpretation due to overlap with XP peaks, their presence can also be of substantial value. As will be discussed in Sect. 7.3.3, the use of these Auger lines can yield valuable chemical shift information.

7.1.3 Surface Specificity

As was discussed in Sect. 1.5.1, the surface sensitivity of electron spectroscopies is due to the low inelastic mean-free path, λ_m, of the electrons within the sample. For XPS, the main region of interest relates to electron energies from 100–1200 eV, which gives rise to a λ_m value of 0.5–2.0 nm (or 5–20Å). However the actual escape depth, λ, of the photoelectrons depends on the direction in which they are travelling within the solid, such that [7.10]

$$\lambda = \lambda_m \cos \theta \tag{7.4}$$

where θ is the angle of emission to the surface normal. Thus electrons emitted perpendicular to the surface ($\theta = 0°$) will arise from the maximum escape depth, whereas electrons emitted nearly parallel to the surface ($\theta \sim 90°$) will be purely from the outermost surface layers. Thus XPS is an extremely surface sensitive technique, and as we will see from Sect. 7.5.3, varying the angle of detection during an experiment, as described above, can enhance this surface sensitivity.

7.2 Instrumentation

As with most techniques, certain components are essential, while others, although desirable, can be optional. In this section we discuss the components used in modern XPS systems, and consider different alternatives available for many of these components.

7.2.1 Essential Components

Figure 7.4 is a schematic diagram indicating the essential components necessary for performing XPS. These consist of an X-ray source, a sample/support system, an electron energy analyzer and an electron detector/multiplier, all maintained under ultra-high vacuum (UHV), and suitable electronics to convert the detected current into a readable spectrum. These components will now be discussed individually in more detail.

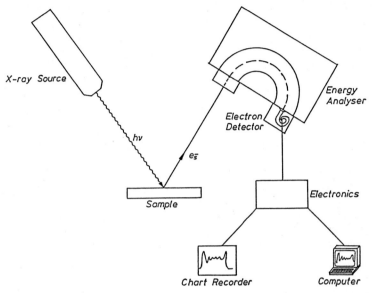

Fig. 7.4. Schematic representation of the components necessary for performing XPS

a) X-Ray Source

An ideal photon source must be sufficiently energetic to access core levels, intense enough to produce a detectable electron flux, have a narrow line width and be simple to use and maintain. Listed in Table 7.1 are some of the more common sources, together with their energies and line widths. The full-width-half-maximum (FWHM) of an XPS peak depends on several factors, but the major contribution comes from the line width of the X-ray line. From Table 7.1, the only lines which are sufficiently energetic and narrow are the Na $K\alpha$, Mg $K\alpha$ and Al $K\alpha$ lines. However it is very difficult to design a suitable, stable sodium anode, thus the Mg and Al sources are most commonly used. Many manufacturers now provide X-ray sources with dual Mg/Al anodes as standard. Other dual-anode combinations are also available, such as Mg/Zr and Al/Zr, since the Zr $L\alpha$ line has a particularly high sensitivity to elements such as aluminum and silicon [7.11].

The sources described above do not, however, produce single X-ray lines, but a series of lines superimposed on the Bremsstrahlung continuum. One method

Table 7.1. Common x-ray sources

X-ray line	Energy [eV]	Line Width [eV]	Comments
Y $M\zeta$	132. 3	0.47	Very low energy; very narrow line
Zr $M\zeta$	151. 4	0.77	Very low energy; narrow line
Na $K\alpha$	1041. 0	0.70	Very difficult to design
Mg $K\alpha$	1253. 6	0.70	Requires 15 keV; narrow line; stable
Al $K\alpha$	1486. 6	0.85	Requires 15 keV; narrow line; stable
Zr $L\alpha$	2042. 4	1.7	Wide line
Ti $K\alpha$	4510. 0	2.0	Requires 20 keV source; wide line
Cu $K\alpha$	8048. 0	2.6	Requires 30 keV source; wide line

of removing the unwanted components, and eliminating the continuum, is to monochromatize the radiation. This is most conveniently achieved by using a diffraction grating, with the line spacings for selecting the correct component determined from the Bragg relation [7.12]. Of course the resulting monochromatized radiation will be greatly reduced in intensity, due to dispersion.

All the sources discussed thus far are discrete line sources. Synchrotron radiation provides a continuous source of photons from 10 eV to 10 keV, which can allow the radiation to be tuned to attain the ideal photoionization cross-section for a particular set of core levels, however synchrotron time is extremely limited and very expensive, hence not available to most people.

b) Electron Energy Analyzer and Detector

Once the photoelectrons have been produced, they must be separated according to their energy and subsequently converted into a spectrum. The electron energy analyzer is thus the heart of any XPS system and is the critical component in determining sensitivity and resolution. There are several types of analyzer that have been used in electron spectrometers, but the simplest is the Retarding Field Analyzer (RFA) which although used mainly in LEED/AES systems, was occasionally used in UPS/XPS [7.13–15]. Of the electrostatic analyzers, the simplest is the 127° cylindrical analyzer, which consists of two concentric cylindrical deflectors focusing the electrons, which travel through an arc of 127° ($\pi/\sqrt{2}$), in one dimension. This type of analyzer is relatively easy to construct, and hence appears often in home-built systems [7.16]. The two most common analyzers encountered in XPS are the Cylindrical Mirror Analyzer (CMA) and the Concentric Hemispherical Analyzer (CHA).

The CHA is also known as the spherical deflection analyzer, and the basic operating principle is depicted in Fig. 7.5 [7.17]. Two hemispherical surfaces of inner radius R_1 and outer radius R_2 are positioned concentrically, with a potential ΔV applied such that the outer sphere is negative and the inner positive (with respect to ΔV). The electrons of kinetic energy eV travel along the median radius, R_0, where

$$R_0 = (R_1 + R_2)/2 \qquad (7.5)$$

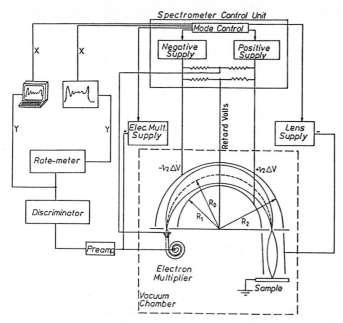

Fig. 7.5. Schematic diagram showing the operating principles of a concentric hemispherical analyzer (CHA) (see text for explanation), together with the electronics used to produce XP spectra

and

$$eV = E_k(R_2/R_1 - R_1/R_2) \tag{7.6}$$

where the entrance and exit slits are separated by an angle of 180°. This results in focusing in two dimensions, i.e. at a point.

The resolution of an analyzer can be defined in two ways. One is the *absolute* resolution ΔE measured as the full width at half-maximum height (FWHM) of a peak. The other is the *relative* resolution, ϱ, of a peak at KE E_0

$$\varrho = E_0/\Delta E . \tag{7.7}$$

Thus the absolute resolution is independent of peak position, but the relative resolution must be referred to the kinetic energy of the peak.

The current actually reaching the analyzer exit slit following photoionization is typically in the region of 10^{-16}–10^{-14} A, which is well below conventional current-measuring techniques. Thus pulse counting is required, and an electron multiplier is used as the detector. There are two types of electron multiplier currently in use in modern spectrometers: the traditional discrete-dynode type and the more common channel electron multiplier.

The dynode multiplier functions by allowing the electrons to strike the first electrode (or dynode), and then achieving current amplification via secondary electron multiplication of many stages (typically 10–20). A resistor chain is used

171

to establish the separate dynode potentials. Multipliers of this type can achieve amplification of 10^4–10^7, depending on their design.

A channel electron multiplier consists of a small curved glass tube, the inside wall of which is coated with high resistance ($\sim 10^9 \, \Omega$ full length) material. When a potential ($\sim 3\,\text{kV}$) is applied between the ends of the tube the resistive surface becomes a continuous dynode. An electron entering the low potential (typically +500 V) end of the multiplier generates secondary electrons on collision with the wall of the tube. These are accelerated until they strike the wall again, giving an avalanche effect. The gain of these devices is of the order of 10^7–10^8.

As seen in Fig. 7.5, the multiplier output is normally taken through a pre-amplifier, an amplifier, a discriminator and a rate-meter system, with the spectrum being displayed on an X-Y recorder. Alternatively, a computer interface is situated between the discriminator and rate-meter, with the spectra subsequently digitized and stored for future reference. Also shown in Fig. 7.5 is a schematic of the electronics required to operate the analyzer/detector system.

7.2.2 Optional Components

There are many experiments performed via XPS that require further components, other than those discussed above. For example, the addition of an electron gun allows AES to be performed with the same analyzer/electronics. The addition of a rare-gas ion gun allows samples to be cleaned in vacuum, as well as permitting depth-profiling studies to be performed. The ability to heat and cool a sample in situ can also be extremely useful, especially in adsorption/desorption and catalysis studies. While it is ideal to have a sample manipulator with these facilities, one can provide these options in a separate part of the instrument, e.g. a sample preparation chamber. It may also be necessary to install a low-energy electron flood gun for neutralizing the charge on insulating materials, such as ceramics or polymers. A mass spectrometer is also extremely valuable for residual gas analysis and, if the correct geometry is chosen, for thermal desorption spectroscopy. Many of these options are discussed elsewhere in this book, and in review articles [7.18, 19].

7.3 Spectral Information

Much important information can be gleaned from an XP spectrum, beyond simple BE data and elemental identification. In this section we look at what additional information the spectra can reveal.

7.3.1 Spin-Orbit Splitting

If an unpaired electron is in a degenerate orbital (i.e. $p, d, f, \ldots$) the spin angular momentum, S, and the orbital angular momentum, L, can combine in different

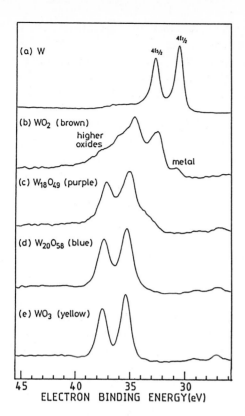

Fig. 7.6. W $4f$ XP spectra of tungsten and four tungsten oxides. (Reproduced with permission from [7.20])

(a) W

$4f_{5/2}$ $4f_{7/2}$

(b) WO$_2$ (brown)

higher oxides

metal

(c) W$_{18}$O$_{49}$ (purple)

(d) W$_{20}$O$_{58}$ (blue)

(e) WO$_3$ (yellow)

45   40   35   30
ELECTRON BINDING ENERGY (eV)

ways and produce new states that are characterized by the total electronic angular momentum, J:

$$J = |L \pm S| \tag{7.8}$$

where $L = 0, 1, 2, \ldots$; $S = 1/2$; $J = 1/2, 3/2, 5/2, \ldots$. The energies of these new states are thus different because the magnetic moments due to the electron spin and orbital motion may oppose or reinforce each other. The degeneracies of these states is $2J + 1$, and the relative intensities of these split peaks is given by the ratio of these degeneracies.

For example, for the Tungsten $4f$ orbital ($L = 3$) shown in Fig. 7.6a [7.20],

$$J = |3 \pm 1/2| = 7/2, 5/2 .$$

Thus we have two components, viz. $4f_{7/2}$ and $4f_{5/2}$. The relative peak intensities in this example are therefore $(2 \times 7/2 + 1) : (2 \times 5/2 + 1)$, which equates to 4 : 3.

7.3.2 Chemical Shifts

As with a technique such as nuclear magnetic resonance (NMR), XPS can distinguish between a particular element in different environments. This is due to

173

the fact that placing a particular atom in a different chemical environment, or a different oxidation state, or in a different lattice site, etc., gives rise to a change in BEs of the core-level electrons. This BE variation is called the chemical shift, and appears as a definite "movement" in BE of the elemental peak in the XP spectrum. As an example, returning to Fig. 7.6, which shows the $4f$ peaks for Tungsten (W) and four common oxides of W [7.20], we can observe the BE of these peaks moving to higher binding energy as the oxidation number of the W increases. A great many of these shifts have been documented for instant reference, and an example for W can be found in the Perkin–Elmer Handbook [7.8]. These can be of substantial aid in determining the chemical environment of an element from a given XP spectrum.

7.3.3 Auger Chemical Shifts in XPS

Chemical information is also available from AES, although the effects on spectral features are not as well understood as in XPS. However since many core-type Auger lines are generated via Mg $K\alpha$ and Al $K\alpha$ X-rays, it is convenient to consider the combined chemical shifts in both the photoelectron and Auger lines. The difference between Auger and photoelectron chemical shifts results from the difference in final-state relaxation energies between chemical states. It is thus possible to define a parameter, α, called the Auger parameter, such that [7.21]

$$\alpha = E_k(A) - E_k(P) \tag{7.9}$$

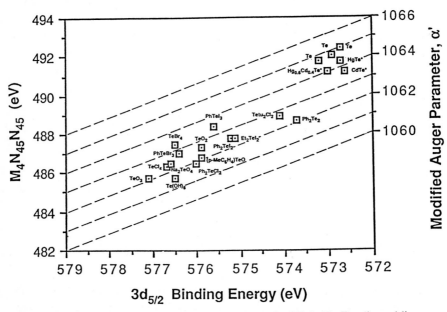

Fig. 7.7. Plot of Auger $M_4 M_{45} M_{45}$ kinetic energy versus $3d_{5/2}$ BE for Te. The diagonal lines represent the sum of these two quantities, and hence yields the modified Auger parameter, (see (7.11). (Data obtained from [7.7, 8])

174

where $E_k(A)$ = Auger electron KE, and $E_k(P)$ = photoelectron KE. One of the main advantages of α is that it is independent of static charging, and is indicative of a particular chemical state. Unfortunately, for some systems α can have negative values, however (7.11) can be modified by using (7.2) as follows:

$$\alpha = E_k(A) + E_b(P) - h\nu \tag{7.10}$$

or, the "modified" Auger parameter,

$$\alpha' = \alpha + h\nu = E_k(A) + E_b(P) . \tag{7.11}$$

Thus a graph $E_k(A)$ vs $E_b(P)$ becomes independent of photon energy. Such two-dimensional plots have been generated for many elements, one of which is tellurium, Te shown in Fig. 7.7 [7.8]. Many other examples of the use of the Auger parameter can also be found in the literature [7.23–25].

7.3.4 X-Ray Line Satellites

As mentioned in Sect. 7.2.1(a), X-ray sources do not produce single lines, but a series consisting of a main line with some minor components at higher photon energies. Those minor lines can cause small satellite peaks to occur for each major photoelectron peak, the intensities and spacings of which are characteristic of the particular X-ray anode material. Examples of these can be seen in Fig. 7.3, where the $K\alpha_{1,2}$ peaks for the Hg $4f$, Cd $3d$ and Te $3d$ orbitals have accompanying $K\alpha_3$ satellite peaks (arrowed), the latter peaks being displaced by 8.4 eV. Table 7.2 lists the X-ray satellite energies and intensities for Mg and Al radiation.

Table 7.2. X-ray satellite energies and intensities

		$\alpha_{1,2}$	α_3	α_4	α_5	α_6	β
	relative intensity	100	8.0	4.1	0.55	0.45	0.5
Mg		67, 33					
	displacement, eV	0	8.4	10.2	17.5	20.0	48.5
	relative intensity	100	6.4	3.2	0.4	0.3	0.55
Al		67, 33					
	displacement, eV	0	9.8	11.8	20.1	23.4	69.7

7.3.5 "Shake-up" Lines

Following photoionization there is often a finite probability that the resultant ion will remain in an excited state, a few electron volts above ground state. The resultant photoelectron thus suffers a loss in KE, corresponding to the energy difference between the ground and excited states. In the spectrum this appears as a satellite a few eV higher in BE from the main peak. An example of this

175

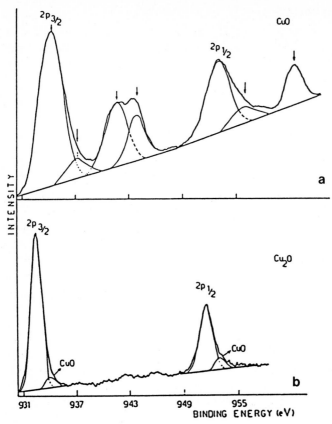

Fig. 7.8. The copper $2p_{1/2}$ and $2p_{3/2}$ XP spectra from (a) CuO, showing very prominent shake-up satellites, and (b) Cu_2O, where such satellites are absent. (Reproduced with permission from [7.26])

phenomenon can be seen in the XP spectrum of CuO, reproduced in Fig. 7.8 [7.26], where several satellites occur. These satellites can often be useful in determining the chemical nature of an element, e.g. distinguishing between CuO and Cu_2O, as seen in Fig. 7.8.

7.3.6 Ghost Lines

Ghost lines are small peaks appearing in an XP spectrum which result from X-radiation from foreign material. For example, common ghost lines arise from Mg impurity in the Al source, or *vice versa* in a dual anode source. Other sources can be Cu from the anode base structure or X-ray photons arising from the foil window. The positions of these ghost lines can be easily calculated, and Table 7.3 shows where such lines are expected to occur.

176

Table 7.3. Displacements of X-ray ghost lines

Contaminant Radiation	Anode material	
	Mg	Al
O ($K\alpha$)	728.7	961.7
Cu ($L\alpha$)	323.9	556.9
Mg ($K\alpha$)	–	233.0
Al ($K\alpha$)	−233.0	–

7.3.7 Plasmon Loss Lines

A plasmon loss occurs following interaction between photoelectrons and surface electrons in a solid, i.e. electrons passing through a solid can excite a collective oscillation of bulk electrons. This is called a bulk plasmon energy loss, and if the frequency of oscillation is ω_b, then the plasmon energy loss is $\hbar\omega_b$. Subsequent plasmon losses can also occur for these photoelectrons, resulting in a series of lines in the spectrum, all equally spaced by $\hbar\omega_b$, but of decreasing intensity.

At the surface a more localized oscillation can occur, and this is called a surface plasmon, with frequency ω_s, and hence energy loss $\hbar\omega_s$. The surface

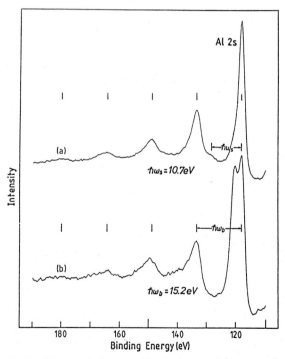

Fig. 7.9. XP spectrum of the Al $2s$ region, showing a succession of sequential bulk plasmon losses ($\hbar\omega_b$ = 15.2 eV). In spectrum (*a*), for clean Al, a surface plasmon loss can also be seen ($\hbar\omega_s$ = 10.7 eV), which disappears on oxidation, as in spectrum (*b*). (After [7.28])

plasmon loss can be related to the bulk plasmon loss as follows [7.27]:

$$\omega_s = \omega_b/\sqrt{2} \, . \tag{7.12}$$

For all elements the fundamental, or first, bulk plasmon will always be observable, with multiple plasmon loss peaks also visible in some cases. Depending on surface conditions, and the energy of the corresponding photoelectron peak, surface plasmons may also be seen, especially in the case of clean metal surfaces. An example of this can be seen in Fig. 7.9, for aluminium, showing plasmon loss peaks, both bulk and surface, for the $2s$ photoelectrons [7.28].

7.4 Quantitative Analysis

One of the major advantages of XPS is the ease with which quantitative data can be routinely obtained. This is usually performed by determining the area under the peaks in question and applying previously-determined sensitivity factors. For a homogeneous sample, the number of photoelectrons per second in a given peak, assuming constant photon flux and fixed geometry, is given by [7.29]:

$$I = KN\sigma\lambda AT \, , \tag{7.13}$$

where K = const, N = number of atoms of the element per cm^3, σ = photoionization cross section for element, λ = inelastic mean-free path length for photoelectrons, A = area of the sample from which the photoelectrons eminate and T = analyzer transmission function. If we define the sensitivity factor for element x as

$$S_x = K\sigma\lambda AT \, , \tag{7.14}$$

then

$$I_x = N_x S_x \tag{7.15}$$

or

$$N_x = I_x/S_x \, . \tag{7.16}$$

If we wish to know the relative amounts of elements in a sample, it is thus necessary to know the sensitivity factors for the elements and to measure their intensities (i.e. peak areas)

$$N_1 = I_1/S_1 \, ; \quad N_2 = I_2/S_2 \, . \tag{7.17}$$

This approach will provide semi-quantitative results for most situations, except where heterogeneous samples are involved, or where serious contamination layers obscure the underlying elements. Also, any peak interference (e.g. overlapping Auger lines or another XPS peak) must be avoided. Ideally elemental sensitivity factors should be determined specifically for each instrument.

178

A simple application of this type of quantitative analysis has been provided by *Leech* et al. in a study of the effect of etchants on II-VI semiconductor materials [7.30]. Their work used XPS to determine the relative amounts of Hg, Cd and Te in the samples before and after etching, and hence were able to follow the changes which the etchants induced.

7.5 Experimental Techniques

7.5.1 Variation of X-Ray Sources

As mentioned in Sect. 7.2.1(a), many X-ray sources are available for use in XPS, with the most common being a dual Mg/Al source. What is the value of the availability of two X-ray lines of different energy? Firstly, the Mg $K\alpha$ line is narrower than the Al $K\alpha$, thus allowing a better resolution to be achieved.

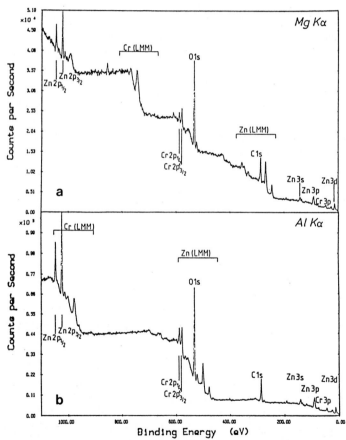

Fig. 7.10. XP spectra of galvanized steel obtained using (a) Mg $K\alpha$ and (b) Al $K\alpha$ X-ray sources. Note the relative positions of the Auger and Photoelectron lines

Secondly, some core levels may not be accessible with the Mg source, and require the slightly higher (by 233 eV) energy Al line. Thirdly, often in an XP spectrum Auger lines and XP peaks may overlap, and changing to another source can alleviate this problem, since the Auger electrons always have the same KE, whereas the photoelectron KE depends on the ionizing radiation as the BE is constant. Thus the Auger lines appear to move relative to the XP peaks as the X-ray energy is varied. An example of this application is shown in Fig. 7.10, where spectra have been recorded for galvanized iron. In the Al $K\alpha$ spectrum (b), the Zn(LMM) lines overlap with the O $1s$ and Cr $2p$ orbitals, while the Cr(LMM) lines interfere with the Zn $2p$ peaks. Changing to Mg $K\alpha$ (a) solves this problem, with the Cr(LMM) lines clear of any other features, and the Zn(LMM) lines now overlapping with the C $1s$ peak.

7.5.2 Depth Profiles

This topic has been broached previously in this book, both generally and with specific application to AES (Chap. 6). Depth profiling has applications in XPS also, although it has been less utilized than in AES. The process again involves etching, or sputtering, away the material using a rare-gas ion beam (usually Ar) and recording spectra as a function of depth. However for an XPS depth profile a relatively large surface area must be etched. This is necessary since the X-ray beam cannot easily be focused to a small spot in most instruments.

Although ion sputtering can change the chemical states in a material, much information is still available from an XPS depth profile. An excellent example is the variation of the oxidation state of an element in a material. Figure 7.11 shows a simple depth profile for a CdTe film on a nickel substrate, where the tellurium has oxidized [7.25]. The first thing to notice is that Cd seems to be preferentially concentrated at the surface, whereas the Te near the surface occurs largely as the oxide TeO_2, which disappears quickly on sputtering. The sequence

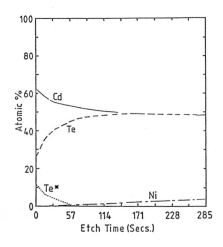

Fig. 7.11. XPS depth profile for a CdTe electrodeposited film. "Te*" represents Te arising from TeO_2

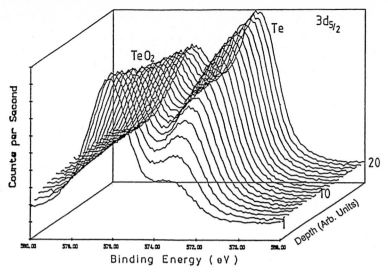

Fig. 7.12. Montage showing the variation of Te peaks at the TeO₂/Te interface as a function of depth

of XP spectra recorded as a function of depth can be presented as a montage, as in Fig. 7.12, which highlights the relative variation of the peaks.

7.5.3 Angular Variations

While discussing surface specificity in Sect. 7.1.3, it was pointed out that the depth from which the photoelectrons emanated depended on the angle of detection, i.e. the angle of emission to the surface normal, θ. This is shown diagramatically in Fig. 7.13, where it can be seen that detection close to the normal enhances the signal from the bulk relative to the surface, while detection close to the surface plane enhances the signal from the surface relative to the bulk. Thus varying the angle of detection can yield non-destructive depth information, an example of which is presented in Fig. 7.14, showing data for a thin film of SiO₂ on Si [7.31]. For small values of θ the main contribution to the spectrum is

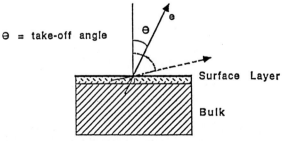

Fig. 7.13. Schematic showing surface sensitivity as a function of emission angle. Small θ enhances the signal from the bulk, while large θ enhances the signal from the surface

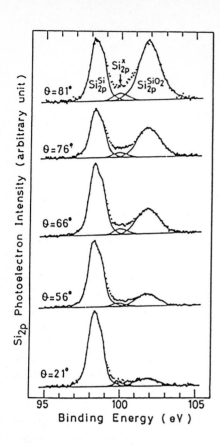

Fig. 7.14. Study by XPS of the interface between silicon and a thin (0.91 nm) film of SiO_2 on its surface. θ varies from 21° to 81° away from the normal to the surface

from the bulk Si, while at larger values of θ the contribution from the oxide layer becomes substantial. This approach is obviously preferable to the destructive ion etching, but is limited to very thin layers.

7.5.4 Sample Charging

When insulating materials are irradiated by X-rays they quite often develop a static surface charge due to their inability to replace the photoemitted electrons. This charging can be substantial, say 2–5 eV, but is always positive and quite small in comparison to electron-induced AES where charging can be of the order of several hundred eV. By using reference elements, such as a small amount of gold or silver (or, in some cases, adventitious carbon) the BE shifts can easily be calculated. In some instances, partial charging is possible, e.g. insulating domains on a conducting substrate. The most common approach to this problem is to use a neutralizing source, such as a low energy electron flood gun, to compensate for the charge. Using a reference peak to observe the amount of BE shift, the flood gun can be tuned to provide just the right amount of current to shift the peaks back to their "uncharged" binding energies [7.20].

7.6 Comparison with Other Techniques

Obviously no one technique can ever solve every surface problem, and for most studies two or more techniques are combined. Compromises must be made in the use of a particular method, between, for example, resolution or sensitivity, or lateral resolution or chemical information. In order to place XPS in perspective with other surface-science tools, we will compare its characteristic properties with those of two other frequently used techniques, viz. AES and secondary ion mass spectrometry (SIMS). (Refer to Tables 1.3, 4 for a list of the specific properties of these surface analytical techniques.)

All three spectroscopies have similar element detection capabilities, although only SIMS can detect hydrogen. XPS can detect all other elements, while AES can detect all but hydrogen and helium. Thus, except for special applications, the range of elemental detection is fairly comparable. Quantitative data can be obtained from all the methods, although standards are needed, however AES and XPS can provide semi-quantitative information directly from peak heights. In some situations SIMS can detect elements down to concentrations of less than 0.01 ppm (10^{-6} Atomic %), whereas XPS and AES are virtually identical in their detection ability of only 0.1 Atomic %. All three techniques are comparable when it comes to depth resolution, however AES and SIMS can achieve high lateral resolution unlike XPS. This arises because it is much easier to focus an electron or ion beam than an X-ray flux.

One of the major advantages of XPS is its ability to provide consistent chemical information. SIMS also provides chemical data, although more difficult to interpret, and AES chemical information is both difficult to obtain routinely and to interpret. It is difficult to study organic materials or many adsorbate systems with AES, due mainly to electron-beam induced damage, although this is minor compared to the destructive nature of SIMS. XPS is by far the least destructive of the three techniques considered here, and X-ray damage is not often a problem. Finally, it is much simpler to eliminate static charging during XPS than when using AES or SIMS, since the latter methods rely on beams of charged particles.

In summary, the main advantages of XPS (in comparison to AES and SIMS) are the ability to provide vital chemical information, simple quantitative information, low sample damage and the fact that it can be used to study insulating materials. The only major disadvantage of XPS is its very poor lateral resolution.

7.7 Applications

While there are a plethora of applications in which XPS has played a major role, the list is far too exhaustive to consider in any detail here. Instead the main catagories in which XPS has been used are tabulated in Table 7.4 along with the information yielded by the technique in each case, and references to some of the original work.

Table 7.4. Applications of XPS

Area of Application	Information available	References
Corrosion and oxidation	Elemental Identification; chemical (oxidation) state of corrosion products; change of chemical composition of surface (or bulk: depth profiling) during process	[7.32–38]
Metallurgy/ tribology	Elemental identification; analysis of alloys: effects of added lubricants; study of surface protective coatings	[7.10, 39–43]
Adhesion	Measure of film purity and thickness; details of bonding between film and substrate; chemical changes observed during adhesion	[7.44–48]
Catalysis	Identification of intermediate species formed; oxidation state of active species; study of modification of catalyst and support material during reaction	[7.20, 29, 49–51]
Semiconductors	Characterization of thin film coatings; identification of native oxides; interface characterization	[7.52–57]
Fibres and Polymers	Elemental composition data without serious charging; information available on typical polymer groupings; shake-up satellites for indicating aromaticity; contamination identification	[7.58–64]
Chemisorption	Variation of chemical state of substrate and adsorbate during chemisorption; uptake curves	[7.65–70]
Superconductors	Valence states; stoichiometries; electron structure	[7.71–76]

7.8 Conclusion

X-ray photoelectron spectroscopy has been used extensively over the past ten years in many areas of surface and materials analysis. The disadvantages of the technique, such as the need for UHV and the poor spatial resolution of the X-ray source, are more than adequately outweighed by the advantages, viz the ease of interpretation of spectra and the ability to derive chemical-state and simple quantitative information. During the next ten years the technical aspects of XPS will improve substantially, and the number of areas of application for XPS will grow considerably, making this technique almost mandatory in modern analytical laboratories.

Acknowledgement. I would like to thank C.G. Kelly for valuable suggestions relating to this work. The permission of the Executive General Manager, Telecom Research Laboratories, to publish this chapter is also acknowledged.

References

7.1   J.G. Jenkin, R.C.G. Leckey, J. Liesegang: J. Electron Spectrosc. **12**, 1 (1977)
7.2   J.G. Jenkin, J.D. Riley, J. Liesegang, R.C.G. Leckey: J. Electron Spectrosc. **14**, 477 (1978)
7.3   J.G. Jenkin, J. Liesegang, R.C.G. Leckey, J.D. Riley: J. Electron Spectrosc. **15**, 307 (1979)

7.4   J.G. Jenkin: J. Electron Spectrosc. **23**, 187 (1981)
7.5   K. Siegbahn, C. Nordling, A. Fahlman, R. Nordberg, K. Hamrin, J. Hedman, G. Johansson, T. Bergmark, S.-E. Karlsson, I. Lindgren, B. Lindgren: *ESCA: Atomic, Molecular and Solid State Structure Studied by Means of Electron Spectroscopy* (Almqvist & Wiksells, Stockholm 1967)
7.6   K. Siegbahn, G. Nordling, G. Johansson, J. Hedman, P.F. Heden, K. Hamrin, U. Gelius, T. Bergmark, L.O. Werme, R. Manne, Y. Baer: *ESCA Applied to Free Molecules* (North-Holland, Amsterdam 1969)
7.7   M.H. Kibel: *X-Ray Spectrometry* **19**, 73 (1990)
7.8   G.E. Muilenburg (Ed.): *Handbook of X-Ray Photoelectron Spectroscopy* (Perkin-Elmer Corp., 1979)
7.9   D. Briggs, M.P. Seah (Eds): *Practical Surface Analysis by Auger and Photoelectron Spectroscopy* (Wiley, New York 1983) Appendix 6
7.10   J.B. Lumsden: In *Metals Handbook*, 9th edn. Vol. **10** (1986) pp. 568–580
7.11   J.E. Castle, L.B. Hazell, R.H. West: J. Electron Spectrosc. **16**, 97 (1979)
7.12   M.A. Kelly, C.E. Tyler: Hewlett-Packard Journal **24**, 2 (1972)
7.13   D.E. Golden, A. Zecca: Rev. Sci. Instr. **42**, 210 (1971)
7.14   J.D. Lee: Rev. Sci. Instr. **43**, 1291 (1972)
7.15   F.J. Leng, G.L. Nyberg: J. Phys. E. **10**, 686 (1977)
7.16   R.G. Dromey, J.B. Peel: Aust. J. Chem. **28**, 2353 (1975)
7.17   D. Roy, J.-D. Carette: Canadian J. Phys. **49**, 2138 (1971)
7.18   J.C. Riviere: In *Practical Surface Analysis by Auger and X-Ray Photoelectron Spectroscopy*, ed. by D. Briggs, M.P. Seah (Wiley, New York 1983) Chap. 2
7.19   A. Barrie: In *Handbook of X-Ray and and Ultraviolet Photoelectron Spectroscopy* ed. by D. Briggs (Heyden & Sons, London 1977) Chap. 2
7.20   P.J.C. Chappell, M.H. Kibel, B.G. Baker: J. Catalysis **110**, 139 (1988)
7.21   C.D. Wagner: Anal. Chem. **47**, 1201 (1975)
7.22   C.D. Wagner, L.H. Gale, R.H. Raymond: Anal. Chem. **51**, 466 (1979)
7.23   R.H. West, J.E. Castle: Surf. Interface Anal. **4**, 68 (1982)
7.24   J.C. Riviere, J.A. Crossley, B.A. Sexton: J. Appl. Phys. **64**, 4585 (1986)
7.25   M.H. Kibel, C.G. Kelly: Materials Aust. **19** (10), 15 (1987)
7.26   M. Scrocco: Chem. Phys. Lett. **63**, 52 (1979)
7.27   J.M. Ritchie: Phys. Rev. **106**, 874 (1957)
7.28   A. Barrie: Chem. Phys. Lett. **19**, 109 (1973)
7.29   B.A. Sexton: Materials Forum **10**, 134 (1987)
7.30   P.W. Leech, P.J. Gwynn, M.H. Kibel: Appl. Surf. Sci. **37**, 291 (1989)
7.31   A. Ishizaka, S. Iwata: Appl. Phys. Lett. **36**, 71 (1980)
7.32   B.L. Maschhoff, K.R. Zavadil, K.W. Nebesny, N.R. Armstrong: J. Vac. Sci. Technol. A **6**, 907 (1988)
7.33   D.E. Fowler, J. Rogozik: J. Vac. Sci. Technol. A **6**, 928 (1988)
7.34   T. Czeppe, P. Nowak: Materials Sci. Forum **25/26**, 525 (1988)
7.35   T.M.H. Saber, A.A. El Warraky: J. Mater Sci. **23**, 1496 (1988)
7.36   E. Paparazzo: Surf. Interface Anal. **12**, 115 (1988)
7.37   E.W.A. Young, J.C. Riviere, L.S. Welch: Appl. Surf. Sci. **28**, 71 (1987)
7.38   N.S. McIntyre: In *Practical Surface Analysis by Auger and Photoelectron Spectroscopy*, ed. by D. Briggs, M.P. Seah (Wiley, New York 1983) Chap. 10, p. 397 and references therein
7.39   M. Ben-Haim, U. Atzmony, N. Shamir: Corrosion Sci. **44**, 461 (1988)
7.40   K.L. Rhodes, P.C. Stair: J. Vac. Sci. Technol. A **6**, 971 (1988)
7.41   J.C. Langevoort, I. Sutherland, L.J. Hanekamp, P.J. Gellings: Appl. Surf. Sci. **28**, 167 (1987)
7.42   F.-M. Pan, P.C. Stair: J. Vac. Sci. Technol. A **5**, 1036 (1987)
7.43   R.J. Bird: Metal Sci. J. **7**, 109 (1973)
7.44   S. Akhter, X.-L. Zhon, J.M. White: Appl. Surf. Sci. **37**, 201 (1989)
7.45   J.G. Clabes, M.J. Goldberg, A. Viehbeck, C.A. Kovac: J. Vac. Sci. Technol. A **6**, 985 (1988)
7.46   H.M Meyer III, S.G. Anderson, Lj. Atanasoska, J.H. Weaver: J. Vac. Sci. Technol. A **6**, 1002 (1988)
7.47   J.L. Jordan, C.A. Kovac, J.F. Morar, R.A. Pollack: Phys. Rev. B **36**, 1369 (1987)
7.48   H. Ohno, T. Ichikawa, N. Shiokawa, S. Ino, H. Iwasaki: J. Mater Sci. **16**, 1381 (1981)
7.49   G. Moretti, G. Fierro, M. Lo Jacono, P. Porta: Surf. Interface Anal. **6**, 188 (1980)
7.50   T.L. Barr: In *Practical Surface Analysis by Auger and X-Ray Photoelectron Spectroscopy*, ed. by D. Briggs, M.P. Seah (Wiley, New York 1983) Chap. 8 p. 283 and refences therein

7.51   J.S. Brinen: In *Applied Surface Analysis* ed. by T.L. Barr, L.E. Davis (ASTM, Philadelphia 1978) p. 24
7.52   J.H. Thomas III, G. Kaganowicaz, J.W. Robinson: J. Electrochem. Soc. **135**, 1201 (1988)
7.53   T. Wada: Appl. Phys. Lett. **52**, 1056 (1988)
7.54   R.N.S. Sodhi, W.M. Lan, S.I.J. Ingrey: Surf. Interface Anal. **12**, 321 (1988)
7.55   C.F. Yu, M.T. Schmidt, D.V. Podlesnik, R.M. Osgood, Jr.: J. Vac. Sci. Technol. B **5**, 1087 (1987)
7.56   M.D. Biedenbender, V.J. Kapoor, W.D. Williams: J. Vac. Sci. Technol. A **5**, 1437 (1987)
7.57   H.J. Kim, R.F. Davis, X.P. Cox, R.W. Linton: J. Electrochem. Soc. **134**, 2269 (1987)
7.58   A. Nelson, S. Glenin, A. Frank: J. Vac. Sci. Technol. A **6**, 954 (1988)
7.59   B. Mutel, O. Dessaux, P. Goudmand, J. Grimblot, A. Carpentier, A. Szarzbajnski: Rev. Phys. Appl. **23**, 1253 (1988)
7.60   D.T. Clark, D.R. Hutton: J. Polym. Sci. A **25**, 2643 (1987)
7.61   T.J. Hook, J.A. Cardella Jr., L. Salvati, Jr.: J. Mater Res. **2**, 117 (1987)
7.62   G. Gilberg: J. Adhes. **21**, 129 (1987)
7.63   D. Briggs: In *Practical Surface Analysis by Auger and Photoelectron Spectroscopy* ed. by D. Briggs, M.P. Seah (Wiley, New York 1983) Chap. 9, p. 359 and references therein
7.64   M.M. Millard: In *Industrial Applications of Surface Analysis*, ACS Symposium Series, ed. by L.A. Casper, C.J. Powell (Am. Chem. Soc. 1982) Chap. 8, p. 143
7.65   M.A. Barteau, E.I. Koh, R.J. Madix: Surf. Sci. **102**, 99 (1981)
7.66   G.A. Sormorjai: *Chemistry in Two Dimensions: Surfaces* (Cornell University Press, Ithaca 1981)
7.67   M.W. Roberts, C.S. McKee: *Chemistry of the Metal–Gas Interface* (Clarendon, Oxford 1981)
7.68   M.W. Roberts: Adv. Catal. **29**, 55 (1980)
7.69   R. Gomer (Ed.): *Interactions on Metal Surfaces* (Springer, Berlin, Heidelberg 1975)
7.70   C.R. Brundle, C.J. Todd (Eds.): *Adsorption at Solid Surfaces* (North-Holland, Amsterdam 1975)
7.71   H. Ihara, M. Jo, N. Terada, M. Hirabayashi, H. Oyanagi, K. Murata, Y. Kimura, R. Sugise, I. Hayashida: Physica C **153–155**, 131 (1988)
7.72   P.C. Healy, S. Myhra, A.M. Stewart: Philos. Mag. B **58**, 257 (1988)
7.73   A. Mogro-Campero, L.G. Turner, E.L. Hall, M.C. Burrell: Appl. Phys. Lett. **52**, 2068 (1988)
7.74   G.G. Peterson, B.R. Weinberger, L. Lynds, H.A. Krasinski: J. Mater. Res. **3**, 605 (1988)
7.75   S. Horn, J. Cai, S. Shaheen, C.L. Chang, M.L. Den Boer: Rev. Solid State Sci. **1**, 411 (1987)
7.76   P. Steiner, R. Courths, V. Kinsinger, I. Sander, B. Siegivart, S. Hüfner, C. Politis: Appl. Phys. A **44**, 75 (1987)

8. Fourier Transform Infrared Spectroscopy of Surfaces

N.K. Roberts

With 12 Figures

8.1 Introduction to Fourier Transform Infrared Spectroscopy

Infrared spectroscopy has been a widely used technique in industry for the structural and compositional analysis of organic, inorganic and polymeric samples and for quality control of raw materials and commercial products. With the advent of Fourier transform infrared spectroscopy (FTIR) the range of applications and the materials amenable to study has increased enormously, owing to its increased sensitivity, speed, wavenumber accuracy and stability. Oils, coal, shale, polymers, paints, catalysts, pharmaceuticals and industrial gases have been successfully analyzed [8.1]. More recently FTIR has been developed for the quality control of minerals and in exploration [8.2]. Cellulosic materials, wood, paper and cotton have also proved amenable [8.3]. It is also now possible to study electrode processes using FTIR [8.4, 5]. Biochemical processes once excluded from the field of infrared spectroscopy owing to the presence of water now form one of the fastest growing areas of research in FTIR. Specific examples are discussed in the text.

Conventional infrared spectroscopy relies on the dispersion of an infrared beam via a grating into its monochromatic components, and slowly scanning through the entire spectral region of interest. When a sample is placed in the beam, various wavelengths of infrared radiation are absorbed by the sample as the beam is scanned, and the result recorded as the infrared spectrum of the sample. To maintain a resolution of 4 wavenumbers, this typically requires around 2–3 minutes for a single scan of the spectrum.

Fourier transform infrared (FTIR) spectroscopy relies on a totally different principle to record the same information – that of interferometry. A Michelson or Genzel interferometer forms the basis of the FTIR spectrometer. Figure 8.1 shows a comparison of the optics for dispersive and Michelson interferometric instruments. The FTIR instrument consists of a standard infrared source, collimation mirrors, a beamsplitter, a fixed mirror and moving mirror. 50% of the beam is passed through the beamsplitter and 50% is reflected both on the initial pass and from reflections from the two mirrors. This results in the potential for interference patterns to be generated by varying the path length that one portion of the beam travels before recombination at the beamsplitter.

When $L2 = L1 + n\lambda$, constructive interference occurs, while when $L2 = L1 + n\lambda/4$, destructive interference occurs. For a monochromatic light source this

DISPERSIVE OPTICS

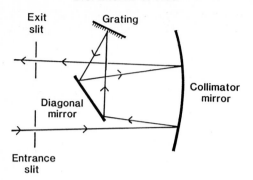

Exit slit

Grating

Diagonal mirror

Collimator mirror

Entrance slit

Fig. 8.1. Comparison of optics for dispersive and Michelson interferometric instruments

INTERFEROMETRIC OPTICS

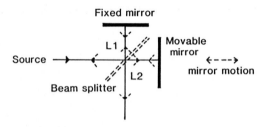

Fixed mirror

Source

L1

L2

Beam splitter

Movable mirror

← - - - →
mirror motion

results in an "interferogram" of intensity versus distance travelled in the form of a cosine wave. A normal infrared source covers a wide range of wavelengths and the interferogram obtained is a combination of all the individual interference patterns. Only at the point $L1 = L2$ are all waves in phase together, giving a strong "centreburst". The further away from this point one travels in either direction the various wavelengths tend to cancel giving a reduced signal.

The interferogram is detected as a function of mirror travel at the detector. For example, two optical velocities, $0.31 \, \mathrm{cm \, s^{-1}}$ and $1.2 \, \mathrm{cm \, s^{-1}}$, are possible with Digilab FTS-2OE instrument. A choice of optical velocities is necessary for the photoacoustic technique described later, usually within this range. Any substance which absorbs infrared radiation modifies the interferogram according to the wavelengths that are absorbed. Thus all the absorption information is contained within a single interferogram (the "Fellgett" advantage) which at a resolution of 4 wavenumbers/cm can be obtained in about 2 seconds (corresponding to a moving mirror travel of 0.125 cm). However, this only becomes useful when the wavelength information is "unscrambled" from the interferogram by the mathematics of the Fourier transformation resulting in a "single beam" spectrum of intensity versus wavenumber (or wavelength if desired). The "single beam" spectrum represents the combination of the source and beamsplitter characteristics and any absorbances due to components in the air inside the spectrometer. In the FTS-2OE a purge of dry air is maintained to minimize the contribution from water vapour while carbon dioxide is not removed and hence a large absorbance is observed. The spectral information is not usually presented in this format. Instead, a "reference" spectrum is taken of the system without the sample, and the subsequent sample spectrum is ratioed to this producing either a transmission spectrum or an absorption spectrum.

One of the important features of FTIR spectroscopy is the ability to signal average a large number of scans in a relatively short amount of time. Thus spectra on very small quantities of material, or on highly absorbing materials, in which the signal to noise ratio of individual scans is very poor, can be achieved. The signal to noise ratio improves as the square root of the number of scans taken. It is thus possible to acquire 10 000 scans overnight to achieve a 100-fold improvement in the signal to noise ratio. In practice, the interferograms are signal averaged first, and a single Fourier transform then done, as the Fourier transform can take up quite significant amounts of computing time.

The second major advantage of FTIR over conventional dispersive instruments is the high throughput of infrared radiation, since narrow slits are no longer necessary to achieve resolution. This is the so-called "Jacquinot" advantage. In a dispersive instrument measuring in the range 4000–400 cm^{-1} at a resolution of 8 cm^{-1} only 8/3600 or 0.2% of the incident radiation reaches the detector. For 1 cm^{-1} resolution this is reduced to 0.25%. The FTIR spectrometer is also capable of very high resolution of absorption bands. Higher resolution is achieved by moving the mirror further while maintaining the same starting point. By allowing mirror movement of 5 cm, resolution of 0.1 wavenumbers is possible, taking 30 s per scan on the FTS-2OE, and over 30 minutes to compute the Fourier transform. More recent FTIR spectrometers are faster. Moreover, the interferometric derived frequencies are internally calibrated by a laser giving greater frequency reproducibility. This advantage is very important in multiple scanning needed for low energy signals. FTIR spectroscopy, therefore, has two main advantages over conventional IR spectroscopy, improved sensitivity and improved computational ability from having the data in digital form in a powerful computer [8.6, 7].

8.2 Surface Techniques

The main techniques for studying surfaces with FTIR spectroscopy are,

a) Specular and grazing angle reflectance
b) Diffuse reflectance (DRIFT)
c) Attenuated total reflectance (ATR)
d) Photoacoustic spectroscopy (PAS)

Of these techniques (a), (b) and (c) have been used successfully with dispersive IR instruments on solid samples. FTIR has extended infrared spectroscopy to aqueous solutions using (c) and enabled (d) to be carried out in the IR region; previously PAS was only possible in the UV-visible region. Both ATR and PAS allow depth profiling of the surface.

a) Specular and Grazing Angle Reflectance. This technique has rarely been used because of the usual requirement for a collimated beam, the strict requirement for sample flatness and the difficulty of interpreting spectra dominated by the effect of the absorbing species on the complex refractive index. Its main application is in the area of surface coatings, such as lubricants, adhesives and paints.

b) Diffuse Reflectance is used for powdered samples. Although little or no sample preparation is required, the technique has the disadvantage that specular reflection may give rise to spurious spectral features. It is important that the samples be highly scattering. Coal, pharmaceutical products, foodstuffs and mineral samples are amenable to this technique.

c) Attenuated total reflectance may be used to study
  i) the surface of solid samples, provided good contact can be achieved between the sample and the ATR element. Depth profiling is also possible by varying the angle of incidence provided it does not exceed the critical angle. This technique is ideal for solid samples that can make good contact with the ATR element, in this class fall rubber, polymer films fabrics, coated and painted surfaces.
  ii) ATR has been used with excellent results to investigate the solid water interface. The application of FTIR-ATR Spectroscopy to aqueous systems has opened up a whole new area of study, from the clotting mechanism of blood on foreign surfaces to the mode of action of flotation collectors on mineral surfaces.

d) Photoacoustic Spectroscopy has the great advantage of requiring very little sample preparation and in contrast to the ATR method the sample does not have to be in contact with an element. Higher surface/volume ratios of the sample increase the signal intensity. Thus powdered samples and rough surface morphologies are more favorable in PAS. The technique has been used with great success in the depth profiling of solid surfaces. Previously variable angle ATR was the only method for depth profiling and the maximum depth was of the order of a micron. PAS allows maximum depths of many microns by using low mirror velocities. As with the ATR method the sampling depth is wavelength dependent, making quantification of surface concentrations difficult. PAS is best suited to intractable solid samples. Of the four techniques only ATR has found any application in aqueous systems. Biological applications are now dominating the field of FTIR-ATR. Previously dispersive instruments make aqueous systems inaccessible to IR spectroscopy. A more detailed description of these techniques follows with some specific applications.

8.2.1 Diffuse Reflectance Infrared Fourier Transform (DRIFT)

The theory of diffuse reflectance at scattering layers within powdered samples has been developed by *Kubelka* and *Munk* [8.14, 15]. For an "infinitely thick" layer, the Kubelka–Munk equation may be written as

$$f(R_\infty) = \frac{(1 - R_\infty)^2}{2R_\infty} = \frac{k}{s}$$

where R_∞ is the absolute reflectance of the layer, s is a scattering coefficient, and k is the molar absorption coefficient. In practice there is no perfect diffuse reflectance standard and R_∞ is replaced by R'_∞ where

$$R'_\infty = \frac{R'_\infty \text{ (sample)}}{R'_\infty \text{ (standard)}} .$$

R'_∞ (sample) represents the single-beam reflectance spectrum of the sample and R'_∞ (standard) is the single-beam reflectance spectrum of a selected nonabsorbing standard exhibiting high diffuse reflectance throughout the wavelength regions being studied. Potassium chloride or potassium bromide with a particle size of less than $10\,\mu\text{m}$ is often used as a standard. The depth of the KCl or KBr matrix required to yield samples of "infinite thickness", i.e where further increasing the thickness causes no change in the spectrum, is less than 5 mm. The Kubelka–Munk theory predicts a linear relationship between the molar absorption coefficient, k, and the peak value of $f(R_\infty)$ for each band, provided s remains constant. Since s is dependent on particle size and range, these parameters should be made as consistent as possible if quantitative data are needed.

Diffuse reflectance can be used with dispersive IR instruments, however the low signal level and the poor sensitivity limits the range of applications. The advent of FTIR spectroscopy has made the technique more accessible. The samples must be highly scattering, that is, the optical depth, $\mu\beta$, must be larger than the geometrically average path length, μD, through the sample. For scattering samples μD is a function of particle size. Light impinging on an opaque, nonscattering sample is completely absorbed, except for a small reflected component ($\sim 7\%$ at normal incidence). If the surface is rough, a diffuse reflectance accessory ($\sim 20\%$ efficient) is required to collect this small amount of light. The great advantage of DRIFT is the short preparation time and the ability to obtain a strong spectrum with less than $100\,\mu\text{g}$ of sample in a suitable matrix. Diffuse Reflectance has been successfully applied to finely powdered samples, however it is not suitable for samples in the form of coarse powders, granules or intractable lumps. High pressure and temperature DRIFT cells are now available for studying in situ catalytic reactions at surfaces. Samples may also be used without being mixed with a matrix powder, but considerable distortions in relative peak heights within the spectrum may occur if the diffuse reflectance accessory does not completely eliminate the specularly reflected radiation.

A successful application of this technique is ex-situ investigation of the adsorption of the flotation collector, amylxanthate, on the mineral pyrite [8.8]. For the more interesting and relevant in-situ investigations this method cannot be used and the ATR method is the method of choice (see next section). Figure 8.2 shows the DRIFT spectra of pyrite after wet grinding at natural pH (5.1)

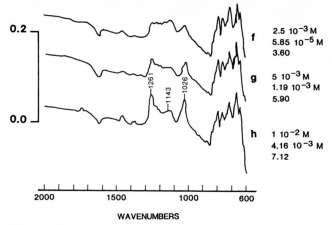

Fig. 8.2. Diffuse reflectance spetra of pyrite after wet grinding at natural pH (5.1) and adsorption of amylxanthate with a conditioning time of 10 min. For each spectrum are given the initial and final concentration, and the statistical surface coverage

and adsorption of the collector amylxanthate. The sample, of which 50% was less than 7 μm, was dispersed in KBr. Above monolayer coverage bands appear at 1026 cm^{-1} (CS stretching mode) 1259–1261 cm^{-1} (C-O-C stretching in dixanthogen) and bands due the hydrocarbon chain at 1347 and 1465 cm^{-1}. The bands at 1746 and 1716 cm^{-1} indicate the presence in the adsorbed layer of the dimer of amylmonothiocarbonate from an oxydegradation reaction. The bands between 600 and 900 cm^{-1} are due to the oxidation products on pyrite surface after grinding and before contact with the xanthate solutions.

8.2.2 Attenuated Total Reflectance Spectroscopy (ATR)

Attenuated total reflectance (ATR) spectroscopy, sometimes known as internal reflection spectroscopy (IRS), was developed independently by *Harrick* and *Fahrenfort* [8.9a] in 1960. The theory and practice of the technique is described extensively by *Harrick* (1967) [8.9b]. If a beam of radiation strikes the interface of two media of refractive indices n_1 and n_2 from the side of an IRS crystal (n_1) at any angle greater than the critical angle, it is totally reflected and a standing wave is established at the crystal-sample interface. The intensity of the radiation that penetrates into the rarer medium (n_2) falls off exponentially from its value I_0 at the surface, and the total reflection is attenuated (Fig. 8.3). The depth of penetration of the sample is given by the relation,

$$d_p = \frac{\lambda}{2\pi}(\sin^2\theta - n_{21}^2)^{1/2}$$

where d_p = depth at which the intensity is reduced to 1/e of its original value. θ is the angle of incidence and n_{21} the ratio of the refractive indices of the sample (n_2) and the internal reflection element (n_1). Consequently the penetration is

192

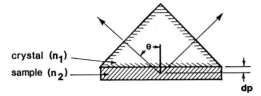

crystal (n₁)
sample (n₂)

dp

Fig. 8.3. Schematic of the attenuated total internal reflection process

greater at longer wavelengths. Over the mid-IR range (2.5–25 μm) there is a factor of 10 difference in sampling depth. Furthermore as θ gets larger, d_p gets smaller and approaches 0.1 λ at grazing incidence ($\theta = 90°$) for high index media. Theoretically d_p is infinitely large at the critical angle. It is not possible to vary the angle continuously and several elements of fixed angle, usually 30°, 45° and 60°, are employed, provided they do not exceed the critical angles. The sensitivity of the ATR technique can be markedly increased by using multiple internal reflection elements. A typical crystal 50 mm long with its end faces cut at the crystal angle can produce between 20 and 30 reflections. If l is the crystal length and d its thickness, the number of reflections is given by

$$N = l/d \cot \theta .$$

The most common materials employed as ATR elements are thallium bromo-iodide-5, germanium, zinc selenide, silicon, silver bromide, synthetic sapphire and diamond (used mainly in the far infrared region). Table 8.1 lists typical depths of penetration of the infrared beam for ATR elements of germanium and KRS-5 of different crystal angles in contact with polyethylene [8.1].

Table 8.1. Penetration depths (d_p) in μm for polyethylene for various ATR elements

ATR element	Angle	Wavenumber [cm^{-1}]			
		2000	1700	1000	500
KRS-5	45	1.00	1.18	2.00	4.00
KRS-5	60	0.55	0.65	1.11	2.22
Ge	30	0.60	0.71	1.20	2.40
Ge	45	0.33	0.39	0.66	1.32
Ge	60	0.25	0.30	0.51	1.02

Thus one of the major differences between transmission and ATR spectra is the dependence of the intensity of the spectral bands on wavelength in ATR spectra since d_p is larger at longer wavelengths. There is also distortion on the longer wavelength side. Programs are available for correcting for these effects.

ATR spectroscopy can therefore study the surfaces of samples in the range of a fraction of a μm to several μm provided intimate and uniform contact is maintained with the ATR element. It is particularly important to maintain uniformity of contact if further data processing such as spectral subtraction is to be made. A torque wrench to press the sample against the ATR crystal element should be used to obtain reproducible contact. The surfaces of the ATR element should be clean and highly polished. For the best results the sample should be pliable and have a smooth surface. In practice, a spectrum is run of the ATR element alone then in contact with the sample. The latter spectrum is then ratioed to the former, since FTIR instruments are single beam instruments, producing the spectrum of the surface of the sample alone. Depth profiling of surfaces is possible by varying the crystal material and the angle of incidence. An interesting comparison of ATR and PAS depth profiling of some biocompatible polymer surfaces showed that PAS was more sensitive to surface impurities and segregation [8.10]. The outstanding advantage of the FTIR-ATR spectroscopy is the ability to study the solid/aqueous solution interface, previously inaccessible to dispersive IR spectrometers because water is a strong IR absorber. Two examples will suffice to show its capability.

The first example is protein adsorption on foreign surfaces from whole blood. Interactions between flowing blood and a foreign surface are of great practical significance when medical polymers such as those used in heart valves, indwelling catheters, dialysis membranes, vessel grafts, and other artificial organs are implanted in the body. Most typical polymers in contact with blood give rise to a number of complex reactions culminating in a blood clot (thrombus). FTIR-ATR spectroscopy has contributed significantly to our understanding of clot formation [8.11]. Since FTIR spectra can be collected in fractions of a second, the kinetics of the process can be readily studied. Figure 8.4 contains the spectrum of flowing blood, a reference saline blank, and the spectrum obtained by subtracting the latter from the former. Figure 8.5 is a continuation of Fig. 8.4, showing additional expansion of the subtraction spectrum. These spectra were obtained in 0.8 s. The presence of the amide I ($1650\,cm^{-1}$) and amide II ($1550\,cm^{-1}$) peaks due to proteins so close to the strong OH binding vibration of water ($1640\,cm^{-1}$) shows that successful spectral subtraction can be obtained in less than 1 s close to strong interfering peaks. Figure 8.6 shows the total amount of protein adsorbed, as measured by the amide II peak ($1550\,cm^{-1}$), as a function of time. Careful analysis of the spectrum below $1550\,cm^{-1}$ allows a more detailed eludication of the proteins adsorbed. The ATR crystal may be bare, as in this case, or coated with a very thin film of polymer (100–1000Å) for a study of its biocompatibility.

The second application of FTIR–ATR spectroscopy is the adsorption of flotation collectors on mineral surfaces. FTIR–ATR spectroscopy is particularly suitable for studying the in situ adsorption of flotation collectors on mineral surfaces. Previously only ex situ methods were available for studying the process, i.e. the mineral with adsorbed collector was removed from the aqueous environment, dried and the spectrum recorded. However, virtually nothing is known about the effect of removing the water from the sample. FTIR–ATR spectroscopy allows ready subtraction of the water and mineral spectra to leave only the spectrum of the adsorbate in situ [8.12]. The experimental procedure is as follows. Firstly, the mineral is vacuum evaporated onto an ATR element to produce a thin film

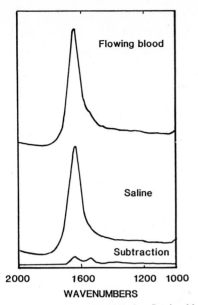

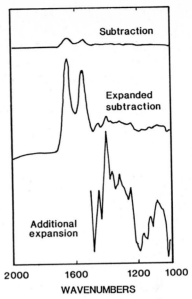

Fig. 8.4. Spectra collected during flowing blood experiment: *top:* representative spectrum collected during run *middle:* spectrum of saline blank *bottom:* subtraction of flowing blood spectrum minus saline spectrum

Fig. 8.5. Expansion of difference spectrum Fig. 8.4

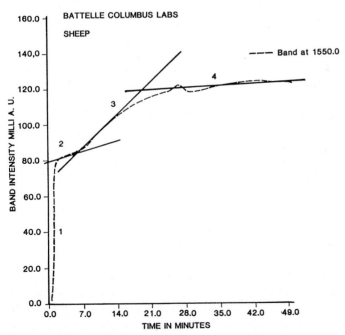

Fig. 8.6. Plot of amide II (1550 cm^{-1}) band intensity versus time of blood flow showing total amount of protein adsorbed in the time period indicated. Spectra collected every 5 s

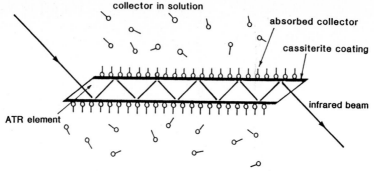

Fig. 8.7. Schematic representation of the use of ATR method to study collector adsorption

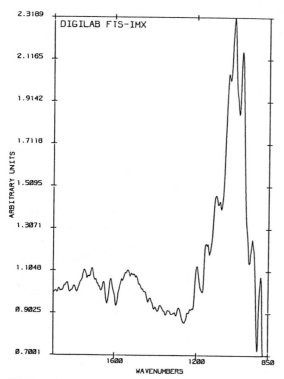

Fig. 8.8. FTIR-ATR spectrum of 300 ppm SPA adsorbed on a thin film ($\sim 0.05\ \mu m$) of tin (IV) oxide at pH 4.5 (100 scans)

($\sim 0.05\ \mu m$). A word of caution is required here. It is important to be sure that the mineral is unchanged during evaporation. The term "thin film" is a misnomer, as it is known that the "thin film" consists of very small crystallites. Evaporation of tin (IV) oxide produces a variety of oxides. Fortunately in this case the lower oxides may be converted to tin IV) oxide by annealing in oxygen. Fluorite and other minerals have been successfully produced as "thin films". If

196

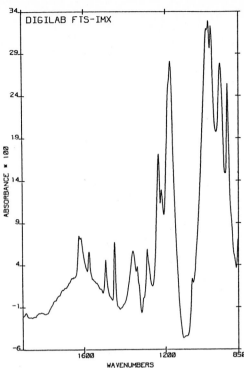

Fig. 8.9. FTIR spectrum of SPA in KBr disc

thin films cannot be obtained it may be possible to obtain spectra on suspensions of the mineral, e.g. Goethite, or use a transmission technique if the mineral is infrared transparent in the appropriate region, for example sphalerite. Secondly, a spectrum is run of the ATR element coated with the mineral film in the presence of water. The experimental set-up is shown in Fig. 8.7. The water is then pumped out and replaced with the collector solution. Spectra may then be run at appropriate intervals and difference spectra obtained showing the adsorbate on the mineral surface. Figure 8.8 shows the result of the adsorption of styrene phosphoric acid (SPA) on cassiterite. Figure 8.9 is the spectrum of styrene phosphoric acid. Comparison with the corresponding SPA complexes shows a close resemblance between the adsorbate species and the corresponding tin (IV) and titanium (IV) complexes. On adsorption of the collector the bands due to the νP=O, ν_{as}P–O and ν_sP–O vibrations of the phosphonic acid group are replaced by a complex broad band due to the resonance stabilized phosphonate group. The kinetics of the adsorption process may be followed in situ along with the effect of such variables as pH, temperature and ionic composition. At the same time spectra of the solution species may be obtained for comparison with the adsorbed species.

In the pH range studied the mechanism of adsorption is as follows:

monoanion   cassiterite
of SPA   surface

slow →

fast ↓

bidentate surface
complex of SPA

8.2.3 Photoacoustic Spectroscopy (PAS)

UV-visible photoacoustic spectroscopy bas been used for some time to investigate solid samples; however, PAS was not applied to the IR region until the advent of FTIR. Only when photoacoustic detection was combined with an interferometer and the large capacity data system in the FTIR instrument could PAS be used in the IR region. The signal to noise (S/N) ratio in dispersive instruments with PAS is far too low to make it a useful technique.

The solid sample is placed in a sealed cell of small volume containing an infrared transparent gas such as helium to carry the photoacoustic signals. The IR radiation absorbed by the solid sample is converted to heat by a radiationless transfer process. When the heat propagates to the sample surface and is transferred at the sample–gas interface into the surrounding gas, pressure variations of the gas are generated because the intensity of the IR radiation is modulated by the interferometer at an audio frequency region. The Michelson interferometer modulates the IR beam as does a chopper in a dispersive IR instrument. The gas expands and contracts at the modulation frequency resulting in a pressure wave within the sealed cell. The photoacoustic signal is detected by a sensitive microphone and preamplified in the cell. Subsequently the signal is fourier-transformed to yield a single beam FTIR/PAS spectrum. The single beam spectrum is referenced to that of carbon black since it can absorb all IR radiation and convert it to photoacoustic signals. A schematic represention of the PAS sample cell is shown in Fig. 8.10.

The strength of the signal I_{PAS} given by the relation,

I_{PAS} = number of photons × energy per photon × thermal diffusivity of sample

The thermal diffusivity depends on the particular modulated frequency of the interferometer incident on the sample. The thermal diffusion length μ_s, in cm

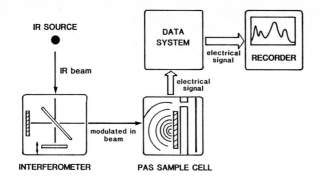

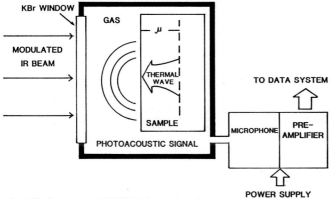

Fig. 8.10. Schematic of FTIR/PAS sample cell

can be calculated from the following equation,

$$\mu_s = (2k/\rho c\omega)^{1/2} \, ,$$

where k is the sample's thermal conductivity in cal. cm s^{1-} C^{-1}, ρ is the density of the sample in g cm^{-3}, and ω is the angular modulation frequency in radian s^{-1}. The angular modulation frequency can be calculated from the modulation frequency, f, as follows;

$$\omega = 2\pi f \, , \quad f = u\bar{\nu}$$

where u is the optical velocity of the interferometer in cm s^{-1} and $\bar{\nu}$ is the infrared frequency in wavenumbers (cm^{-1}). For a Michelson interferometer a mirror displacement of x results in an optical path difference of $2x$, i.e. the "optical velocity" is twice the "mirror velocity". For FTIR instruments using a Genzel interferometer the "optical velocity" is four times the "mirror velocity", so they achieve the same resolution for half the mirror travel. For example if the "mirror velocity" is 0.155 cm s^{-1} (equivalent to an "optical velocity" of 0.31 cm s^{-1} with a Michelson interferometer) then the modulation frequencies at 1000 cm s^{-1} and 4000 cm s^{-1} will be 320 and 1280 Hz respectively, i.e. in the

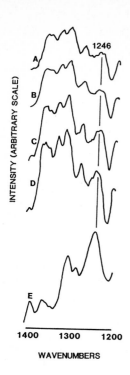

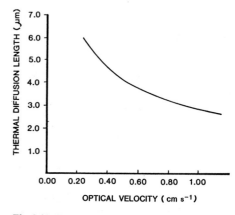

Fig. 8.11. Photoacoustic infrared spectra of cotton yarn sized with polyurethane collected at (A) 0.235 cm s^{-1}, (B) 0.396 cm s^{-1}, (C) 0.665 cm s^{-1}, (D) 1.119 cm s^{-1}, (E) polyurethane sizing agent

Fig. 8.12. Thermal diffusion length of cellulose at 1730 cm^{-1} versus the optical velocity of the interferometer

audio frequency range. That is, as far as the FTIR instrument is concerned, each frequency component of the infrared source contributes a frequency component to the interferogram. In principle one can obtain the spectra of different layers of the sample by varying the mirror velocity of the Michelson interferometer. The lower the mirror velocity the greater the depth of penetration. However if the radiation is absorbed within a length that is smaller than the thermal diffusion length of the sample, photoacoustic saturation sets in. Therefore one must be extremely careful interpreting the results of PAS experiments on depth profiling particularly at low modulation frequencies as varying photoacoustic saturation effects of different bands in the spectrum may occur.

An interesting application of PAS is the non-destructive depth profiling of cellulose materials and polymers to determine the penetration of chemical additives. In the textile industry, sizing agents are applied to warp yarns to increase their resistance to abrasion during weaving. After weaving the sizing agents are removed through a desizing process which produces desired properties. Figure 8.11 shows the PAS-FTIR spectra of cotton yarn sized with a polyurethane at different interferometer velocities [8.13].

The intensity of the peak at 1246 cm^{-1}, due to the polyurethane sizing agent, increases dramatically as the optical velocity changes from 0.235 cm s^{-1} to 1.119 cm s^{-1}, indicating that the agent is concentrated at the surface. Figure 8.12 shows the thermal diffusion length of cellulose at 1730 cm s^{-1} for different optical velocities. Hence it may be concluded that the polyurethane sizing agent

200

is more concentrated in the top 2.8 μm and gradually decreases to a depth of 6.8 μm.

Acknowledgement. I am grateful for the assistance of Mr. K. Kuys in obtaining and checking references.

References

8.1   T. Theophanides (ed.): *Fourier Transform Infrared Spectroscopy* (Reidl, Dordrecht, 1984)
8.2   P.M. Fredericks, J.B. Lee, P.R. Osborn, D.A.J. Swinkels: Appl. Spectrosc. **39**, 311 (1985)
8.3   C.Q. Yang, R.R. Bresee, W.G. Fateley, T.A. Perenich: *The Structure of Cellulose*, ACS Symposium Series Vol. 340 (1987)
8.4   A. Bewick, S. Pons: *Advances in Infrared and Raman Spectroscopy*, Vol. 12 (Wiley, New York 1985)
8.5   S.I. Yaniger, D.W. Vidrine: Appl. Spectrosc. **40**, 174 (1986)
8.6   J.R. Ferraro, L.J. Basile (eds.): *Fourier Transform Infrared Spectroscopy* Vols. 1–3 (Academic, New York 1978–82)
8.7   P.R. Griffiths, J.A. Haseth: *Fourier Transform Infrared Spectrometry* (Wiley, New York 1986)
8.8   J.M. Cases, P. de Donato, M. Kongolo, L. Michot: Colloids and Surfaces **36**, 323 (1989)
8.9a   J. Fahrenfort: Spectrochim. Acta **17**, 689 (1961)
8.9b   N.J. Harrick: *Internal Reflection Spectroscopy* (Wiley, New York 1967)
8.10   J.A. Gardella, G.L. Grobe, W.L. Hopson, E.M. Eyring: Anal. Chem. **56**, 1169 (1984)
8.11a   S. Winters, R.M. Grendreau, R.I. Leininger, R.J. Jacobsen: Appl. Spectrosc. **36**, 404 (1982)
8.11b   R.I. Leininger, T.B. Hutson, R.J. Jacobsen: Annal. New York Acad. Sci. **516**, 173 (1987)
8.12   K.J. Kuys, N.K. Roberts: Colloids and Surfaces **24**, 1 (1987)
8.13   C.Q. Yang, R.R. Brisee, W.G. Fateley: Appl. Spectrosc. **41**, 889 (1987)
8.14   P. Kubelka, F. Munk: Z. Tech. Phys. **12**, 593 (1931)
8.15   P. Kubelka: J. Opt. Soc. Am. **38**, 448 (1948)

9. Rutherford Backscattering Spectrometry and Nuclear Reaction Analysis

S.H. Sie

With 8 Figures

The RBS (Rutherford backscattering spectrometry) and NRA (nuclear reaction analysis) are a subset of what is generally known as ion beam analysis (IBA) methods, performed with energetic (typically a few hundred keV to a few MeV) ion beam from accelerators. The energies involved render the measurements insensitive to molecular or atomic shell effects, and thus chemical effects. RBS is based on elastic Coulomb scattering between the projectile and the target nuclei, and is usually applied to obtain data for most if not all elements present in the specimen. In contrast, NRA is based on nuclear reactions which are element specific. The most commonly used beam in RBS is He^4 (alpha particles) with 1–4 MeV energies. Protons are also used for RBS, typically with energies between 100 keV and 2 MeV, but the beam is more suited for NRA applications. The methods are non-destructive, and provide the elemental composition and/or structure, namely the depth profiles from the surface region spanning the first few hundred layers of atoms to up to 10 μm depth. Beyond this, complications due to straggling effects result in loss of sensitivity and/or of depth resolution. At very low energies, the effect of screening of the atomic electrons must be taken into account and this is covered in another section. The main advantages of the methods are the rapidity of analysis (few minutes), and the direct and simple way the information can be obtained from the data. The methods are amenable to simple calibration procedures to facilitate quantitative, standardless analysis.

The methods are most suitable for characterization of mainly inorganic compounds as thin films or bulk samples, giving thickness, composition variation or impurity distribution as a function of depth with resolution of few tens of Angstrom (NRA) to a few hundreds Angstrom (RBS). Applications to organic or biological samples are limited by radiation and thermal damage to the specimen. In RBS, the element identification is poor for high Z due to limits in the detector resolution, and the sensitivity for light element detection is also limited, especially in the presence of high Z matrix. NRA is limited to only those elements exhibiting strong nuclear (resonant) reactions, and these are mainly light nuclei. Interpretation of data can be ambiguous when the sample is not homogeneous, for instance when precipitation, segregation or clustering of impurities in the matrix occur.

Table 9.1 summarizes the salient features of the methods to be used as a rough guide. Special cases can extend the general limits shown. The usefulness of the methods is enhanced with microbeams, which for energetic beams are typically limited to a few microns in diameter for viable beam intensities.

Table 9.1. Important parameters and capabilities of RBS and NRA

	RBS	NRA
Typical probing beam	H, He^4	H, He^4, ^{15}N, ^{19}F
Beam energies	0.2–4 MeV	0.5–20 MeV
Beam currents	1–100 nA	10–1000 nA
Sample structure: layers	yes	yes
bulk	yes	yes
Depth resolution [μm]	> 0.03	> 0.01
Range [μm]	> 10	< 3
Sensitivity [ppm]	> 500	> 100
Element range: Z	> 1	< 30
Z discrimination: high Z	poor	–
low Z	good	good

For more detailed treatment of the subject, and all other IBA methods, the reader can refer to a number of excellent handbooks, the foremost being 'The Ion Beam Handbook for Material Analysis' edited by *Mayer* and *Rimini* [9.1]. Proceedings of conferences based on the theme of ion beam analysis are also excellent resource material and usually contain the repertoire of the applications in various disciplines. One of the best known series is the biennial International Ion Beam Analysis conference, now approaching its 20th year. The latest proceedings published as an issue of Nuclear Instruments and Methods is given in [9.2].

9.1 Principles

Both RBS and NRA involve measurements of energy and flux of the radiations resulting from the interaction of the ions and the atoms of the material under analysis. In RBS, they are the beam projectiles themselves after being scattered by the target; and in NRA they could be gamma-rays or fragments of the reacting nuclei. Unless stated otherwise the general nomenclature is as follow:

Z_1, m_2 – ion beam projectile atomic number and mass (in amu)
Z_2, m_2 – target atomic number and mass
E – incident ion beam energy (MeV)
q – charge state of the incident ions
e – electronic charge (1.6×10^{-19} Coulomb)
I – ion beam current (Amperes)
F – ion beam flux = $\int I \, dt / eq$
N – target molecular density (molecules/cm^3)
s – solid angle of the detector (steradians) subtended at the target.

The surface texture of the sample must be smooth for accurate measurements. With microbeams surfaces with rough texture can still be analyzed if the undulation is larger than the microbeam size. Surface roughness degrades the depth resolution at the surface, and non-flatness results in error due to angular distribution effects. A conductive coating has to be applied on the surface to prevent charge build-up, which can affect the result. Typically a thin layer of carbon or Au are applied, selected to minimize interference with the analyzed material. In NRA measurements the coating may not be desirable.

The measurements are usually carried out in high vacuum (pressure in the 10^{-5} to 10^{-6} mbar range). An ultra high vacuum may be required for prolonged measurement (> 10 min) on the same spot to prevent carbon build-up on the target due to break-up of residual hydrocarbons in the chamber under beam bombardment. This would result in shifting the beam energy for NRA, or in interference with light elements (e.g. O) for RBS. Measurements with external beam are possible, where the beam is brought out of the accelerator through a thin window or through a differentially pumped aperture. In such case the RBS detector is usually operated in an inert gas atmosphere to minimize background problems. At any rate the resolution of the detector suffers, and this kind of measurement is usually conducted only for special cases, e.g. in situ measurements of large objects.

Analysis of the spectrum requires a knowledge of the stopping power and the cross section of the reaction. The stopping power is defined as the rate with which the beam particles lose energy in traversing the target material. The cross-section is defined as the probability of an incident beam particle producing the reaction product of interest for individual atoms in the target, expressed as an effective area presented to the beam by the target atoms. The yield from a layer of the target where the beam of energy E loses E is given by:

$$dY = N\frac{d\sigma}{d\omega}E \Big/ \frac{dE}{dx} \tag{9.1}$$

where:

$\dfrac{d\sigma}{d\omega}$ — differential cross section of the reaction product (cm^2)

$\dfrac{dE}{dx}$ — stopping power of the target matrix for beam at energy E(MeV/cm).

The count rate, dN/dt, in the detector depends on the beam current, on the detector and on absorption effects. The last of these arise from self-absorption processes or additional absorbers used with the detectors. Thus the count write can be expressed as

$$\frac{dN}{dt} = dY\, s e_f f_1 f_2 I / eq \tag{9.2}$$

where:

e_f – efficiency of the detector
f_1 – transmission the reaction product through the target matrix
f_2 – transmission of detector absorbers.

To obtain the yield for analysis of the data, these basic formulae can be integrated.

9.1.1 Stopping Power

Energetic ions lose energy mainly through electronic excitation of the stopping medium atoms, which does not involve change in direction, referred to as electronic stopping. At low velocities ($< c/137$), nuclear collisions become more frequent, giving rise to an additional energy loss mechanism known as nuclear stopping, usually involving many changes of directions [9.3]. However, this occurs towards the end of the ion trajectory, and thus does not usually affect, and is not usually observed in the data obtained.

Extensive measurements over the past two decades have established the values for the stopping power of protons and alpha particles in various elemental material. The data are not as extensive for heavy ions, but these can be related to the stopping power of protons by a scaling factor based on the so-called effective charge. The current usage of stopping power is based on an empirical parametrization, first presented by *Northcliffe* and *Schilling* [9.4]. Subsequently *Ziegler* and *Andersen* [9.5] refined the parametrization for protons and He^4 incorporating many new experimental data, and extended them to all ions. These values agree with available experimental data to within 3–10%.

Molecular stopping power can be derived from the elemental values according to *Bragg* and *Kleeman* [9.6], generally known as Bragg's rule. For a general compound $A_x B_y \ldots$ the stopping power is given by:

$$\frac{dE}{dx} = f_A \left(\frac{dE}{dx}\right) A + f_B \left(\frac{dE}{dx}\right) B + \ldots \tag{9.3}$$

where

$$f_A = x m_A / M, \quad f_B = y m_B / M \ldots$$
$$M = x m_A + y m_B + \ldots .$$

Deviations from this rule have been observed for gaseous organic compounds as well as oxides and nitrides, but they are generally less than 20%. The rule applies well for higher velocities, especially for protons [9.7].

9.1.2 Straggling

The statistical nature of the energy loss at low velocities involving many binary collisions gives rise to fluctuations in the energy loss known as the straggling effect. This progressive broadening of the energy distribution of the incident particles as they lose energy degrades the depth resolution in RBS and the sensitivity in NRA. A theoretical estimate of this effect was first given by *Bohr* [9.8]:

$$\Omega_B^2 = 4\pi Z_2 Z_1^2 e^4 N \, dR \qquad\qquad (9.4)$$

where Ω_B is the width of the energy distribution at one standard deviation (corresponding to FWHM/2.355), and dR is the target thickness. Another theoretical estimate [9.9] incidated a smaller effect than Bohr's theory, particularly in the low energy region and for heavier elements. It is instructive to use Bohr's estimate to determine whether straggling should be considered when information at depths of a few microns is required. For instance for Sn, at $1\,mg/cm^2$ depth ($1.74\,\mu m$) the Bohr estimate gives $7.5\,keV$ FWHM. The total effect on backscattered particles is the sum in quadrature of this and another for the outgoing path, and the detector resolution. For a detector resolution of $16\,keV$, the resulting effect is a broadening of the resolution to $19\,keV$.

9.2 Rutherford Backscattering Spectrometry

In a typical RBS analysis the scattered beam particles are usually detected at the "back angles" ($> 90°$ with respect to the beam direction) using a surface barrier detector. The energy distribution and the yield contain respectively the identity and of the concentration of the target nuclei. In this method, f_1 and f_2 in (9.2) are unity, as is the detection efficiency e_f. However the energy of the scattered particle is attenuated by the target matrix. The particles scattered from a layer thus have an energy uniquely determined by the depth of the layer, and the species of the scattering atom. The yield is determined by the Rutherford scattering cross section:

$$\frac{d\sigma}{d\omega} = \left(\frac{Z_1 Z_2 e^2}{4E}\right)^2 \frac{4[\cos\theta + \sqrt{1 - (m_1 \sin\theta/m_2)^2}]^2}{\sin^4\theta \sqrt{1 - (m_1 \sin\theta/m_2)^2}} \qquad (9.5)$$

where θ is the scattering angle in the laboratory system.

The strong quadratic dependence on Z_2 can be seen from this formula, and the angular dependence is strongly peaked towards the forward angle, and at the back angles the cross section is virtually constant.

The scattered beam energy E' is determined from the kinematics of an elastic collision; its ratio to the incident beam energy is the kinematic factor:

$$K = \frac{E'}{E} = \left(\frac{m_1 \cos\theta + \sqrt{m_2 - m_1 \sin^2\theta}}{m_1 + m_2}\right)^2 . \qquad (9.6)$$

The angular dependence is strongest at the back angle and the contrast for different m_2 is higher for larger m_1, as can be seen in Fig. 9.1, calculated for H and ^{4}He.

Consider now a scattering from a thick target set at an angle θ_1 with respect to the beam, and detection at angle θ_2 (Fig. 9.2). At a depth dt, the incident en-

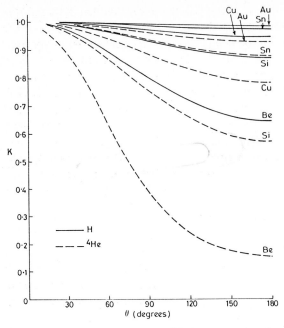

Fig. 9.1. Kinematic factors for H and ^{4}He beams on selected targets as a function of angle

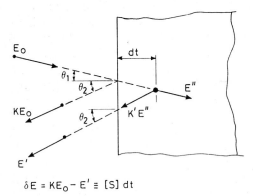

$$\delta E = KE_0 - E' \equiv [S]\, dt$$

Fig. 9.2. Kinematics of Rutherford scattering from a near surface layer

ergy E_0 is attenuated by $dE = (dE/dx)E_0\, dt/\cos\theta_1$, and the scattered energy is $K'E'' = K'(E_0 - dE)$ where K' is the kinematic factor at E''. The scattered particle energy is further attenuated on the way out by $dE' = (dE/dx)E''\, dt/\cos(\theta_2)$. This particle is separated in energy from that scattered from the surface by δE. For infinitesimal dt a quantity S can be defined:

$$\delta E = S\, dt \tag{9.7}$$

where

$$S = K(dE/dx)E_0/\cos\theta_1 + (dE/dx)E''/\cos\theta_1 \ .$$

208

The quantity S reflects the total stopping power for incident and scattered particles, and can be used to define the depth scale. This general expression is the basis for manual analysis of the spectrum, which is practical for thin targets.

For thick and complex targets numerical simulation of spectrum is necessary in the analysis. While many laboratories tend to produce their own RBS simulation programs, some are available from a number of authors, e.g. [9.10, 11]. In calculating the simulated spectrum the response of the surface barrier detector must be taken into account. The response of the detector to monoenergetic particles appears as a Gaussian distribution in an energy dispersive pulse height analyzer with the width dominated mainly by the electronic noise of the system. Incomplete charge collection in the detector produces a continuum extending from the full energy peak to low energies, which could amount to as much as 10% of the total counts.

9.2.1 Experimental Considerations

Most RBS measurements are carried out with H or ^{4}He ions. Heavier ions such as Li or B can be used to gain better mass resolution but due to poor detector resolution for such ions this is applied only in special cases. Typical conditions for RBS measurements are as follows:

1. The particles are detected in surface barrier detectors set between 90 and 170°, typically subtending a few msr solid angle at the target. For maximum mass resolution, the angle must be close to 180°, which can be achieved with annular detectors. The resolution for this detector however is not as good as that for standard detectors (12–16 keV). Larger solid angles can degrade the depth resolution due to kinematic effects. The count rate in the detector is usually maintained below 10 000 counts/s to maintain good resolution and reduce electronic pulse pile-up effects.

2. Targets are usually set normal to the incident beam, except when 90° detection is required, in which case the target is set at some angle sufficient to prevent blocking of the detector by the target frame and to minimize the straggling effect.

3. The energy of the beam is usually chosen to avoid nuclear reactions, which result in deviations from the Rutherford cross-section. However, these deviations can be corrected by means of relative measurements against suitable standards, since the other quantities governing the yield are unaffected by nuclear effects.

4. Low beam currents (below 1 nA) are desirable to minimize damage to samples but may be difficult to measure accurately due to secondary electron emission from the target and leakage currents in the beam integration circuit.

9.2.2 Examples

Figure 9.3 shows a simulation spectrum for a target consisting of a thin layer of Sn on a thick Cu substrate, analyzed using a 1.7 MeV proton beam. The detector is a surface barrier detector set at 143°, subtending a 2 msr solid angle and its

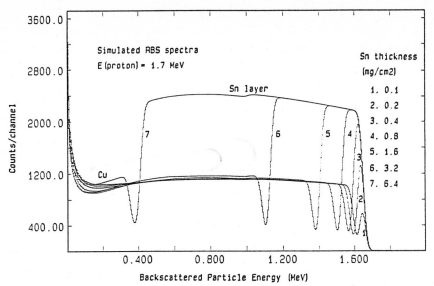

Fig. 9.3. Calculated RBS spectra from a series of targets consisting of a Cu substrate with a thin surface layer of Sn of various thicknesses. The spectra are computed for 1.7 MeV proton beam and a detector at 143° with respect to the beam direction subtending a solid angle of 2 msr at the target

response simulated by a Gaussian (16 keV FWHM) with a constant continuum tail amounting to 10% of the total counts. The Sn layer appears as a Gaussian with a total area (counts) A, for thin layers, i.e. when the energy loss through it is less than the detector resolution. The thickness of the layer, as areal density, is given by:

$$W(\text{g/cm}^2) = \frac{A}{F s \frac{d\sigma}{d\omega}} \ .$$

The peak height would increase with target thickness until the maximum height H is reached. Further increase results in the increase of the width of the peak with a trapezoidal shape. The Sn peak is partially separated from the spectrum from Cu through kinematic effect. The Cu spectrum shows a high energy edge corresponding to the interface layer, and extends to the low energy region showing the characteristic shape of the spectrum from a thick target. The depth scale for the Sn peak can be determined from the dispersion of the spectrum. If dE is the energy per channel of the spectrum, then the height of the spectrum H at the edge, corresponding to the surface is given by:

$$H = F \frac{d\sigma}{d\omega}\bigg|_{E_0} \frac{s N dE}{S_0} \ .$$

At greater depth, dE would correspond to a different depth interval corresponding to the appropriate energy at that depth. More conveniently, this depth scale can be established from the simulated spectra calculated for the various target thickness greater than the detector resolution.

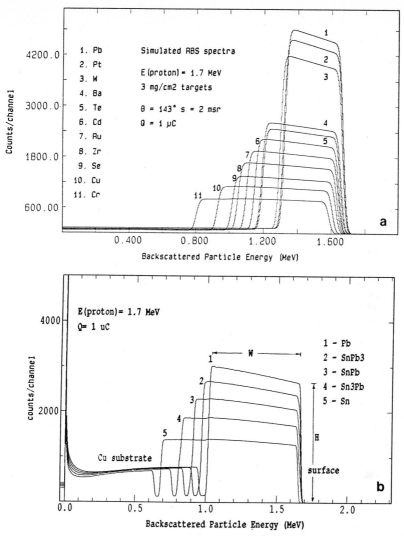

Fig. 9.4. (a) Calculated RBS spectra from a series of targets of 3 mg/cm² layer of various elemental material. The spectra are computed for the same conditions as in Fig. 9.3. The width of the spectra generally increases with lower Z. This is mainly due to kinematic and stopping power effect. Occasionally this trend does not hold, e.g. in the case of Ba. (b) Calculated RBS spectra from a series of targets consisting of a Cu substrate with a 5 mg/cm² surface layer of Sn, Pb and their various compounds. Note the variation of the spectrum height H and width W of the surface peak as a function of the material

Figure 9.4a shows the calculated RBS spectra for a series of targets consisting of a 3 mg/cm² layer of various elemental material, for the same experimental condition. The "width" W of the spectra generally increases with lower atomic number, although occasionally this does not hold, for instance in the case of Ba. The general trend is primarily due the kinematic factor effect and stopping power

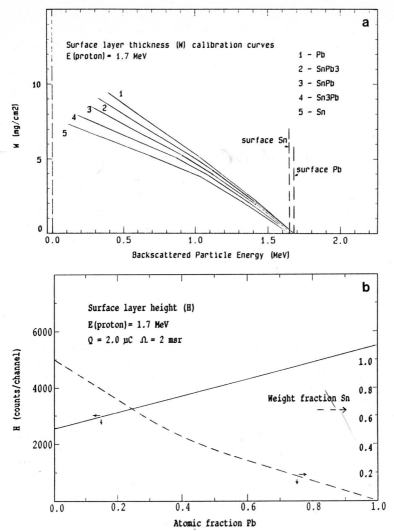

Fig. 9.5. (a) Depth scale for a thick layer of material (W) corresponding to those in Fig. 9.4b. The non-linearity of the scale is due to the energy dependence of the stopping power. **(b)** The composition of the Sn/Pb mixture can be deduced from the spectrum height (H)

(9.6), obscuring effects of the different atomic density. Figure 9.4b shows the calculated spectra for an assortment of materials on a Cu substrate. The materials range from pure Sn to pure Pb with an assortment of their compounds in between. Each spectrum shows similar features: a trapezoidal peak corresponding to the layer at higher energy and a lower one corresponding to the substrate. The features observed can be related to (9.5, 6); the energy term in (9.5) results in the increase of yield towards lower energy, and the "height" H of the spectrum (the yield per unit energy interval), is proportional to the atomic number. The separation between the surface layer and the substrate is due mainly to kinematic effects (9.6). Figure 9.5a shows the corresponding general depth scale (W) for the different surface layers for Fig. 9.4b. The non-linearity of the scale is due to the energy dependence of the stopping power. For a binary mix where the components are not resolved kinematically such as the present case, the composition can be obtained by a graphical method from the height H. Figure 9.5b shows the graph for H for the Sn/Pb system.

These examples illustrate that RBS is an ideal tool for analysis of thin layers of a few μm thickness for stoichiometric information. However, its applicability is limited to analysis of relatively simple structures and compositions. Unless the components are resolved kinematically, compositions of ternary and more complex mixtures are difficult to obtain unambiguously. The method is also limited to depth ranges where the scattered particle can still escape the target. This varies between 5 and 10 μm for most materials for proton and alpha beam energies below the Coulomb barrier, i.e. where nuclear reactions start.

9.2.3 Special Cases

The following special cases are worthy of mention:

– Channeling effect: when the target is a single crystal, the backscatter yield will be reduced when the incident beam is aligned with the major (low index) axes [9.12]. This phenomenon can be used to probe the site of an impurity atom. For interstitial sites, the backscatter yield will increase again towards the random direction value.

– Resonant scattering: for a number of light nuclei, e.g. ^{16}O, ^{12}C, a number of resonances (marked by increase of yield) occur for certain energies. Application of this is similar to that which will be discussed further in the NRA section below.

– Forward recoil: in this technique, under bombardment with heavy ions, the recoiling nuclei are detected in the forward direction. This provides some depth profile information but it is usually limited to a depth of around 1 μm due to straggling effects [9.13].

– Non-resonant reaction: nuclear reactions can have a much higher cross-section than Rutherford scattering, and can thus be used to enhance the detection of selected elements.

9.3 Nuclear Reaction Analysis

A certain class of nuclear reactions is known as resonant reactions, defined as those where the reaction cross-section increases dramatically, as much as few hundred fold over the Rutherford cross section, for a narrow range of energy (res-

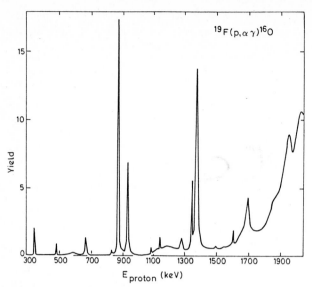

Fig. 9.6. Resonances in the reaction between a proton beam and a fluorine target, appearing as peaks in the reaction yield, e.g. gamma rays, as a function of the incident beam energy

onance width). Resonances are common for reactions induced by protons, and to a lesser extent by alpha particles, on light nuclei (up to $Z_2 \sim 30$). These resonances can occur below the Coulomb barrier energy ($E_c \sim Z_1 Z_2 / (m_1^{1/3} + m_2^{1/3})$ MeV in the centre of mass system) – a phenomenon known as sub-Coulomb transfer reactions. The most common reactions are capture reactions which show resonances corresponding to discrete high energy states in the product (compound) nucleus. In proton capture reactions, alpha particles are sometimes emitted, and can be used to detect the resonance. These reactions occur at specific energies for specific elements, and this is the feature exploited in NRA. At the resonant energy, the reaction occurs at the surface, sampling the surface region for which the energy loss of the beam corresponds to the resonance width. For higher beam energies the reaction will occur deeper in the target where the resonance energy is reached. Hence this method can be used to determine the concentrations of the particular atom involved in the resonance as a function of depth, giving the depth profile of the specific element. With typical resonance widths in the range of a few keV, depth resolutions of a few nm are obtained.

For example, the resonances in the reaction between a proton beam and a fluorine target can be seen in Fig. 9.6 as peaks in the reaction yield against the incident beam energy. The criteria for selection of the reactions for NRA applications are that the cross section should be large and that the radiation can be detected readily. Gamma rays are commonly detected in this reaction.

9.3.1 Formalism

In the NRA method the yield is obtained only at the resonance energy, and from a layer corresponding to the natural width of the resonance. The correction factor f_1 (9.2) depends on the nature of the radiation; it is 1 for particles as long as they are not stopped within the target. For gamma rays, the usual attenuation must be computed. The intensity of a gamma ray traversing a medium with a linear absorption coefficient μ' is given by:

$$I = I_0 e^{-\mu' x} \tag{9.8}$$

where:

I_0 = initial intensity (no of counts)
μ' = linear attenuation coefficient of the medium for gamma ray of energy E_g, in cm^{-1} unit
x = thickness of the material traversed in cm.

A more convenient variable is the areal density (g/cm^2), and the corresponding coeffcient is referred to as mass attenuation coefficient μ (= μ'/ϱ, where ϱ is the density). For a compound medium, Bragg's rule of additivity applies. Thus for a compound $AxBy\ldots$ the mass attenuation is given by:

$$\mu = \left(m_A^* x^* \mu_A + m_B^* y^* \mu_B + \ldots \right)/mt \tag{9.9}$$
$$mt = m_A^* x + m_B^* y + \ldots .$$

Extensive tabulation of μ is available, based on both theoretical calculation and empirical data [9.14]. Parametrization of the coefficients are given by *Theisen* and *Vollath* [9.15].

For high energy gamma rays this absorption could be negligible, hence $f_1 \sim 1$. The absolute value of the cross-section can only be determined empirically, but in most applications measurements are carried out against a standard. The "raw" profile, i.e. the yield as a function of incident beam energy are further unfolded to take into account the effect of straggling, to give the true profile. The Bohr estimate can be used to determine whether this effect is significant for the range of energy (depth) involved in the measurement.

9.3.2 Experimental Considerations

In NRA measurements, the following points should be considered:

1. The cross-section in (9.1) must contain the angular distribution effect of the detected radiation. Most resonant reactions show a very strong angular distribution, and care must be taken in choosing the angle of the detector with respect to the beam direction.

2. When the resonant reaction produces discrete, relatively low energy gamma rays (below $\sim 2\,MeV$) usually emitted in the last stages of the product nucleus deexcitation, Ge detectors are used. The high resolution of this detector (typically $< 2\,keV$ at $1.33\,MeV$) results in a good signal to background ratio.

3. In measurements involving high energy gamma rays ($> 2\,\mathrm{MeV}$), NaI detectors are commonly used for better efficiency. The poor energy resolution (typically 6% of the full energy peak) results in poorer signal to background ratio. The main sources of background are gamma rays from $^{40}\mathrm{K}$ ($1.461\,\mathrm{MeV}$) and one of the daughters of $^{232}\mathrm{Th}$ decay ($2.614\,\mathrm{MeV}$ from the decay of $^{208}\mathrm{Tl}$) from building materials, and cosmic rays.

4. To improve the ratio gamma-rays to background, a closed geometry is necessary and can be achieved by appropriate design of the target chamber. Background can be reduced further by absorbers, which however are only effective for the room background and less so for cosmic rays. More active reduction of cosmic rays can be achieved by using anti-coincidence shields.

5. If sharp resonances are used, it is desirable to carry out the measurements in ultra high vacuum conditions to reduce the problem of carbon build-up on the target which shifts the energy of the beam. This can be achieved by means of a cold finger (cooled by liquid nitrogen) around the target.

9.3.3 Examples

Figure 9.7 shows the hydrogen profile from a set of samples of hydrogenated amorphous silicon, obtained using the resonance at $6.412\,\mathrm{MeV}$ fluorine beam energy, corresponding to the $0.340\,\mathrm{MeV}$ resonance (Fig. 9.6) when protons are

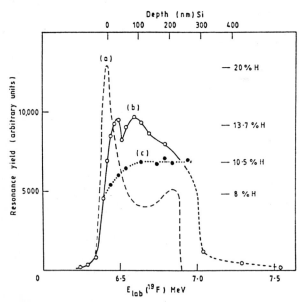

Fig. 9.7. A hydrogen profile of a selection of a-Si:H samples, obtained using the resonance at $6.412\,\mathrm{MeV}$ fluorine bombarding energy. Two of the samples show surface peaks which may arise from adsorbed moisture. The width of the surface peak indicates the depth resolution of $44\,\mathrm{nm}$ obtained with this resonance. Note the non-linearity of the hydrogen content scale due to change in the stopping power with composition

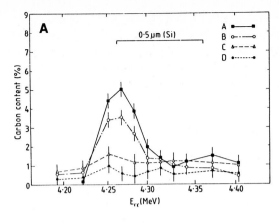

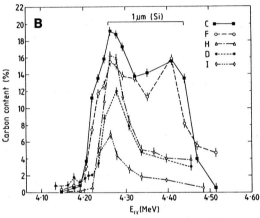

Fig. 9.8. Carbon profiles (A and B) from a series of amorphous hydrogenated Si-C compounds, measured using the alpha resonance at 4.26 MeV. The corresponding backscattered particle spectra for one of the samples are shown in C: part (a) shows the spectrum below the resonances; (b) the onset of the resonance due to oxygen at 3.036 MeV; (c) the spectrum at an energy where the oxygen no longer shows resonance, and (d) the onset of the carbon resonance at 4.26 MeV

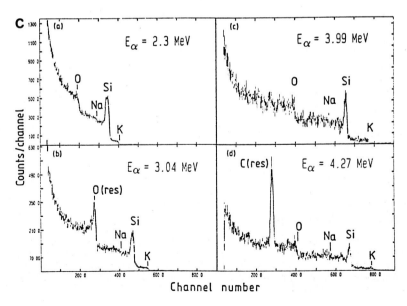

used on fluorine target [9.16]. In this particular case, the inverse of the resonant reaction induced by protons is used, where now one uses the ^{19}F as the beam to detect hydrogen. The 6.13 MeV gamma rays emitted by the reaction are detected in NaI detectors. The surface peaks on two of the samples indicate a depth resolution of 44 nm, corresponding to the resonance width of 3 keV. The range of applicability is usually limited to a few μm, because of increasing diffuseness of the beam with depth due to multiple scattering. In the present case, the onset of another resonance corresponding to the 0.484 MeV in Fig. 9.6, limits the applicability of the reaction above to a depth of only 1 μm. The sensitivity of this method is limited by background to around 100 ppm. This can be improved when anti-coincidence techniques are applied to reduce the cosmic ray background. A detection limit as low as 10 ppm has been reported [9.17].

The resonance yield can be calibrated against a hydrogen-rich target of known composition. Organic polymers (e.g. polyethylene, mylar) are convenient but they are easily damaged by the beam mainly due to thermal effects. Care must be taken in the calibration run to minimize such effects by running very low level and diffuse beam on the target. The damage can be observed by monitoring the yield as a function of time or beam dosage, and the true yield can be obtained by extrapolation to zero time or beam dosage. More stable standards can be obtained using implanted material [9.18] or hydrogenated amorphous silicon.

Sub-Coulomb barrier resonances in scattering of protons and alpha particles on some light nuclei can be utilized in this method, where the detected radiation is the scattered particle. The main advantage are the somewhat longer range of applicability, and higher efficiency of detection afforded by particle detectors [$f_1 = 1$ in (9.2)]. As an example, Fig. 9.8 shows the depth profiles of carbon obtained from several samples of thin layers of hydrogenated amorphous Si-C compound using the 4.26 MeV resonance of alpha scattering on carbon, and the corresponding spectra obtained at various bombarding energies showing the onset of the resonance [9.19]. The figure also shows the resonance of oxygen, from the glass substrate, at 3.036 MeV. These particular resonances are peaked at 180° with respect to the beam direction, hence the ideal measurement geometry is consistent with the standard RBS requirements.

9.4 Summary

Two of the suite of ion beam analysis techniques, namely RBS and NRA, and variations thereof, are versatile and powerful non-destructive methods for characterizing the elemental composition and structure of the surface region, extending from a few hundred atomic layers to several μm. In this respect they complement other surface analytical methods which probe fewer atomic layers of the surface, and their chemical states. The insensitivity of the forces governing the interaction between the probing beam and the sample results in methods which are highly quantitative, being amenable to straightforward calibration procedures.

Acknowledgement. The simulation spectra presented in this article were produced using a computer program by Chris Ryan.

References

9.1   J.W. Mayer, E. Rimini (eds.): *Ion Beam Handbook for Material Analysis* (Academic, New York 1977)
9.2   J.F. Ziegler, P.J. Scanlon, W.A. Lanford, J.L. Duggan (eds.): *Ion Beam Analysis*, Proc. of the 9th Int. Conf., Kingston 1989, Nucl. Instr. Meth. B45 (1990)
9.3   J. Lindhard, M. Scharff, H.E. Schiott: Kgl. Danske Videnskab. Selskab., Mat. Fys. Medd. **33**, 14 (1963)
9.4   L.C. Northcliffe, R.F. Schilling: Nuclear Data Tables A7, no. 3–4 (1970)
9.5   H.H. Andersen, J.F. Ziegler: *Hydrogen Stopping Powers and Ranges in All Elements* (Plenum, New York 1977) and related titles in the series "The Stopping and Ranges of Ions in Matter", ed. by J.F. Ziegler (Pergamon, New York 1980)
9.6   W.H. Bragg, R. Kleeman: Phil. Mag. **10**, 318 (1905)
9.7   J.F. Ziegler, J.M. Manoyan: Nucl. Instr. Meth. in Phys. Res. B35, 215 (1988)
9.8   N. Bohr: Kgl. Danske Videnskab. Selskab., Mat. Fys. Medd. **18** no. 8 (1948)
9.9   W.K. Chu: Phys. Rev. A13, 2057 (1976); also in Ref. [9.1] p. 1
9.10   P.A. Saunders, J.F. Ziegler: Nucl. Instr. Meth. in Phys. Res. **218**, 67 (1983)
9.11   L.R. Doolittle: Nucl. Instr. Meth. B9, 344 (1985)
9.12   B.R. Appleton, G. Foti: In Ref. [9.1] p. 67
9.13   B.L. Doyle, P.S. Peecy: Appl. Phys. Lett. **34**, 811 (1979)
9.14   J.H. Hubbell: Atomic Data **3** no. 3 (1971)
9.15   R. Theisen, D. Vollath: *Tables of X-Ray Mass Attenuation Coefficients* (Stahleisen, Düsseldorf 1967)
9.16   S.H. Sie, D.R. MacKenzie, G.B. Smith, C.G. Ryan: Nucl. Instr. Meth. in Phys. Res. B15, 525 (1986)
9.17   H. Damjantschitsch, M. Weiser, G. Heusser, S. Kalbitzer, H. Mannsperger: Nucl. Instr. Meth. in Phys. Res. **218**, 129 (1983)
9.18   J.F. Ziegler et al.: Nucl. Instr. Meth. **149**, 19 (1978)
9.19   S.H. Sie, D.R. MacKenzie, G.B. Smith, C.G. Ryan: Nucl. Instr. Meth. in Phys. Res. B15, 632 (1986)

10. Scanning Tunnelling Microscopy

B.A. Sexton

With 15 Figures

The scanning tunnelling microscope, or STM, has emerged over the last few years as a fascinating new technique for examining conducting solid surfaces with high resolution [10.1–5]. A sharpened metal wire is brought close enough to the surface so that the electrons "tunnel" across the narrow gap (0.5–1.5 nm). A small bias potential (2 mV–2 V) provides the necessary potential difference for tunnelling to occur. As a result of the exponential dependence of tunnelling current on separation, the tip height above the surface can be kept constant by using a feedback controller. The tunnel current is monitored and applied to a "PI" controller which in turn drives a piezoelectric arm attached to the tip. By scanning another set of piezoelectric arms, the tip can be rastered in an XY plane whilst simultaneously following the surface corrugations. In this way, a three-dimensional image of the surface can be formed by plotting $z(x, y)$. The imaging is non-destructive since the tip does not normally touch the surface during the scans.

In Fig. 10.1 we show a schematic of the tip traversing an idealized, corrugated surface. The normal mode of microscope operation is constant current mode

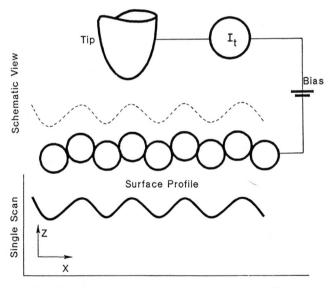

Fig. 10.1. Schematic of the scanning tunnelling microscope. The tracking motion of the tip during a scan is recorded as a three dimensional image

where the tip traverses the surface at a constant tunnelling distance. Tunnel microscope images are measured as a series of linescans and displayed on a computer as a three-dimensional representation of the surface. Both grey scale and false colour representation may be used to enhance the images.

STM experiments may be conducted in air, under liquids or in vacuum. The requirements for successful imaging depend primarily on the conductivity of the sample and the nature of the surface oxide film or contamination layer. In air, the surface should have no oxide at all, or a thin ($< 2\,\mathrm{nm}$) conducting oxide film. Insulating oxide films, such as those found on silicon, gallium arsenide and aluminium are prone to charging under the tip causing excessively noisy images. The problems encountered in imaging some semiconductors in air have led to the development of vacuum – compatible microscopes as offered by several commercial manufacturers. Almost any conducting surface can be studied in these microscopes, since the oxide films can be removed by sputtering or annealing [10.4]. Imaging under liquids is difficult, and is complicated by the presence of Faradaic currents between tip and sample in conducting liquids. Several authors have reported special cells designed to overcome these problems, however [10.6, 7]. In Table 10.1 we summarize the range of materials which can be studied with the STM in different media.

The STM has three main areas of application, as discussed by *Baumeister* [10.8]. These are (a) as an imaging device to reveal surface topography from several μm down to the atomic level, (b) as an analytical tool for, obtaining spectroscopic information and (c) as a micromanipulation device for modifying surfaces on the "nano-scale". For surface topographical measurements the STM can reveal the atomic arrangement of atoms on semiconductor surfaces, or measure surface features up to several μm in size [10.9]. In spectroscopic mode, the current-voltage (I–V) curves at constant height show the conductivity of the tunnel gap, and also the surface electronic structure [10.9].

Table 10.1. Some materials that can be studied in an STM in various media

Air[a]	Vacuum[b]	Under Liquids[c]
Gold	Silicon	Platinum
Platinum	GaAs	Nickel
Silver	HgCdTe	Graphite
Rhodium	Aluminium	Gold
Graphite	Copper	Stainless steels
Nickel	+	
Iron	all	
	air	
PbS	Materials	
PbO$_2$		
Co$_3$O$_4$		

[a] These surfaces have very thin or highly conducting oxide films which do not charge.
[b] Many semiconductor and metal surfaces have insulating surface films which require removal in vacuum before images can be obtained.
[c] Surfaces of interest to electrochemists.

222

Spatial variations of the surface conductivity can also be mapped out at high resolution. Recent measurements include the mapping of spatially resolved electronic structure on GaAs [10.9] and the observation of $p - n$ junctions on silicon on a much larger scale [10.10]. Interest in the micro-manipulation of surfaces stems from a desire to increase the storage density of computer-related devices. Whether the STM will ever contribute to this area is the subject of conjecture. The STM has, however, already successfully imaged sub-μm features on the high density storage surfaces of compact discs [10.11, 12] and provided complementary information to the scanning electron microscope.

Interest in the STM technique from biologists and botanists has been high, but the difficulties of working in this area are indicated by the lack of published results. Thin film sensors and Langmuir-Blodgett layers, both new growth fields, have some potentially interesting applications for the STM and some preliminary results have been reported [10.13, 14]. In Table 10.2 we list some of the areas of application of the STM taken from recent conference proceedings [10.9].

Table 10.2. Some areas of application of the STM

Semiconductor surface structure
Layered materials
Langmuir-Blodgett films
Thin film surface morphology
Ion beam and laser damage
Etching and processing of semiconductors
Molecular clusters
Mineral surface topography
X-ray optics
Machined surfaces
Microfabricated patterns
Optical and compact discs
Conductive ceramics
Gold-coated ceramics
Adsorbates on metals and semiconductors
Alloy surfaces
Surface modification with the STM
Polymers
Biological structures

10.1 History of Development of the STM

For a detailed account of the historical development of the tunnel microscope, the article by *Quate* [10.5] is recommended. In brief, the first instrument similar to an STM was built by *Young* and colleagues at the National Bureau of Standards [10.15] and was operated primarily as a field emission microscope. Later, *Binnig* and colleagues at the IBM Zurich laboratory obtained the first atomic resolution images of surfaces by operating a similar instrument in the tunnelling mode

[10.1–3]. They successfully demonstrated the power of the technique as a new form of microscopy. *Binning* and *Rohrer* shared the Physical Nobel prize in 1986 with *Ruska* for their pioneering work [10.5]. Several international conferences devoted to STM have already been held and the proceedings as well as the earlier issues in IBM Journal of Research and Development are worth reading [10.9, 16, 17].

10.2 Theory of Metal-Vacuum-Metal Tunnelling

Electron tunnelling occurs when two conductors come into near contact, separated by an insulating barrier (Fig. 10.2). With an applied bias, V, the tunnel current I becomes significant when the wavefunctions of the two surfaces overlap, typically at distances less than 1.5 nm. With the tip biased negative with respect to the sample in Fig. 10.2, the electrons tunnel through the trapezoidal barrier to empty

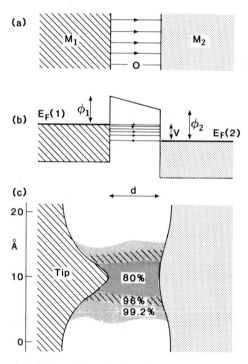

(a) Planar Tunnel Junction

(b) Tunnelling through trapezoidal barrier by application of bias

(c) Localized tunnelling between tip and sample

Fig. 10.2. The metal-vacuum-metal tunnelling junction

states above the Fermi level of the surface. The tunnel current I, is a strong function of the separation, d and also depends on the bias, V. At low bias $V \ll \Phi$, where Φ is the effective barrier height [$\Phi = (\Phi_1 + \Phi_2)/2$], the current density, j can be written as [10.18];

$$ j = \frac{e^2}{h} \frac{k_0}{4\pi^2 d} V \exp(-2k_0 d) \qquad (10.1) $$

where d = tunnel distance

k_0 = inverse decay length of wavefunctions

V = bias (volts

$e^2/h = 2.44 \times 10^{-4} \ \Omega^{-1}$

$2k_0(\text{nm}) = 0.1025 \sqrt{\Phi}$.

At low bias, therefore, the tunnel current depends exponentially on distance, d. In Fig. 10.2c the calculated current density at a tip is shown based on the exponential approximation, showing localization of the current density to the extreme end of the tip. The exponential current dependence also means that the tip height can be easily controlled to a sub-angstrom level, since small tip displacements cause large current fluctuations. The vertical resolution is therefore typically 0.1 Angstroms or better.

The horizontal resolution, however, depends somewhat on the surface roughness and the tip radius, r. For a sharp tip with micro-tips consisting of small atom clusters on the end, imaging a very flat surface, the calculated resolution can be on the order of 0.2 nm or less as calculated by *Hansma* [10.4]. Certainly the recent work on atomic resolution of semiconductors allows individual atoms to be resolved. An average surface, however can look like something approaching the Grand Canyon with large surface hills and valleys with amplitudes up to some μm in height. Since the average tip radius will be 0.1–0.5 μm, the images generated on rough surfaces can be a convolution of the tip radius with the topography. We demonstrate and discuss these effects in Sect. 10.5 on compact discs.

In Fig. 10.2 it can be seen that the tunnelling probability will depend on the availability of electrons in states near the Fermi level of the tip and unoccupied states above the Fermi level of the sample. The tunnel current will therefore be dependent on the bias in a manner more complicated than that described by (10.1). Particularly on semiconductors, the conductivity of the tunnel barrier will be strongly modulated by the surface electronic structure [10.17]. Suffice it to say at this stage that current-voltage measurements form the basis of a second technique called scanning tunnelling spectroscopy (STS) which is extensively used on semiconductors.

10.3 Experimental Aspects

10.3.1 Piezoelectric Drivers

The extremely small displacements required to move the tip over a surface can be achieved by the use of piezoelectric materials. Figure 10.3 describes their operation. A piezoelectric material (e.g. lead zirconate titanate) expands or contracts when voltages are applied across the faces of a block of the material. The displacement is linear with voltage and can be calculated from the manufacturer's data [10.19, 20]. For a rectangular bar, the extension is given by:

$$\delta L = d_{31} V L / t \tag{10.2}$$

where L = length

Piezoelectric Drivers : Mechanisms for Achieving
Small Displacements

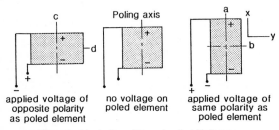

Piezoelectric Actions From Applied Voltages

Two commonly used microscope drivers

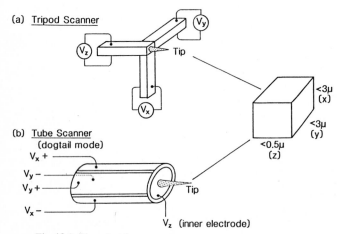

Fig. 10.3. Piezoelectric movements

226

t = thickness parallel to poling direction

d_{31} = piezoelectric charge constant along poling direction

V = applied voltage

For example, for PZT5A ceramic, $d_{31} = -171 \times 10^{-12}$ m/V which is 0.171 nm/V [10.19]. For a bar with a length: thickness ratio of 10, an applied voltage of 500 V will cause an extension of 0.86 μm. For tubes, (10.2) is valid substituting the tube wall thickness for t and the length, L.

In Figure 10.3, two microscope driver types are shown, the XYZ tripod, originally described by *Binnig* [10.1–3], and the tube scanner, by *Binnig* and *Smith* [10.21]. The tripod scanner has independent movements of the three orthogonal axes, whereas in the tube scanner there are four external electrodes. By applying potentials to the various electrodes, a three dimensional scanning function is achieved. Tube scanners can be scanned faster than tripods and are the preferred design for rapid imaging.

10.3.2 The Tunnelling Tip Preparation

Both tungsten wire (< 0.5 mm diameter) and Pt or Pt-Ir wire are currently used for tips. Etching of the sharp radius is carried out in either a solution of KOH (W) or HCl-containing solutions for Pt or Pt-Ir [10.22, 23]. Pt-Ir or Pt-Rh tips can also be cut with wire cutters at an angle to generate tips which work quite well. The apparatus for tungsten etching is shown in Fig. 10.4. An AC voltage of < 10 V is applied between a stainless steel electrode and the tungsten wire in the KOH solution. The wire etches to a fine point characterized by an end radius of $< 0.5\,\mu$m. On the microscopic scale, however, small asperities as small as single atom clusters will potentially become the active tip during tunnelling.

Tungsten gradually corrodes in air and has to be re-etched periodically, whereas in vacuum, the metal is quite stable. For air operation, Pt or Pt-Ir tips are better since they have no surface oxide and are non-corrosive in moist air or solutions. For high resolution atomic imaging, tungsten is the preferred metal for vacuum microscopes [10.9, 16, 17].

10.3.3 A Microscope Design

In Fig. 10.5 is shown the design of a tube scanner built in this laboratory [10.24]. The tube scanner incorporates an inverted configuration with the sample approach from below. The head is temperature compensated with the ceramic tube and legs glued to a quartz base [10.2, 5]. A single micrometer, obtained from a surplus electron microscope moves the sample through a 2–3 mm approach to establish tunnelling. The tube(2) has four equal electrodes cut on the outer surface with a diamond saw. In addition, the sample stage in Fig. 10.5 has piezoelectric offset arms attached to translate the stage during operation, to access new areas.

Preparation of Sharp Tungsten Tips

(<0.5 Micron Radius)

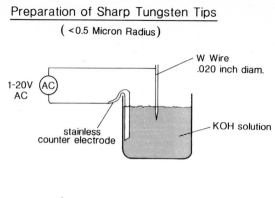

1-20V AC
AC

W Wire
.020 inch diam.

stainless
counter electrode

KOH solution

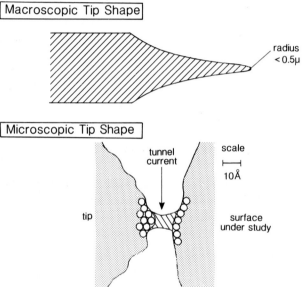

Macroscopic Tip Shape

radius
< 0.5μ

Microscopic Tip Shape

tunnel
current

scale

⊢———⊣

10Å

tip

surface
under study

Fig. 10.4. Preparation of tungsten tips for use in an STM

During operation, the sample is mounted on an aluminium stub and clamped to the sample holder. By manually adjusting the micrometer, the sample can be brought up to the tip (or vice versa) to establish tunnelling. Approaches must be smooth to avoid crashing the tip. Isolation of the microscope in a sound-proof box or bell-jar is necessary to prevent modulation of the tunnel gap. An anti-vibration mounting is also essential to remove the effects of building vibrations. Some important design considerations for tunnel microscopes include: (1) temperature stability and compensation, (2) structural rigidity, (3) low mass of the microscope head and a high resonant frequency, (4) vibrational damping (5) smooth tip approach with no backlash or mechanical drift, and (6) incorporation of x, y and z offsets (piezoelectric) which operate independently of the driver head.

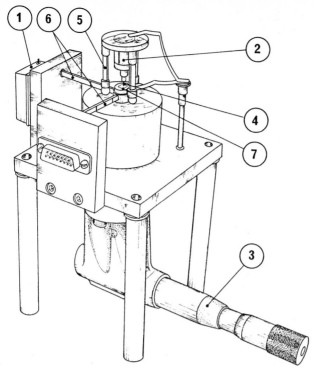

Fig. 10.5. A tube-scanner STM (*1*) *I–V* converter (*2*) scan tube (*3*) approach micrometer (*4*) tip (*5*) *Z*-offset leg (*6*) *XY* offset legs (*7*) sample

10.3.4 Electronics

There are many different electronic designs in current use although almost all have as a common element the PI or PID feedback controller (proportional-integral-differential). Full circuit diagrams of such a system have been published by *Park* and *Quate* [10.25]. A schematic of our electronics, which was designed in our laboratory is shown in Fig. 10.6. The PI controller (no differentiator was found necessary) senses the tunnel current from the *I–V* converter (typically 100 mV/nA) and compares this with a tunnel current "target" or set-point (< 1 nA, typically). An error signal is amplified and sent to an integrator and a high voltage driver (150 V) which drives the *z*-piezo (inner electrode of the tube). A bias of less than 10 V is applied to the tip. The computer sends sweep signals to an *XY* raster generator which provides the bipolar sawtooth voltages for the outer electrodes of the tube. The tube then scans in the *XY* plane with the PI controller keeping the tip at a constant height (current) above the surface.

Images are collected by sampling the *z*-piezo monitor voltage with a fast A/D data acquisition card in synchronism with the *XY* map. The function $z(xy)$ is a three-dimensional representation of the surface and is stored on the computer

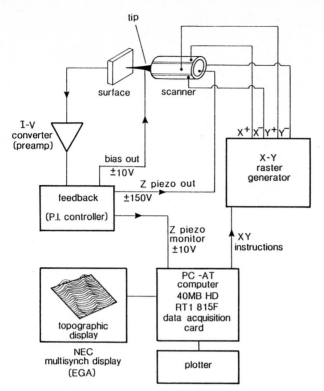

Fig. 10.6. The feedback and control electronics for an STM

as a file. At a later date the file is recalled and image processing and display programs are actuated.

10.4 Imaging Surfaces in Air with the STM

10.4.1 Gold Surfaces

Gold represents an excellent starting surface for STM work in air, since it has no surface oxide and remains clean for long periods of time. We have studied both polycrystalline gold, polished with 0.25 μm diamond paste, and gold films sputtered onto Si(100) wafers with an SEM coating unit. Other work has been published on the morphology of gold single crystal surfaces.

Figure 10.7 shows an STM image taken in air of sputter-evaporated gold, over an area of 130 × 130 nm (0.13 μm). The analysis conditions were tip bias, −100 mV, tunnel current 0.2 nA. This film, evaporated in an SEM coater, consists of a close packed structure of rounded nodules, some 20 nm diameter and 5 nm high. These nodules are derived from the sputtering process: spherical clusters

Sputtered Gold Surface

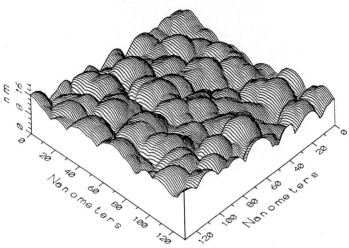

Fig. 10.7. STM image in air of a gold film sputtered onto Si(100) in an SEM coating unit (15 nm thick)

of gold are "splattered" on the substrate and coalesce to form the film. We can therefore see in this first example that the STM can resolve the surface morphology at these high magnifications ($\times 500\,000$). Some of the surface structure in the image will have information from the tip convoluted into the picture, especially near sharp edges. We can estimate the tip radius at less than 5 nm based on the smallest resolvable details in the image of Fig. 10.7.

Gold single crystal faces are useful substrates to search for atomic steps. When single crystal substrates are cut at different angles, both low and high Miller-index planes are exposed. One can tailor a surface to provide different terrace widths and step orientations. The small interplanar spacing on gold (0.25 nm) makes it a useful resolution test for an STM. *Salmeron* et al. [10.26, 27] measured the step heights on gold at between 0.2 and 0.4 nm with terrace widths of 2.2 nm on the Au(334) surface. *Hallmark* et al. measured a 0.25 nm step height on Au(111) and obtained variable lateral resolutions of between 0.5 and 4 nm when tracking across the step edge [10.28]. They also resolved the corrugations of the atoms in the Au(111) plane with a 0.3 nm spacing and a 0.03 nm corrugation height. *Jaklevic* and *Elie* reported some very interesting results for surface diffusion of gold on a Au(111) crystal which had been indented approximately three layers deep. They obtained a time-lapse sequence of images which clearly showed the surface diffusion of gold filling in the shallow craters [10.29]. In Fig. 10.8 we show a 100×100 nm image of epitaxial gold(111) on a mica substrate taken in this laboratory. The 0.25 nm steps are clearly visible in the image.

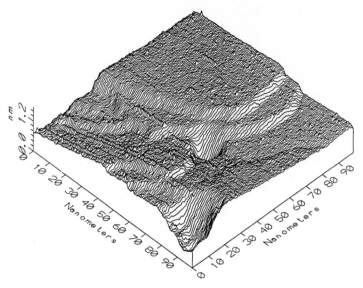

Fig. 10.8. STM image of epitaxial gold (111) on mica in air showing 0.25 nm steps

10.4.2 Graphite

It was recognized at an early stage that atomic resolution on highly oriented pyrolitic graphite could be achieved with the STM, in vacuum [10.30], air [10.31] or under liquids [10.32]. The images showed what appeared to be every second graphite atom on the surface, and the corrugation height depended on the bias setting. At low bias, V approaching zero, corrugation heights as large as 0.5 nm were observed, which is much larger than the surface buckling predicted from crystallographic models. Graphite became a standard surface with which to calibrate an STM, since it provided the unit cell spacing with good signal-to-noise on the corrugations. The surfaces are easily prepared by cleaving to expose a fresh surface. More recently, some groups have used graphite as a reference surface on which to grow metal atom clusters, with the background lattice providing a useful calibration [10.33].

In Fig. 10.9 is shown a constant current map (2 nA current, −50 mV bias) or HOPG over a 4×4 nm area. The corrugation height is 0.5 nm. The image shows six-fold symmetry with the perpendicular spacing between the rows equal to 0.213 nm. A contour map of the same area is shown in Fig. 10.10. The linearity of the rows is a measure of the drift of the microscope, and graphite is generally regarded as a good calibration source. The map in Fig. 10.10 does not match the surface structure of graphite (0001), since only one hill or valley is seen per unit cell, whereas graphite has two atoms per unit cell. *Tersoff* [10.34] has recently clarified the interpretation of these images. The STM image can

232

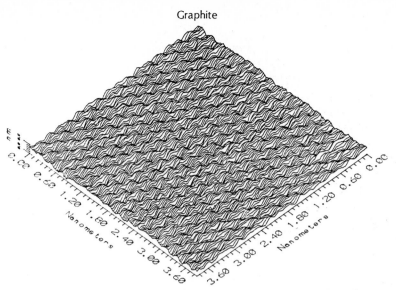

Fig. 10.9. Constant current image in air of a highly oriented pyrolitic graphite surface over a 4 × 4 nm area

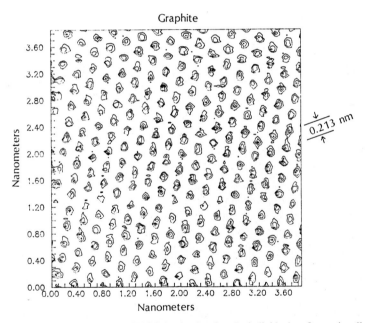

Fig. 10.10. Contour map of HOPG in air, showing the individual surface unit cells and the 0.213 nm row spacing

233

be theoretically understood as being dominated by electronic effects and not just topographical fluctuations. On graphite (and some other two-dimensional compounds, e.g. TaSe₂), the Fermi surface collapses to a point at the corner of the surface Brillouin zone. The STM tip essentially sees only one wavefunction at the surface which has a node at each unit cell. Tersoff also explained the anomalously large corrugation observed on graphite. Graphite is therefore a special case in which the surface unit cells can be easily seen with an STM. It is incorrect to attribute the atom-like features in Fig. 10.10 to "graphite atoms", however.

Beta Lead Dioxide

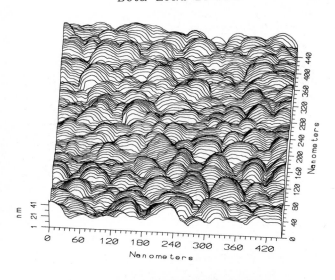

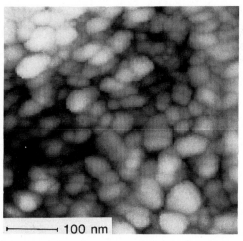

Fig. 10.11. STM images, linescans (*upper*) and greyscale (*lower*) of beta-lead dioxide (PbO₂) surface in air showing the surface microstructure

Clarke and colleagues [10.33] have recently reported imaging gold and silver clusters supported on graphite in ultra-high vacuum. They were able to resolve the two-dimensional packing arrangements of rafts of metal atoms, imaged against the background graphite lattice. There are many other interesting adsorbates which could be studied on graphite.

10.4.3 Lead Dioxide (PbO_2)

In recent work we successfully imaged the surfaces of electrochemically grown alpha- and beta-lead dioxide, which is the material used in lead-acid batteries. The STM resolved the surface microstructure of the material, exposing the surface grains and allowing measurement of the crystallite sizes [10.35]. The image of beta-lead dioxide is shown in Fig. 10.11, and the surface crystallites can be readily seen. There is considerable potential for studying semiconducting mineral surfaces and electrodes in air or under liquids.

10.4.4 Galena (PbS)

Lead sulphide or galena (PbS) is a semiconducting mineral which cleaves readily in air along (100) planes, exposing a fresh surface which tunnels reasonably well in air. STM images of PbS reveal large, flat cleavage planes with (100) cleavage steps ranging from monatomic (0.3 nm) up. Other features such as dislocations can be readily seen, as shown in the image of Fig. 10.12, where two dislocations

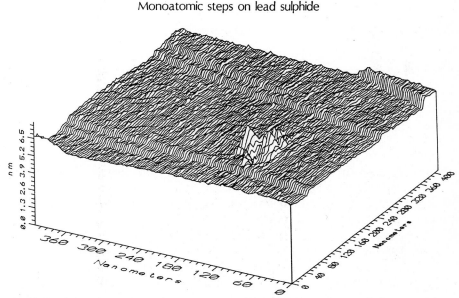

Fig. 10.12. STM image of cleaved lead sulphide (PbS) in air showing 0.3 nm steps of (100) orientation and two surface dislocations

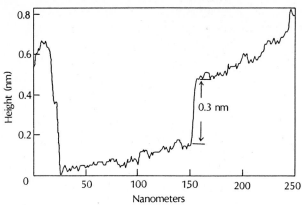

Fig. 10.13. Linescans across a single 0.3 nm step on PbS in air

can be seen emerging from the surface, with a double step in the corner of the frame. A linescan across a single step is shown in Fig. 10.13, and is measured at 0.3 nm, which is half the unit cell size. Other measurements which are useful on air-exposed mineral surfaces include the measurement of oxide growth and distribution of defects in the surface layer. There has been no work done to date on imaging flotation agents attached to minerals, particularly sulphides, but this area could provide useful information on the mechanisms of flotation.

10.5 An Application of Air Microscopy: Compact Disc Technology

Compact discs are currently in popular use as musical recording and computer storage (retrieval) devices. The disc, constructed of polycarbonate plastic has small μm-sized slots impressed in one side by a metal stamper disc. Aluminizing the pits allows light to be reflected off the surfaces by a playback laser in the CD player. Of critical importance in the production of the CD master disc is the requirement for optically flat surfaces with no defects, and the correct height of 0.25 wavelengths of the playback light (120–140 nm). Scanning electron micrographs of compact disc surfaces do not resolve the details below 20 nm size easily or give quantitative measures of the height of surface defects.

We have recently conducted an STM study of compact disc masters obtained from a local manufacturer [10.11]. The aim of the study was to obtain topographic measurements of the surface roughness on the nickel master disc, and to measure the depths of the holes in the plastic replica discs. Light scattering from defects can degrade the signal-to-noise in a CD, and the depth of the impressed holes must be near 130 nm after allowing for some contraction after stamping.

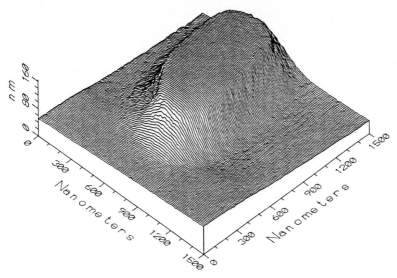

Fig. 10.14. STM image in air of a single pit on a compact disc master (nickel) taken over a 1.5 × 1.5 μm area

Figure 10.14 shows an STM image of a single pit on a nickel compact disc master, taken in air. The area scanned is 1.5 × 1.5 μm. Individual pits are 0.5 μm wide at the top and vary in length up to several μm. Accurate measurements of the surface roughness of the top of the pit and the valley floor were also made by zooming in to each region and taking 0.1 μm square maps. The pit top is extremely smooth with only atomic steps visible across the area scanned. The $p - p$ surface roughness value is < 2 nm. By contrast, the valley floor area is considerably rougher, exhibiting $p - p$ roughness of > 4 nm. Nevertheless, both surfaces fall within the specifications required for optical flatness at the 550 nm laser wavelength. The extra roughness on the valley floor is a consequence of etching of the photoresist surface on which the nickel master is plated. The pit top is in contact with an optically flat glass plate during formation and is not etched [10.11].

Images of the stamped replica discs were made by sputtering gold onto the surface, since plastic is non-conducting [10.11]. By taking linescans across the top of the master pit we were able to calibrate the STM z axis accurately, since the 140 nm step is accurately derived from an ellipsometer. Measurements of the hole depth in the plastic could then be accurately made, and the percentage contraction established [10.11].

In imaging high-density storage media such as compact discs, the problem of convolution of the tip shape with the image becomes apparent. Figure 10.15 illustrates the problem. The macroscopic tip radius (0.1–0.5 μm) is comparable with the width and height of the features to be resolved. Convolution manifests itself as a broadening of the base of a step such as a CD pit in Fig. 10.15, depending on the radius. The tunnelling switches from an end tip to a series of

Compact Disc Master

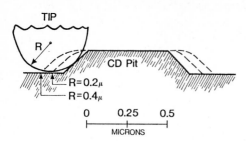

Fig. 10.15. Tip radius effects in imaging submicron surface structures

Compact Disc Replica

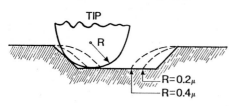

tips on the side of the main tip. In imaging holes, the apparent width of the hole is reduced as the tip radius increases. For very fat tips, cusp-shaped images can be produced, where the bottom of the tip may not touch the base of the hole. Okayama et al. have imaged microfabricated patterns with a 160 nm periodicity and 40–50 nm depth on silicon. They fabricated micro-tips on the end of the main tip to achieve the required resolution, although it was not a reproducible operation [10.36].

10.6 Vacuum Microscopy on Semiconductor Surfaces

10.6.1 Topographic Imaging

Semiconductor surfaces, especially silicon and gallium arsenide require good UHV conditions for imaging at atomic resolution. The native oxides SiO_2 and Ga_2O_3, which form rapidly on the clean semiconductors in air, act as an insulating layer whose thickness exceeds the normal tunneling gap distance (< 1.5 nm). In air, the tip consequently oscillates up and down, striking the oxide layer in its search for the required current and exhibits an anomalously noisy behaviour. Some workers have circumvented the problem on silicon by preparing the hydrogen-terminated form by passivation in HF-ethanol solution and transferring to a nitrogen-filled STM [10.37]. The development of vacuum microscopes has solved the oxidation problems on semiconductors, since clean surfaces can be prepared and maintained in UHV. *Binnig* and colleagues [10.3] were the first group to image the famous Si(111) 7 × 7 reconstruction with the STM. They etched the silicon in HF and heated the wafer to 900°C in UHV to remove the SiO layer. Large biases (+2.9 V) were required to exceed the band gap of silicon and establish stable tunnelling. Corrugation heights for these structures were less than 0.2 nm and extremely stable conditions were necessary to obtain such good data.

There have been a very large number of publications dealing with silicon surfaces. Some of the recent work includes imaging steps on the Si(111) 7 × 7 surface [10.38], observing the structure of Si(112) [10.39] and Si(001) surfaces [10.40], the structure of Ag overlayers on Si(111) [10.41, 42], and the reaction of Si(111) 7 × 7 with ammonia [10.43]. High resolution imaging of silicon surfaces has produced some of the most graphic and spectacular examples of the capabilities of the STM. The recent conference proceedings provide an excellent introduction to the topic [10.9, 16, 17]. Besides silicon, both germanium [10.44] and gallium arsenide [10.45] have also been studied with vacuum microscopes at atomic resolution.

10.6.2 Scanning Tunnelling Spectroscopy (STS)

It was observed at an early stage that the contrast in semiconductor images depended strongly on the bias, V. Tersoff and Hamann predicted that the ratio of the differential conductivity dI/dV to the total conductivity I/V should be proportional to the surface density of states [10.46]. This type of experiment is run by freezing the tip position and rapidly ramping the bias, V, through a preset range whilst simultaneously measuring the tunnel current:

$$\frac{dI/dV}{I/V} \propto \text{Surface DOS} .$$

Results for Ge(001) from the work of Kubby et al. [10.44] show the surface state band gap of 0.9 eV and the presence of filled (valence band) and unfilled (conduction band) states, both below and above the Fermi level (0 V on scale). The peaks in the density of states can be compared with results from photoemission and inverse photoemission experiments. The presence of these strong peaks in the current voltage relationship of the tunnel gap on semiconductors dominates the contrast mechanism. With the tip negative, electrons tunnel from the tip to unoccupied states in the conduction band of the semiconductor. If the density of states of the surface is low at this particular bias, the tip will move closer to the surface to increase the current to the target value (decreasing tip height). If the surface DOS is high at another point on the surface at the same bias, then the tip will move less, or perhaps even away from its present value to control the current. The high and low points in the image of the surface are therefore constant DOS contours and not constant electron density contours.

The technique of STS (scanning tunnelling spectroscopy) has been successfully applied to selectively image Ga and As atoms on GaAs [10.45]. In the reaction of ammonia with Si(111), the reactivity of different surface dangling bond states was investigated [10.47]. This field is extremely active at the time of writing of this book, and is providing a wealth of fundamental information on the relationship between the electronic and physical structure of semiconductor surfaces.

10.7 Atomic Force Microscopy

The atomic force microscope, or AFM was conceived by *Binnig* et al. [10.48] and incorporates a tunnelling junction, or more recently, an optical level to measure the displacement of a diamond stylus tracing across a surface. The AFM can operate on insulators and relies on obtaining a very sharp tip to establish high resolution. A diamond is normally crushed and a sharp fragment glued to a thin lever arm. Essentially this instrument is identical to the conventional "Talysurf" type surface profilometers which operate at much higher displacements. There is increasing interest in this technique for air work, since many surfaces do not tunnel well in air (e.g. Al, Si). There are many applications, such as ceramic surfaces, where surface deformations at high resolution need to be measured accurately, however. The recent conference proceedings are a useful source of references for this topic.

10.8 The Solid-Liquid Interface

The ability of the STM to operate successfully under liquids was realized in 1986 when *Sonnenfeld* and *Hansma* [10.7] reported tunnelling on graphite and gold films immersed in water. The considerable optimism for the future of this field was dampened somewhat by the realization that the immersed surface had to remain relatively free of surface oxide films. In addition, the Faraday current between tip and sample interfered with the feedback current and, more seriously was found to drift with time. *Sonnenfeld* and *Schardt* subsequently reported imaging the initial electrodeposits of silver on graphite [10.49] in 0.05 M silver perchlorate solution.

Despite these problems, at least four different groups have developed tunnelling electrochemical cells where the potential of the tip and sample can be controlled to obtain more stable imaging [10.23, 50–52]. The conditions for successful imaging in solution are similar to those in air. Metals such as platinum, gold, silver, rhodium or iridium are ideal candidates for a liquid study, since surface oxide formation is minimal and easily controlled by varying the surface potential. By manipulating the surface potential, even difficult surfaces such as stainless steels may be freed of the passivating oxide in certain solutions. The area of electrodeposition is ripe for study with the STM. Initial nucleation and growth processes, of great interest to electrochemists, could be monitored at high resolution. Corrosion processes could be studied in situ, provided the corrosion products were soluble. It remains to be seen whether this field of study will expand beyond the preliminary work already reported. The indications are, however, that some very useful results will be obtained.

10.9 Adsorbed Molecules and Clusters

An early report by *Baro* et al. [10.53] demonstrated that individual adsorbates such as oxygen atoms could be observed on a Ni(110) surface under UHV conditions. More recently, in a remarkable piece of work by *Ohtani* et al. [10.54], individual benzene molecules were imaged on a Rh(111) surface in UHV. Under certain bias conditions, the benzene molecule was observed to have a threefold symmetry and a central depression. The potential of the STM for obtaining information about the location of adsorption sites and the adsorbate symmetry makes STM adsorbate studies particularly attractive.

Besides metals, adsorption of gases on semiconductors has received some attention. *Stroscio* et al. observed individual oxygen atom adsorbed states on a GaAs(110) surface [10.55]. The oxygen atom and its associated charge-depleted region surrounding it were observed either as a hole or a hill depending on the polarity of the bias setting. The lateral size of the oxygen defect was measured against the background lattice, giving a real-space view of the local band-bending.

Clusters of silver on graphite have recently been investigated by *Clarke*'s group at Berkeley [10.33]. Large metal-organic clusters adsorbed on substrates such as graphite would be particularly interesting candidates for study with the STM.

10.10 Surface Modification with the STM

The STM may be used as a surface modification tool in several ways. First, the novice student of tunnel microscopy learns to recognize the holes present in many of his early images as being due to inadvertent tip crashes during the initial approach. Impacting a soft metal such as gold produces a perfect replica of the tip end. Besides mechanical deformation effects, the bias voltage and power density at the tunnel gap are known to produce surface effects. In a recent study, *Staufer* et al. [10.56] produced small hillocks ≈ 10 nm diameter on a Rh-Zr metallic glass by pulsing the bias up to several volts and increasing the tunnelling current at the same time. The surface melted under the tip, producing small mounds as the molten metal was drawn towards the tip. *Ehrics* et al. fabricated micro-tracks on metal and semiconductor surfaces by operating the STM in an atmosphere of organometallic gases [10.57].

It is doubtful whether these preliminary attempts at manufacturing "nano-features" on surfaces will ever lead to a useful application. The STM will more likely be used as an analytical probe for sub-μm structures when optical and electron-beam lithographic methods become more refined in semiconductor production.

One potentially interesting area for the STM, however, is the micro-indentation of metals and the measurement of the annealing and restructuring of the surface craters after impact. We have observed quite dramatic changes in gold surfaces impacted with a tunnelling tip as the surface restructures [10.58]. On ceramic surfaces, there is considerable interest in measuring the shape of nano-indentations at surfaces. For this application, the ceramic would have to be coated with a thin conducting overlayer film.

10.11 Biological Applications and Langmuir-Blodgett Films

Recently, *Travaglini* et al. [10.59] and *Amrein* et al. [10.60] have published images of freeze-dried DNA complexes coated with a Pt-Ir-C film on a thicker carbon film substrate. The images show the helical structure of the DNA strands. *Guckenberger* et al. [10.61] have published the design of an STM incorporating an optical microscope stage for visually accessing areas for examination with the STM tip. The extreme difficulties in working in this area are reflected by the lack of publications. No successful imaging of biological structures under liquid environments (producing meaningful results) have yet been published. There is some potential for imaging conducting organic surfaces such as molecular crystals. *Sleator* and *Tycko* [10.62] imaged TTF-TCNQ crystals at the atomic level with an STM in a recent study.

Langmuir-Blodgett films represent another potentially interesting area for STM work. *Wu* and *Lieber* [10.14] imaged a detergent monolayer 2.2–2.6 nm thick on a graphite substrate in aqueous solution. *Lang* et al. [10.13] have obtained images of the external surface of cadmium arachidate and other LB films on graphite in air. The exact conduction mechanism through these films is not clear, since their thickness exceeds the normal tunnelling gap. LB films can be up to 5–10 nm thick. *Wu* and *Lieber* [10.14] were able to obtain an estimate of the film thickness when they observed two regimes of tunnelling – at the graphite interface and on the external part of the LB layer. With the increasing interest in these film systems as biological and inorganic sensors, it is likely that more reports will be seen in the near future.

10.12 Summary

Scanning tunnelling microscopy has been one of the major growth areas in surface science over the last few years. The explosion of interest in this field has subsided somewhat with the realization that only a limited range of conducting surfaces can be studied. Good results are difficult to obtain and only a small percentage of images taken may produce meaningful results. The STM is unlikely to ever rival the conventional SEM as a general purpose surface instrument. On a more positive note, the images obtained so far have generated a rebirth in the concept of a surface as an active interface between solids and fluids. Defects such as steps and terraces can be measured directly, and appear just as the textbook examples illustrate. Atomic arrays and surface reconstructions have been confirmed and compared with earlier LEED studies.

The STM will always remain a specialized research tool in a select number of research laboratories. The results and areas of application will continue to grow, however. In the area of micro-electronics and high density data storage, the STM will be one of the few tools available to measure the dimensions of nanometer-scale surface structures.

References

10.1   G. Binnig, H. Rohrer, Ch. Gerber, E. Weibel: Phys. Rev. lett. **49**, 57 (1982)
10.2   G. Binnig, H. Rohrer: Surf. Sci. **126**, 236 (1983)
10.3   G. Binnig, H. Rohrer, Ch. Gerber, E. Weibel: Phys. Rev. Lett. **50**, 120 (1983)
10.4   P.K. Hansma, J. Tersoff: J. Appl. Phys. **61(2)**, R1 (1986)
10.5   C.F. Quate: Physics Today (August 1986) p. 26
10.6   P. Lustenberger, H. Rohrer, R. Christoph, H. Siegenthaler: J. Electroanal. Chem. **243**, 225 (1988)
10.7   R. Sonnenfeld, P.K. Hansma: Science **232**, 211 (1986)
10.8   W. Baumeister: Ultramicroscopy **25**, 103 (1988)
10.9   Proceedings of the Second International Conference on Scanning Tunnelling Microscopy, ed. by R.M. Feenstra, J. Vac. Sci. Technol. **A6(2)**, 259–556 (1988)
10.10 S. Hosaki, S. Hosoki, K. Takata, K. Horiuchi, N. Natsuaki: Appl. Phys. Lett. **53(6)**, 487 (1988)
10.11 B.A. Sexton, G.F. Cotterill. J. Vac. Sci. Technol. **A7**, 2734 (1989)
10.12 L. Vasquez, A. Bartolome, R. Garcia, A. Buendia, A.M. Baro: Rev. Sci. Instrum. **59**, 1286 (1988)
10.13 C.A. Lang, J.K.H. Horber, T.W. Hansch, W.M. Heckl, H. Mohwald: J. Vac. Sci. Technol. **A6(2)**, 368 (1988)
10.14 Xian-Liang Wu, C.M. Lieber: J. Phys. Chem. **92**, 5556 (1988)
10.15 R. Young, J. Ward, F. Scire: Rev. Sci. Instrum. **47**, 1303 (1976)
10.16 Proceedings of the First International Conference on Scanning Tunnelling Microscopy, ed. by N. Garcia, Surf. Sci. **181**, 1–412 (1987)
10.17 IBM J. Res. Dev. **30**, Nos. 4 and 5, 353–572 (1986)
10.18 J. Simmons: J. Appl. Phys. **41**, 1915 (1970)
10.19 Modern Piezoelectric Ceramics, Vernitron Piezoelectric Division, 232 Forbes Road, Bedford, Ohio 44146, USA
10.20 Piezoelectric Ceramics, J. Van Randeraat, R.E. Setterington (Mullard Ltd., London 1974)
10.21 G. Binnig, D.P.E. Smith: Rev. Sci. Instrum. **57**, 1688 (1986)
10.22 E.W. Muller, T.T. Tsong: *Field Ion Microscopy* (Elsevier, New York 1969)

10.23 H. Liu, F. Fan, C.W. Lin, A.J. Bard: J. Am. Chem. Soc. **108**, 3838 (1986)
10.24 B.A. Sexton: Unpublished results
10.25 S. Park, C.F. Quate: Rev. Sci. Instrum. **58(11)**, 2010 (1987)
10.26 M. Salmeron, D.S. Kaufman, B. Marchon, S. Ferrer: Appl. Surf. Sci. **28**, 279 (1987)
10.27 M. Salmeron, B. Marchon, S. Ferrer, D.S. Kaufman: Phys. Rev. B **35**, 3036 (1987)
10.28 V.M. Hallmark, S. Chiang, J.F. Rabolt, J.D. Swalen, R.J. Wilson: Phys. Rev. Lett. **59**, 2879 (1987)
10.29 R.C. Jaklevic, L. Elie: Phys. Rev. Lett. **60**, 120 (1988)
10.30 G. Binnig, H. Fuchs, Ch. Gerber, E. Stoll, E. Tosatti: Europhys. Lett. **1**, 31 (1986)
10.31 Sang-il Park, C.F. Quate: Appl. Phys. Lett. **48**, 112 (1986)
10.32 R. Sonnenfeld, P.K. Hansma: Science **232**, 211 (1986)
10.33 E. Ganz, K. Sattler, J. Clarke: Phys. Rev. Lett. **60**, 1856 (1988)
10.34 J. Tersoff: Phys. Rev. Lett. **57**, 440 (1986)
10.35 B.A. Sexton, G.F. Cotterill, S. Fletcher, M.D. Horne: J. Vac. Sci. Technol. A8(1), 544 (1990)
10.36 S. Okayama, M. Komuro, W. Mizutani, H. Tokumoto, M. Okano, K. Shimizu, Y. Kobayashi, F. Matsumoto, S. Waikiyama, M. Shigeno, F. Sakai, S. Fujiwara, O. Kitamura, M. Ono, K. Kajimura: J. Vac. Sci. Technol. **A6(2)**, 440 (1988)
10.37 W.J. Kaiser, L.D. Bell, M.H. Hecht, F.J. Grunthaner: J. Vac. Sci. Technol. **A6(2)**, 519 (1988)
10.38 R.S. Becker, J.A. Goluvchenko, E.G. McRae, B.S. Swartentruber: Phys. Rev. Lett. **55**, 2028 (1985)
10.39 Th. Berghaus, A. Brodde, H. Neddermeyer, St. Tosch: Surf. Sci. **184**, 273 (1987)
10.40 R.J. Hamers, R.M. Tromp, J.E. Demuth: Phys. Rev. B **34**, 5343 (1986)
10.41 E.J. Van Loenen, J.E. Demuth, R.M. Tromp, R.J. Hamers: Phys. Rev. Lett. **58(4)**, 373 (1987)
10.42 R.J. Wilson, S. Chiang: Phys. Rev. Lett. **58**, 369 (1987)
10.43 J.A. Stroscio, R.M. Feenstra, A.P. Fein: Phys. Rev. Lett. **57**, 2579 (1986)
10.44 J.A. Kubby, J.E. Griffith, R.S. Becker, J.S. Vickers: Phys. Rev. B **36**, 6079 (1987)
10.45 R.M. Feenstra, J.A. Stroscio, J. Tersoff, A.P. Fein: Phys. Rev. Lett. **58**, 1192 (1987)
10.46 J. Tersoff, D.R. Hamann: Phys. Rev. Lett. **50**, 1998; Phys. Rev. B **31**, 805 (1985)
10.47 R. Wolkow, Ph. Avouris: Phys. Rev. Lett. **60**, 1049 (1988)
10.48 G. Binnig, C.F. Quate, Ch. Gerber: Phys. Rev. Lett. **56**, 930 (1986)
10.49 R. Sonnenfeld, B.C. Schardt: Appl. Phys. Lett. **49**, 1172 (1986)
10.50 P. Lustenberger, H. Rohrer, R. Christoph, H. Siegenthaler: J. Electroanal. Chem. **243**, 225 (1988)
10.51 S. Morita, I. Otsuka, T. Okada, H. Yokoyama, T. Iwasaki, N. Mikoshiba: Japan J. Appl. Phys. **26**, L1853 (1987)
10.52 M.H.J. Hottenhuis, M.A.J. Mickers, J.W. Gerritsen, J.P. Van der Eerden: Surf. Sci. **206**, 259 (1988)
10.53 M.A. Baro, G. Binnig, H. Rohrer, Ch. Gerber, E. Stoll, A. Baratoff, F. Salvan: Phys. Rev. Lett. **52**, 1304 (1984)
10.54 H. Ohtani, P.J. Wilson, S. Chiang, C.M. Mate: Phys. Rev. Lett. **60**, 2398 (1988)
10.55 J.A. Stroscio, R.M. Feenstra, A.P. Fein: Phys. Rev. Lett. **58**, 1668 (1987)
10.56 U. Staufer, R. Wiesendanger, L. Eng, L. Rosenthaler, H.R. Hidber, H.J. Guntherodt, N. Garcia: J. Vac. Sci. Technol. **A6(2)**, 537 (1988)
10.57 E.E. Ehrichs, R.M. Silver, A.L. DeLozanne: J. Vac. Sci. Technol. **A6(2)**, 540 (1988)
10.58 Unpublished results, this laboratory
10.59 G. Travaglini, H. Rohrer, M. Amrein, H. Gross: Surf. Sci. **181**, 380 (1987)
10.60 M. Amrein, A. Stasiak, H. Gross, E. Stoll, G. Travaglini: Science **240**, 514 (1988)
10.61 R. Guckenberger, C. Kosslinger, W. Baumeister: J. Vac. Sci. Technol. **A6(2)**, 383 (1988)
10.62 T. Sleator, R. Tycko: Phys. Rev. Lett. **60**, 1418 (1988)

11. Low Energy Ion Scattering

D.J. O'Connor

With 12 Figures

Low energy ion scattering (LEIS) is the study of the structure and composition of a surface by the detection of low energy (100 eV–10 keV) ions (and atoms) elastically scattered off the surface. This technique is a subset of ion scattering spectrometry which involves the use of incident ions with energies ranging from 200 eV to over 1 MeV. The range of measurements possible over such a large range of energies extends from purely atomic layer resolution at the low energy end to analysis to depths of the order of microns at the high energy end. Some of the high energy effects (> 250 keV) are covered in Chap. 9, while the intermediate energy range (medium energy ion scattering) has been successfully developed as a near surface structure probe mentioned briefly in Chap. 1. The use of low energy ions to measure the surface structure of solids was established by *Smith* [11.1] in 1968. In that study the basic elements of LEIS were established and these have been built on over the past 20 years to develop into a powerful surface atomic layer structure and composition probe. It has been successfully applied to a wide range of practical surface problems which include the surface composition analysis of:

- binary alloys
- catalysts
- cathode surfaces
- polymers
- surface segregation
- adsorbates
- surface structure
- adsorbate site identification

It can answer the following questions to a precision which is target dependent;

- What is on the surface?
- How much is on the surface?
- Where is it located (relative to other atoms)?

11.1 Qualitative Surface Analysis

LEIS involves the bombardment of the surface with either inert gas or alkali ions and measuring the energy distribution of the ions (or less commonly the neutrals) scattered off the surface. As the energy loss to electronic processes (termed inelastic energy loss) is relatively small at low energies it is possible to identify the mass of the target atoms from the energy of the scattered projectiles. By applying the principles of the conservation of energy and momentum the scattered projectile energy, E_1, is given by

$$\frac{E_1}{E_0} = \left(\frac{\cos\theta + \sqrt{\mu^2 - \sin^2\theta}}{1 + \mu} \right)^2 , \tag{11.1}$$

where E_0 is the projectile energy, the scattering angle is θ, and the ratio of the target mass to the projectile mass is represented by μ. From this simple analysis it is possible to determine the mass (and usually the identity) of atoms situated on the outermost layer of the solid. It is exactly this approach which allows the determination of the target masses from the ion scattering spectrum in Fig. 11.1 where (11.1) predicts the E/E_0 values observed for Cu and Au as the surface components. In the absence of recoil peaks (not observed for scattering angles greater than 90°) the heavier the target the higher will be E/E_0.

The ability to resolve different masses on a surface is a function of the ratio of the projectile mass to the target mass. It is easy for He to resolve C ($\mu = 3$) and O ($\mu = 4$) under normal conditions but it would be impossible for He to resolve W ($\mu = 46$) from Au ($\mu = 49.25$). To improve the mass resolution for heavier

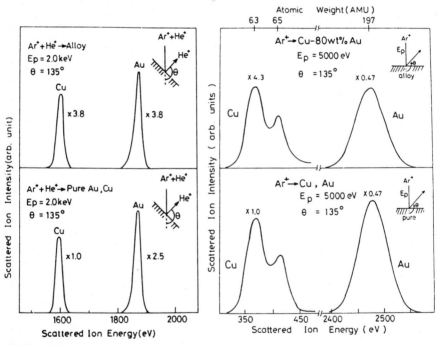

Fig. 11.1. Comparison of the analysis of a Au–Cu(43%) alloy with He and Ar projectiles demonstrating the increased mass resolution attainable with heavier projectiles. From [11.2a]

target masses it is necessary to increase the projectile mass and/or increase the scattering angle. This is more clearly illustrated in Fig. 11.1 where the use of He and Ar as projectiles is compared and it is evident that with He it is possible to resolve Cu and Au, while for Ar it is also possible to resolve the isotopes of Cu. However if the projectile used is heavier than the target atom there is a limiting angle θ_1 beyond which no single scattering is observed. θ_1 is given by

$$\theta_1 = \arcsin(m_2/m_1) .$$ (11.2)

Scattered projectiles can be observed at all scattering angles whenever the projectile mass is less than the target mass. In some cases the recoiling target atom is captured and analyzed, in which case the energy of the recoiling atom is given by

$$\frac{E_2}{E_0} = \frac{4\mu \cos^2 \phi}{(1 + \mu)^2}$$ (11.3)

where E_2 is the energy of the particle recoiling at angle of ϕ to the incident direction. The angle of recoil is limited to 90° so in many systems the detection angle is set at greater than 90° in give the dual benefit of improving the mass resolution and removing the potentially complicating features introduced by recoiling projectiles.

In some cases the identification of the existence of an element on a surface is sufficient to allow some conclusions to be reached. The simplest example of this form of analysis was performed by *Smith* [11.1] who analyzed a clean Ni surface onto which CO had been adsorbed. In the spectrum scattered ions from Ni and from O were observed, but no ion yield was measured from the C atoms. It was concluded from this that the CO molecule was bonded perpendicularly to the Ni surface with the C atom forming the bond hence the C atoms were shadowed from the incident ions by the O atoms and no scattered ion yield was observed off C.

11.2 Advantage of Recoil Detection

The existence of recoils can be exploited in some applications for the specialized tasks of detecting light elements and for the very sensitive detection of electronegative adsorbates. The detection of light elements is extremely difficult directly with ion scattering however the recoils are easily identified and measured [11.2–6] and has been used to identify hydrogen on surfaces. While it is possible to observe recoil peaks in a positive ion spectrum it is often difficult to separate this contribution from other ion scattering processes with inert gas ion yields at the same energy. To overcome this limitation it is possible to take advantage of the fact that some elements will escape the surface in positive, negative and neutral charge states while in general the inert gas projectiles only escape as positive ions or neutrals (Fig. 11.2). In the case of electronegative elements (O,

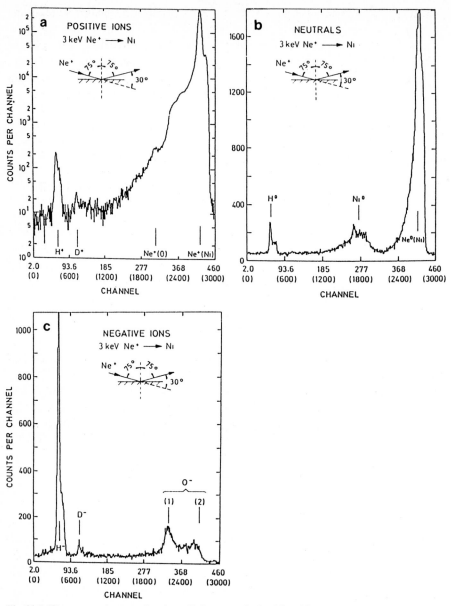

Fig. 11.2. The measured scattered and recoiled spectra obtained for different charge states when a Ni surface is bombarded by 3 keV Ne⁺. From [11.2b]

Cl, etc.) the negative charge fraction can be as large or larger than the positive so by measuring the negative ion spectrum the contribution from the electronegative elements stands out without interferences from the projectiles and as a result it has been estimated that as little as 10^{-5} of a monolayer can be measured [11.6].

11.3 Quantitative Analysis

11.3.1 Scattered Ion Yield

In many analyses it is sufficient to identify the presence of an element on a surface to allow some conclusion to be arrived at, however in some applications the scattered ion yield is used to determine the composition of the surface layer of a sample. The appropriate relationship between these quantities is

$$Y_a^+ = N_0 N_a \left(d\sigma_a / d\Omega \right) \Delta\Omega P^+ T(E) . \tag{11.4}$$

Y_a^+ is the measured ion yield for element a, N_0 in the number of projectiles, N_a is the number of atoms of element a per unit area on the surface, $\Delta\Omega$ is the collection angle the analyzer presents to the target and $T(E)$ is the transmission function of the analyzer and detector. All these terms can be determined to a high degree of precision. The principal uncertainties in the analysis arise from the differential scattering cross section $(d\sigma_a / d\Omega)$ and the charge exchange factor (P^+). These two aspects will be dealt with in the following sections.

11.3.2 Differential Scattering Cross Section

The differential scattering cross section represents the cross-sectional area that each atom presents to the beam for a particular scattering event and it can be determined by integration if the scattering potential is known. In LEIS the projectile energy is sufficiently large that only the repulsive part of the interatomic potential needs to be considered in determining the cross section and a screened coulomb approximation to the repulsive potential of the form given in (11.5) is normally used to represent the interaction,

$$V(r) = \frac{Z_1 Z_2 e^2}{4\pi\varepsilon_0 r} U(r/a) . \tag{11.5}$$

Here Z_1 and Z_2 are the atomic numbers of the interacting particles, r is the interatomic separation and a is the screening length. While the Moliere approximation [11.7] (11.6) has been used extensively and successfully for many years it has recently been surpassed in accuracy by the ZBL potential [11.8] (11.8) which is based on the wavefunctions of a large range of elements.

$$U_M(x) = 0.35 \exp(-0.3x) + 0.55 \exp(-1.2x) + 0.10 \exp(-6x) . \tag{11.6}$$

The screening length a is related to the Bohr radius, a_0, by

$$a = 0.88534 a_0 / \left(\sqrt{Z_1} + \sqrt{Z_2} \right)^{3/2} , \tag{11.7}$$

$$U_Z(x) = 0.1818 \exp(-3.2x) + 0.5099 \exp(-0.9423x)$$
$$+ 0.2802 \exp(-0.4029x) + 0.02817 \exp(-0.2016x) . \tag{11.8}$$

One of the reasons for the success of the ZBL potential is the different expression used for the screening length, a, given by

$$a = \frac{a_0}{Z_1^{0.23} + Z_2^{0.23}} \cdot \tag{11.9}$$

This potential has been shown [11.9] to be the best fit to the range of currently available measurements of the interatomic potential and in absolute terms it may be accurate to a precision of the order of 10%, however the ratio of the cross sections of two atoms will be of higher precision and may be better than 1%.

A more detailed description of the numerical determination of the scattering cross section has been given by *Torrens* [11.10], however as a general guide the cross section increases with

- increasing projectile and target atomic number
- decreasing projectile energy
- decreasing scattering angle.

11.3.3 Charge Exchange

When an inert gas ion is in the vicinity of a surface there is a significant probability that it will be neutralized as the ground state of the inert gas atom is a lower energy state than the conduction electrons of the surface (see Fig. 11.3). Thus a proportion of scattered inert gas projectiles will suffer neutralization and that proportion will be a function of the time spent in the vicinity of the surface. Typically only 0.5–5% of the projectiles scattered off the surface layer are ions while the ion fraction for projectiles scattered off subsurface layers is typically less than 0.1%.

When alkali ions are used as projectiles the escaping ion fraction is significantly different. As the ionization potential is comparable to the work function

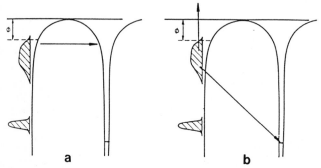

Fig. 11.3a,b. The principal charge exchange processes responsible for the neutralization of low energy ions near surfaces. Part (**a**) depicts the resonance process in which there is a mtach between the energy level of an excited state in the projectile and a filled state in the band structure of the solid. The Auger process (**b**) involves an electron filling the ground state of the projectile and a second electron in the band acquiring the excess energy

of most materials, the projectiles come to a charge equilibrium with the surface and the ion fraction is both large and independent of the time spent near the surface at these energies. In this case the ion fraction for projectiles scattered from subsurface layers is little different from those scattered from the surface layer.

The transfer of an electron from the solid to the projectile can occur by two principal processes. The simplest is the resonance process in which the electron transfers directly from a level in the solid to a vacant level in the projectile at the same energy (Fig. 11.3a). The second major process is the Auger process (discussed in Chap. 6) in which an electron from the solid transfers to a lower energy state in the projectile and the excess energy is transferred to another electron of the solid (Fig. 11.3b). A complete description of the charge exchange process is much more complicated with the inclusion of the possibility of surface electronic states, the changing energy levels of the projectile due to the interaction with its image charge and with the electrons of the solid. As well there are currently two unresolved views of the charge exchange process which favour either an interaction of the projectile with the distributed electron distribution of the solid or a description which favours a discrete interaction with each atom of the solid.

While a completely predictive theory for charge exchange is not yet available recent intense effort, both experimental and theoretical, has established a basis for the description of charge exchange which is most appropriate to inert gas ions. *Hagstrum* [11.11, 12] developed a model for very low energy ions (1–10 eV), and while this is for much lower energies than encountered in LEIS and some approximations were not valid at these energies, it proved to successfully describe the transition rates. Recent, more rigourous, theoretical studies have yielded the same basic equations given by the Hagstrum description. The Hagstrum model considers a transition rate which varies exponentially with distance from the surface and can be integrated over the trajectory of the projectile. The resulting expression for the ion fraction P^+ is given by

$$P^+ = e^{-v_c/v_\perp} \tag{11.10}$$

where v_c is a characteristic velocity (composed of a transition rate and a screening length) which should be dependent on the identity of the projectile and the target, while the second term, $v_\perp$, is the perpendicular component of velocity of the projectile and is a measure of the time the projectile spends in the vicinity of the surface. While in the simplest model v_c is expected to be energy independent it is found experimentally to depend on the projectile velocity [11.13–15]. While it is not possible to yet predict the value of v_c from the properties of the solid and the target, the most recent experimental results reveal that the detailed electronic structure is not significant and that the most important parameter may be the free electron density [11.15]. This implies that the quantification of LEIS is reliable as all scattered projectiles off a surface will experience the same transition rates hence all have a predictable probability of neutralization.

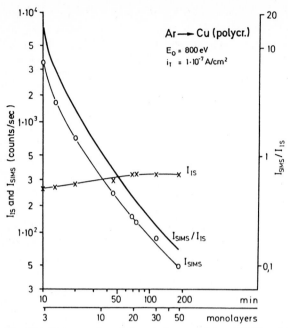

Fig. 11.4. Demonstration of the relative sensitivity of LEIS and SIMS to the existence of contamination on a Cu surface. While the Cu ion yield in SIMS decays by two orders of magnitude as the surface concentration of Cu increases during sputter cleaning, the ISS Cu yield increases only 10%. From [11.16]

An alternative approach to charge exchange treats the transition rates as discrete interactions (11.11) with individual atoms and the net ion yield is then the integration of these independent interactions over the trajectory of the projectile,

$$P^+ = e^{-ar} \tag{11.11}$$

where a is the screening distance and r is the distance between the projectile and the target atom. This model would suggest that the ion fraction depends on the path followed in a mixed surface and that it would be difficult to estimate the charge exchange probability in practical applications. Further work is in progress to establish which of the two models of charge exchange best describe all observations.

Despite this uncertainty it is possible to demonstrate that the role of charge exchange in LEIS is less significant than in SIMS where similar processes occur and it is well known that the ion fraction of secondary ions is very sensitive to low levels of contamination on the surface. The relative sensitivities of LEIS and SIMS to surface contamination were demonstrated by *Grundner* et al. [11.16] as a Cu surface was cleaned by ion bombardment and the SIMS signal for Cu decreased by a factor of 100 while the LEIS signal for Ar scattered off Cu increased by approximately 10% in keeping with the increased concentration of Cu on the surface as contaminants were removed (see Fig. 11.4). Thus LEIS, while not as sensitive to low levels of contamination as SIMS, will more accurately describe the changing composition of a surface.

While the scattering cross section and the neutralization probability are major uncertainties in the quantification of LEIS, it is nevertheless possible to perform quantitative analysis despite them. The form of these measurements fall into the following categories:

1. Relative measurements
2. Standards

11.3.4 Relative Measurements

A number of early studies concentrated on the linearity of the LEIS yield as a function of surface concentration. The principal problem is the lack of a reliable set of calibrated surface composition standards, so comparison was made to the Auger electron spectrometry (AES) signal (or in one case the Rutherford backscattering spectrometry [11.17], RBS, yield) to establish whether a linear relationship existed between the LEIS yields and sub-monolayer concentrations when an impurity was adsorbed onto a clean surface. Early [11.18–22] studies certainly identified a direct relationship between the LEIS and the AES yields under conditions of up to one monolayer of adsorbate (Fig. 11.5). Implicit in this result is the assumption that the ion fraction of the scattered projectiles is independent of the composition of the surface of the solid.

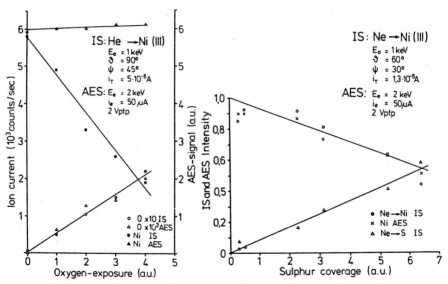

Fig. 11.5. Demonstration of the linearity of LEIS yield response with adsorbed material for O and S on a clean Ni surface. The enhanced surface sensitivity of LEIS is evident in the case of the O adsorption as the LEIS yield from Ni decreases with increasing dose while the AES yield from Ni increases marginally. From [11.19]

This is not a conclusive test as the AES signal has an escape depth which is larger than the one or two atomic layers that LEIS probes so it is proof only if it is assumed that all the adsorbate remains on the surface. Examples can be found where a linear relationship between the LEIS and AES yields breaks down [11.23], in which case the interpretation of this difference may be that the adsorbate moves below the surface so that it continues to be detected by AES and not by LEIS, or that there is a change to the electronic properties of the surface which affects the neutralization rate of the ions used in LEIS.

In other studies LEIS has been applied to compare the surface composition of clean and contaminated catalysts to identify the contaminate. In such studies only the existence of an unexpected element is sufficient to allow a conclusion without the need for quantitative compositional analysis. In the following list of studies of surfaces it was not necessary to establish an absolute scale of concentration but instead the identification of a particular element on the surface or a measurement of the relative change in concentration of an element as some change or profiling is undertaken was sufficient.

- Pb impurity in AgBr [11.24]
- Ni catalysts on alumina and silica [11.25]
- $BaTiO_3$ and $Gd_2(MoO_4)_3$ [11.26]
- Ba surface enrichment in Tungsten cathodes [11.27, 28]
- Be and Sn segregation in Cu [11.29]
- $BaO:CaO:Al_2O_3$ [11.30]
- $Ni-Mo-Al_2O_3$ catalysts [11.31]
- Iron based glasses [11.32]
- Surface composition of $Pt_{10}Ni_{90}(111)$ alloy [11.33]
- Preferential sputtering of TiC by Ar^+ [11.34]
- Segregation of Pb on Yttrium-Iron-Garnet [11.53]

In these cases the assumption that the ion fraction is constant was sufficient to establish a relative concentration scale.

11.3.5 Standards

The analysis of surfaces using standards has been performed with both elemental and molecular standards. In the case of alloy analysis the most convenient reference materials are targets of the pure elements, thus to analyze a compound or alloy [11.35] of two components A and B which have surface concentrations of N_a and N_b (in atoms cm^{-2}) respectively, it is first necessary to establish the measured ion yield I_a^0 and I_b^0 off pure surfaces of A and B which have surface concentrations of N_a^0 and N_b^0 respectively. From the measured ion yields for the alloy surface, I_a and I_b respectively, the surface concentrations can be determined.

$$N_a = \left(I_a / I_a^0 \right) N_a^0 \tag{11.12}$$

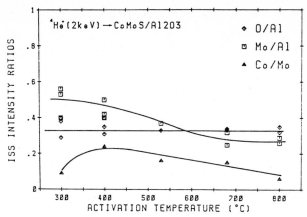

Fig. 11.6. The evolution of the relative concentrations of principal elements in the surface of a CoMoS/γ-Al$_2$O$_3$ catalyst measured by LEIS as a function of temperature. From [11.40]

and

$$N_b = \left(I_b/I_b^0\right) N_b^0 \ . \tag{11.13}$$

If only A and B are present, then the relative concentrations can be determined from the total number of surface sites N_0 (= $N_a + N_b$) as

$$N_a/N_0 = \frac{SI_{a<}}{SI_a + I_b} \tag{11.14}$$

where

$$S = \frac{I_b^0}{I_a^0} \frac{N_a^0}{N_b^0} = \frac{I_b^0}{I_a^0} \left(\frac{d_b}{d_a}\right)^2 \ . \tag{11.15}$$

The sensitivity factor S has been measured [11.36] for Cr and Fe off clean and oxide surfaces and to within experimental accuracy the same value (1.5±0.1) was obtained. In the same study it was established that the relative sensitivity factors for O/Fe was 0.11 and for O/Cr was 0.08, highlighting the difficulty in detecting O with LEIS. There have been variations on this approach [11.37, 38] with similar findings which support the assumption that the ion fraction is little affected by the nature of the surface or the environment of the scattering elements.

The use of the standards approach is complicated for O (and similar adsorbates) as a pure O target is not available. One approach to overcome this difficulty has been employed in analyzing catalytic surfaces involving oxides. To analysis a BiPO$_4$-MoO$_3$ surface samples of BiPO$_4$ and MoO$_3$ as pure samples were used as standards [11.39]. In a later study [11.40] of CoMoS/Al$_2$O$_3$ catalytic surfaces spectra from Co$_9$S$_8$, MoS$_2$ and Al$_2$O$_3$ were used as standards and the relative concentrations of the principal elements were monitored as a function of activation temperature (Fig. 11.6).

In an analysis of magnesium aluminate and magnesium silicate surfaces *Mc-Cune* [11.41] used surfaces of MgO, Al$_2$O$_3$ and SiO$_2$ as the standards and by

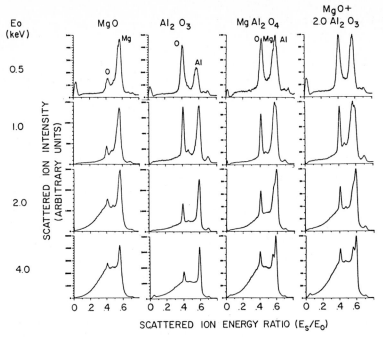

Fig. 11.7. A demonstration of surface analysis of insulators by spectrum synthesis using standards. The columns demonstrate the effect of projectile energy on the background observed and the relative peak amplitudes from different elements. The first two columns are measurements of standards and the third is for a sample of $MgAl_2O_4$ which is compared in the fourth column with the best fit synthesis of the standards as $MgO + 2(Al_2O_3)$. The nonstoichiometric result may reflect the true state at the surface or may result from systematic variations in the specimen surface, crystal orientation effects or density differences between the component materials and the compound. From [11.54]

spectrum synthesis derived a surface composition of the surfaces as $MgO(Al_2O_3)_2$ and $(MgO)_{1.5}SiO_2$ rather than the expected bulk stoichiometries of $MgOAl_2O_3$ and $(MgO)_2SiO_2$ respectively. In this analysis the projectile energy was varied over the range 0.5–4 keV with good agreement at all energies between the synthesized data and the real surface data (Fig. 11.7). The departures from bulk stoichiometry were attributed to systematic variations in the specimen surfaces, crystal orientation effects and differences between the densities of the component materials and the compounds. Without an independent analysis of the surface composition of these surfaces to the same depth resolution it is not possible to attribute the different compositions to a misleading LEIS analysis or to a true surface compositional variation.

11.4 Surface Structural Analysis

The use of LEIS in surface structural analysis most commonly involves one of two basic principles – multiple scattering, or shadowing and blocking of the ions. These methods have been successfully used to determined atomic positions to an accuracy of ± 0.1 Å.

11.4.1 Multiple Scattering

As the projectile energy and the scattering angle decrease the probability of multiple scattering increases and it is often significant in LEIS. An ion can be scattered through an angle of θ by a single scattering event as described above or by two weaker collisions (double scattering) whose total scattering angle is also $\theta(\theta - \theta_1 + \theta_2)$, see Fig. 11.8. Other multiple scattering processes are possible with the next most common the 'zig-zag' sequence [11.42] which is a double scattering process where the two target atoms do not lie in the same plane as the incident and exit trajectories. The projectile energy after two weak collisions is greater than that observed for a single scattering event which results in a clearly defined multiple scattering peak. The relative magnitude of the double scattering peak is inversely dependent on the interatomic spacing of the target atoms and this can be used in a straightforward manner to determine some basic structural information about a surface.

In Fig. 11.9 the ion yield from single (large peaks) and multiple scattering peaks (small peaks) is shown as a function of the azimuthal angle (relative to the [100] direction) of a W(110) surface. The peaks in the multiple scattering at 55° and 125° represent the influence of the [111] directions on multiple scattering yields, while the multiple scattering from the [100] (0°), the [110] (90°) and [113] ($\pm25°$) surface directions are less pronounced as expected by their interatomic spacing.

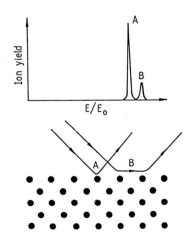

Fig. 11.8. At low energies both single and double scattering sequences as illustrated here can be observed

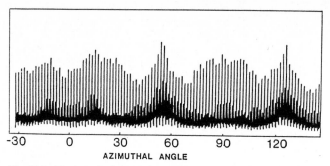

AZIMUTHAL ANGLE

Fig. 11.9. The ion yield from single (large peaks) and multiple scattering peaks (small peaks) is shown as a function of the azimuthal angle (relative to the [100] direction) of the target for 6 keV Kr incident at 25° to a W(110) surface. The peaks in the multiple scattering at 55° and 125° are evidence of the [111] directions, while the influence of multiple scattering at 0°-[100], the 90°-[110] and ±25°-[113] can be observed

11.4.2 Impact Collision Ion Surface Scattering (ICISS)

Perhaps the most direct use of shadowing and the most easily interpreted is the ICISS developed by *Aono* [11.43]. To explain its use it is first necessary to understand the concept of a shadow cone in ion scattering (Fig. 11.10). If a projectile is incident upon a target atom (which we will assume has a greater mass than the projectile) and it strikes 'head on', i.e. with zero impact parameter, then it will be scattered straight back along its incident trajectory. (The impact parameter is the perpendicular distance between the initial undeviated trajectory and the initial target position.) If instead it is incident with a small impact parameter it will be scattered through a large scattering angle, and as the impact parameter becomes larger the scattering angle becomes smaller. Behind the target atom there is a shadowed region, or excluded zone into which no projectiles can penetrate.

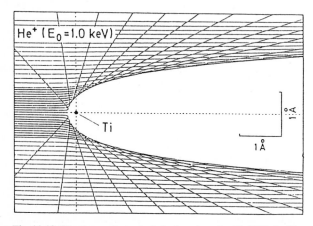

Fig. 11.10. The shape of a typical shadow cone for 1 keV He incident upon a Ti atom. The clear area behind the atom is the excluded region or shadow cone. From [11.43]

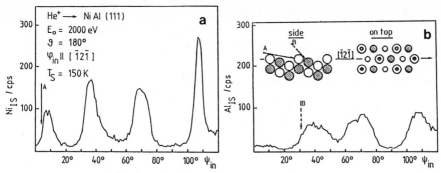

Fig. 11.11. Results of He scattering yield from Ni (**a**) and Al (**b**) on a NiAl(111) surface both demonstrating that there is no Al terminated surface layer and allowing the determination of the relative positions of the surface atoms. From [11.52]

The shape of this shadow cone can be predicted [11.44] with a suitable knowledge of the interatomic potential and the general features are that the shadow cone radius will increase with increasing atomic number of the collision partners and with decreasing projectile energy. The flux at the edge of the shadow cone is enhance over the incident ion flux as most projectiles pass close to the edge of the cone. By suitable choice of geometry this shadow cone can be used to locate atoms above, in, and below the surface layer. This technique has been applied to scattered inert gas ions, scattered alkali ions and scattered neutrals with equal success in different applications [11.45–50].

If an ion beam is incident at a shallow angle to a surface (see [11.43]) and the detector is at a large scattering angle (which means the projectile must have a near zero impact parameter collision to be detected) then no scattering will be observed as each atom in the surface lies in the shadow of the preceding atom. As the angle of incidence of the ion beam is increased the shadow cone rotates about each scattering centre and at some critical angle of incidence the edge of the shadow cone will intersect the neighbouring atom. The scattered ion yield will be zero below the critical angle and beyond this angle it will peak (from the flux enhancement) then fall to an intermediate value. The measurement of the interatomic spacing comes from the critical angle determination and a knowledge of the shape of the shadow cone. An early application of this technique was the location of C under the surface layer of Ti in a TiC(111) surface [11.51]. In this application the C sits asymmetrically between the surface Ti atoms and hence there are two critical angles depending on from which direction the ion beam is incident. The position of the C is 87 ± 8 pm below the Ti surface layer.

In a more recent demonstration of the power of this form of analysis [11.52] a variation of this technique was applied to the NiAl(111) (Fig. 11.11) surface in conjunction with LEED and STM to determine the termination structure of this binary ordered alloy. In Fig. 11.11a the yield from the Ni scattered projectiles yields information on the relative positions of surface Ni atoms while that in Fig. 11.11b reveals that the Al is confined to the second layer.

11.5 Experimental Apparatus

The apparatus used in ion scattering is common to other surface analysis techniques. It is first essential to have a mass analyzed monoenergetic ion beam of inert gas ions or in some applications alkali ions. While it is possible to achieve fine beam spots with ion beams, space charge limitations ensure that the current density decreases with energy and compounded with that the scattered ion yield is sufficiently small that beam spot sizes less than $100\,\mu$m are impractical. In most common applications 0.5–5.0 keV He or Ne ions are used for analysis and the beam size and current depends on the available ion source or sample damage considerations but it is most common to use a 0.1–0.5 μA ion beam with a diameter of approximately 1 mm.

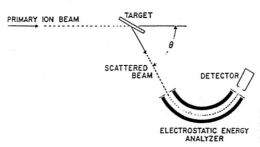

Fig. 11.12. Schematic representation of the layout of a basic LEIS experimental system

LEIS is an extremely surface sensitive technique so the sample must be analyzed in a UHV environment otherwise contamination will build up before practical analysis can be performed. The normal layout of the experimental apparatus is shown in Fig. 11.12. The electrostatic energy analyzer is the same construction as the analyzers used in the electron spectrometries but with the opposite polarities applied to all lens and analyzer elements. The natural line widths in LEIS are principally made up of the kinematic broadening and thermal vibration broadening. The kinematic broadening results from the energy analyzer accepting particles which have suffered a range of scattering angles as a consequence of its finite acceptance angle and this term becomes increasingly significant as the mass ratio (μ) decreases. The magnitude can be estimated from (11.16)

$$\Delta E_1 = E_0 \frac{dK}{d\theta} \Delta\theta .$$
(11.16)

The second contribution arises because the target atom is not stationary and thus the projectile collides with a moving target. As a result the energy width of the scattered ions is broadening with a width given by

$$\Delta E_2 = \sqrt{E_{th} E_0}$$
(11.17)

where E_{th} is the mean thermal vibration energy. At room temperature E_{th} is

0.025 eV so the thermal energy contribution for a 1000 eV projectile is 5 eV or 0.5%. As the natural widths of LEIS peaks are greater than those encountered in electron spectrometries the energy resolution ($\Delta E/E$) need be no better than 0.5% in most practical circumstances. In some instruments the solid angle is increased to increase sensitivity but this results in a greater kinematic broadening and diminished mass resolution. The kinematic broadening increases with increasing projectile mass and with increasing collection solid angle.

References

11.1   D.P. Smith: J. Appl. Phys. **38**, 340 (1967)
11.2a H.J. Kang, R. Shimizu, T. Okutan: Surf. Sci. **116**, L173 (1982)
11.2b P.J. Schneidner, W. Eckstein, H. Verbeek: Nucl. Instrum. Meth. **218**, 713 (1983)
11.3   B.J.J. Koeleman, S.T. de Zwart, A.L. Boers, B. Poelsema, L.K. Verhey: Nucl. Instrum. Meth. **218**, 225 (1983)
11.4   J.A. Schultz, R. Kumar, J.W. Rabalais: Chem. Phys. Lett. **100**, 214 (1983)
11.5   B.J.J. Koeleman, S.T. de Zwart, A.L. Boers, B. Poelsema, L.K. Verhey: Phys. Rev. Lett. **56**, 1152 (1986)
11.6   D.J. O'Connor: Surf. Sci. **173**, 593 (1986)
11.7   G. Moliere: Z. Naturforsch. **2A**, 133 (1947)
11.8   J.P. Biersack, J.F. Ziegler: In *Ion Implantation Techniques*, ed. by H. Ryssel, H. Glawischnig, Springer Ser. Electrophys. Vol. 10 (Springer, Berlin, Heidelberg 1982) pp. 122–156
11.9   D.J. O'Connor, J.P. Biersack: Nucl. Instrum. Meth. **B15**, 14 (1986)
11.10 I.M. Torrens: *Interatomic Potentials* (Academic, New York 1972)
11.11 H.D. Hagstrum: Phys. Rev. **96**, 336 (1954)
11.12 H.D. Hagstrum: In *Inelastic Ion-Surface Collisions*, ed. by N.H. Tolk, J.C. Tully, W. Heiland, C.W. White (Academic, New York 1977)
11.13 R.J. MacDonald, D.J. O'Connor: Surf. Sci. **124**, 423 (1983)
11.14 R.J. MacDonald, D.J. O'Connor, P.R. Higginbottom: Nucl. Instrum. Meth. Phys. Res. **B2**, 418 (1984)
11.15 D.J. O'Connor, Y.G. Shen, J.M. Wilson, R.J. MacDonald: Surf. Sci. **197**, 277 (1988)
11.16 M. Grundner, W. Heiland, E. Taglauer: Appl. Phys. **4**, 243 (1974)
11.17 H. Verbeek: In *Materials Characterisation Using Ion Beams*, ed. by J.P. Thomas, A. Cachard (Plenum, New York 1978)
11.18 E. Taglauer, W. Heiland: Appl. Phys. Lett. **24**, 437 (1974)
11.19 E. Taglauer, W. Heiland: Surf. Sci. **47**, 234 (1975)
11.20 P.J. Martin, C.M. Loxton, R.F. Garrett, R.J. MacDonald, W.O. Hofer: Nucl. Instrum. Meth. **191**, 275 (1981)
11.21 A. Sagara, K. Akaishi, K. Kamada, A. Miyahara: J. Nucl. Mater. **93/94**, 847 (1980)
11.22 H.H. Brongersma, G.C.J. Van der Ligt, G. Rouweler: Philips J. Res. **36**, 1 (1981)
11.23 H. Niehus, E. Bauer: Surf. Sci. **47**, 222 (1975)
11.24 Y.T. Tan: Surf. Sci. **61**, 1 (1976)
11.25 M. Wu, D.M. Hercules: J. Phys. Chem. **83**, 2003 (1979)
11.26 L.L. Tongson, A.S. Bhaila, I.E. Cross, B.E. Knox: Appl. Surf. Sci. **4**, 263 (1980)
11.27 W.L. Baun: Appl. Surf. Sci. **4**, 374 (1980)
11.28 C.R.K. Marrian, A. Shih, G.A. Haas: Appl. Surf. Sci. **24**, 372 (1985)
11.29 C. Creemers, H. Van Hove, A. Neyeens: Appl. Surf. Sci. **7**, 402 (1981)
11.30 W.V. Lampert, W.L. Baun, B.C. Lamartine, T.W. Haas: Appl. Surf. Sci. **9**, 165 (1981)
11.31 H. Jeziorowski, H. Knozinger, E. Taglauer, C. Vogdt: J. Catal. **80**, 286 (1983)
11.32 Th. Berghaus, H. Neddermeyer, W. Radlik, V. Rogge: Phsica Scripta **T4**, 194 (1983)
11.33 J.C. Bertolini, J. Massarddier, P. Delichere, B. Tardy, B. Imojik, Y. Jugnet, Tran Minh Duc, L. Temmerman, C. Creemers, H. Van Hove, A. Neyens: Surf. Sci. **119**, 95 (1982)
11.34 H.J. Kang, Y. Matsuda, R. Shimizu: Surf. Sci. **134**, L500 (1983)
11.35 T.A. Flaim: Research Publication GMR-1942, 1975, Research Laboratories, General Motors, Warren, Michigan

11.36 R.P. Frankenthal, D.L. Malm: J. Electrochem. Soc. **123**, 186 (1976)
11.37 D.G. Swartzfager: Anal. Chem. **56**, 55 (1984)
11.38 M.A. Wheeler: Anal. Chem. **47**, 146 (1975)
11.39 P. Bertrand, J.-M. Beuken, M. Delvaux: Nucl. Instrum. Meth. **218**, 249 (1983)
11.40 J.-M. Beuken, P. Bertrand: Surf. Sci. **162**, 329 (1985)
11.41 R.C. McCune: Anal. Chem. **51**, 1249 (1979)
11.42 D.J. O'Connor, R.J. MacDonald: Radiat. Eff. **45**, 205 (1980)
11.43 M. Aono: Nucl. Instrum. Meth. **B2**, 374 (1984)
11.44 O. Oen: Surf. Sci. **131**, L407 (1983)
11.45 M. Aono, Y. Hou, R. Souda, C. Oshima, S. Otani, Y. Ishizawa: Phys. Rev. Lett. **50**, 1293 (1983)
11.46 J. Moller, H. Niehus, W. Heiland: Surf. Sci. Lett. **166**, L111 (1986)
11.47 H. Niehus: Surf. Sci. Lett. **166**, L107 (1986)
11.48 J.A. Yarmoff, R.S. Williams: Surf. Sci. Lett. **165**, L73 (1986)
11.49 M. Aono, Y. Hou, C. Oshima, Y. Ishizawa: Phys. Rev. Lett. **49**, 567 (1982)
11.50 R. Souda, M. Aono, C. Oshima, S. Otani, Y. Ishizawa: Surf. Sci. Lett. **128**, L236 (1983)
11.51 M. Aono, C. Oshima, S. Zaima, S. Otani, Y. Ishizawa: Japan. J. Appl. Phys. **20**, L829 (1981)
11.52 H. Niehus, W. Raunau, K. Besocke, R. Spitzl, G. Comsa: Surf. Sci. Lett. **225**, L8 (1990)
11.53 S. Priggemeyer, A. Brockmeyer, H. Dotsch, H. Koschmeider, D.J. O'Connor, W. Heiland: Appl. Surf. Sci. **44**, 255 (1990)
11.54 R.C. McCune: Anal. Chem. **51**, 1249 (1979)

12. Reflection High Energy Electron Diffraction

G.L. Price

With 11 Figures

Reflection high energy electron diffraction (RHEED) was first used in the study of a cleaved calcite crystal by *Nishikawa* and *Kikuchi* in 1928 [12.1]. They observed diffraction spots, and lines attributed to diffuse scattering. Other early workers included *Germer* (1936) [12.2] who took diffraction patterns from galena and *Miyake* (1936) [12.3] who examined oxide surfaces. *Uyeda* et al. [12.4] used RHEED with metal films in 1940 and with adsorbed organic molecules in 1950. Commercial RHEED equipment was developed through the 1950s but this operated at 10^{-4} to 10^{-6} Torr and hence only on dirty surfaces. However its application was as an alternative to X-ray diffraction: RHEED's forward scattering nature and comparatively high scattering cross section (10^8:1) made it competitive. As ultra high vacuum equipment became common in the 1960s, systems were equipped with RHEED guns, but good LEED was then possible, and LEED largely displaced RHEED as a diffraction technique for clean single crystal surfaces. The main reasons for this neglect was that RHEED gives quantitative results only on extremely flat surfaces which were not easily prepared and offered no theoretical advantages over LEED in the central surface science question of the surface atomic structure. Also the LEED apparatus was easily constructed and much cheaper than the traditional magnetically focussed RHEED guns. A good review of RHEED and LEED before 1970 is given by *Bauer* [12.5]. The growth technique of molecular beam epitaxy (MBE) brought about a renaissance in RHEED in the 1980s. For this reason the applications of RHEED discussed in this chapter will be drawn from MBE.

The RHEED geometry is shown in Fig. 12.1. An electron beam, of energy 5 to 30 keV, is directed at a glancing angle of $\sim 1°$ at a single crystal held in the centre of an ultra-high vacuum chamber. Diffracted electrons fall on a phosphor screen on the other side of the chamber giving a typically streaked pattern shown in Fig. 12.2. The simplicity and robustness of the technique is immediately apparent. For flat surfaces and energies ≤ 10 keV, only a simple TV electron gun and a phosphor screen are necessary. No accelerating voltages need be applied to the screen and grids are not required. Precautions against stray magnetic and electric fields are minimal compared with LEED because of the high electron energies. The glancing nature keeps the apparatus well clear of other techniques in the chamber.

MBE and its variants such as metalorganic and gas source MBE, have wide application throughout material science for the growth of single crystals [12.6]. It is used to grow metals, semiconductors and ceramics and is a very general

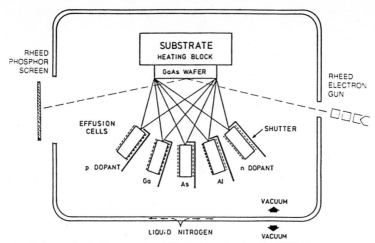

Fig. 12.1. Schematic diagram of an MBE system showing the RHEED geometry

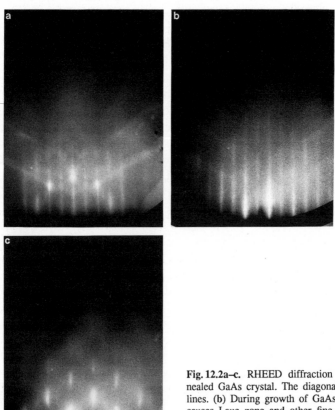

Fig. 12.2a–c. RHEED diffraction patterns for (a) annealed GaAs crystal. The diagonal streaks are Kikuchi lines. (b) During growth of GaAs: the rougher surface causes Laue zone and other fine detail to be lost. (c) Three dimensional growth of InGaAs on GaAs. Much of the background variation is due to adsorbed arsenic on the phosphor screen

method of crystal growth with particular application where extreme purity and precision are required. MBE is simultaneously a mass production and an advanced research technique. It is a unique combination of surface science and production technology. As RHEED is the essential surface science tool for routine MBE, it has been raised to new prominence and very detailed studies have been done in an endeavour to exploit its potential. In this chapter, the examples will be confined to III–V semiconductors as these were the prototype materials for MBE. However the technique applies generally to all other MBE growths.

In an MBE apparatus as shown in Fig. 12.1, the single crystal substrate faces an array of eight or more ovens, each containing elements such as gallium, indium, silicon, aluminium and arsenic. If as an example, a layer of $n-$type GaAs is required, a single crystal GaAs substrate is held at $\sim 580°$ C; the Ga, As and Si dopant shutters are opened and the growth commences. The Ga and Si has unit sticking probability but the As simply supplies an overpressure and is incorporated as needed. The crystal alloy layers are grown epitaxially at about one atomic layer per second. These materials are subsequently transformed into electronic and optoelectronic devices. Generally the ovens are $\leq 1200°$ C, the substrate is $\leq 700°$ C and the chamber may contain 10^{-5} Torr of highly corrosive arsenic vapour. The quality of the semiconductor is critically dependent on the exact surface reconstruction. RHEED is uniquely fitted to this role because of its robustness and geometry. It is one of the few techniques of any kind that can monitor crystal growth *in situ.*

12.1 Theory

The relation between the surface reciprocal lattice and the diffraction pattern is shown in Fig. 12.3a. The RHEED Ewald sphere takes a section through the surface rods and streaks are observed on the phosphor screen rather than the spots of the LEED case. To obtain the full reciprocal lattice map which is given directly by LEED, the crystal substrate must be rotated about its normal; the Ewald sphere then sections each plane in turn. The LEED pattern shown in Fig. 12.3b is a complex gallium rich reconstruction referred to as the centred 8×2. The three main azimuthes which should be observed in RHEED are also drawn. The fine detail of the eighth order is often not observed by RHEED and the pattern is referred to as a 4×2. Practically this is of little consequence as the growth conditions are established to avoid this reconstruction by decreasing the metal/arsenic ratio and obtaining the arsenic rich 2×4 (LEED $c(2 \times 8)$): the fine detail is not required. When the transition from the metal to arsenic rich reconstruction occurs, the four streaks of the [110] change to two. If the surface roughens, then transmission rather than reflection patterns are obtained as shown in Fig. 12.2c. The beam, skimming over the surface, penetrates peaks and ridges. The streaks are replaced with points since the surface reciprocal lattice rods are now replaced by the reciprocal lattice itself. Roughness of order of an atomic

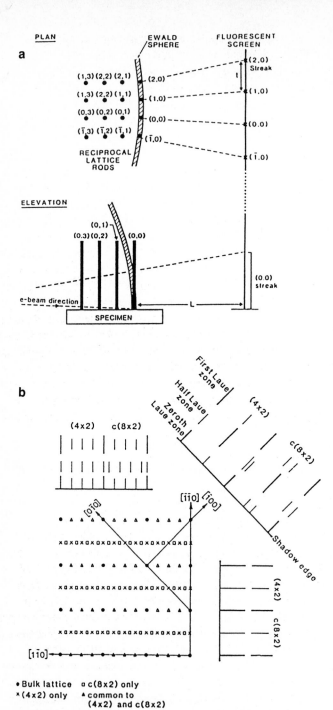

Fig. 12.3. (a) Sectioning of the two-dimensional reciprocal lattice rods give the observed RHEED streaks (after [12.6]). (b) Comparison of the LEED and RHEED patterns for the GaAs $c(8 \times 2)$ reconstruction (after [12.7])

layer can be detected by RHEED. This is extremely useful when monitoring crystal growth and it cannot be easily done by LEED.

To estimate the depth sensitivity of RHEED refer to the mean free path (λ) curve of Fig. 1.8. At 10 keV, the mean free path normal to the surface is $\lambda \sin \gamma$ where $\gamma \sim 1°$ is the glancing angle. The result is about one to two atomic layers. The electron wavelength is ~ 0.1 Å which is an order of magnitude smaller than an atomic layer. Thus the diffraction is sensitive to the surface and the electrons are easily scattered by surface steps and terraces. The diffraction, like LEED, is dominated by multiple scattering and simple kinematic arguments cannot be applied to the streak intensities.

Both LEED and RHEED are surface sensitive, but they differ in their properties parallel to the surface. There are three aspects: the area which can be examined, the coherence length of the diffraction and the ratio of ordered to diffuse scattering. LEED is usually used at normal incidence with a beam size of $\sim 1 \, \text{mm}^2$. The glancing nature of RHEED means that it is imprecise in the direction of the beam, of order of a mm with a simple electrostatic gun, while accurately positioned perpendicular to the beam. More sophisticated high energy and magnetically focussed guns can improve the accuracy by an order of magnitude or more. The coherence length is a measure of the sensitivity of the diffraction to the long range order of the surface. It is the maximum distance between reflected electrons which are able to interfere. By the uncertainty principle, $\delta x = 2\pi / \delta k$. The electron momentum uncertainty has two contributions, δk_α due to the beam convergence half angle α, and dk_E due to the finite energy spread δE. Using $E = \hbar^2 k^2 / 2m$ and assuming α very small, it can be shown that [12.8]

$$\delta k_\alpha = 2k\alpha \sin \gamma \qquad (12.1a)$$

and

$$\delta k_E = (k\delta E/2E) \cos(\gamma + \alpha) \qquad (12.1b)$$

where γ is the glancing angle. In LEED, $\alpha \sim 10^{-2}$ rad, $\delta E \sim 0.5$ eV, $E \sim 100$ eV, $\gamma = \pi/2$, and $\delta x \sim 100$ Å. In RHEED, $\alpha \sim 10^{-4}$ rad, $\delta E \sim 0.5$ eV, $E \sim 10^4$ eV, $\gamma \sim 10^{-2}$ rad, and $\delta x \sim 1000$ Å. Thus RHEED is more sensitive to long range order. Variations in RHEED streaks have been attributed to monatomic steps 2000 Å apart on a GaAs crystal in an MBE system [12.9]. A large coherence length does not necessarily make the diffraction pattern more susceptible to disorder. Disorder gives a diffuse background spread evenly over a large solid angle. Large increases in this background have little affect on the sharp, high intensity diffraction features. Thus the diffraction process tends to select out the ordered parts of the surface. For this reason and because of the beam spread across the crystal, and the glancing angle geometry, RHEED can often give a better diffraction pattern then LEED off a patchy or dirty surface.

12.2 Applications

The main traditional application of RHEED is the monitoring of surface reconstructions and surface roughness which has been mentioned with reference to Fig. 12.2. The importance of this knowledge lies in the primary nature of RHEED. For example, in an MBE growth of al III–V alloy such as $Al_xGa_{1-x}As$ on GaAs, the quality of the resulting material depends on the substrate temperature, the Al and Ga fluxes, the arsenic flux and minute (sometimes in the case of oxygen leaks or source contamination, immeasurable) background contamination. The coupling between these parameters and what is actually measured – the oven thermocouple readings, the pyrometer measurement of substrate temperature, daily ion-gauge flux calibrations, is somewhat loose and can alter day to day. It is often difficult to distinguish between the effects of altering the III/V ratio, changing the substrate temperature or lowering the total growth rate. Over a range of parameter values which provide epitaxial growth there are a number of different reconstructions, but only one of which is optimum. Even with the simple growth of GaAs on GaAs, the chief reconstructions with either increasing temperature or increasing III/V ratio are $c(4 \times 4)$, 2×4, 3×1 and 4×2 with optimum growth in the 2×4. In more complex alloys such as AlGaAs on GaAs, slight contamination which causes high trap densities in the resulting epilayer, can be immediately identified by the behaviour of the reconstructions. Optimum growth areas are found within a reconstruction and these can be identified by the oscillation techniques described below.

Before going into the detail which can be extracted during crystal growth with RHEED, an account of the different growth processes will be useful. Growths which give RHEED streaks, not spots, have a two dimensional nucleation and growth behaviour. This means that the adatoms remain in the sample plane. At the initial states of growth on an atomically flat surface, the adatoms migrate over distances measured by a migration length $l = l_0 \exp(-E_l/kT)$ where E_l is the migration energy. The adatoms are scavenged by a number of processes including: nucleation, where they form an island with other adatoms; step propagation, binding to a step which can be the side of an island or a substrate terrace; and desorption. Three dimensional growth involves adatoms jumping up on top of islands and forming three dimensional crystals which give rise to RHEED spots.

RHEED has recently been found able to calibrate these two-dimensional growths and to give detailed information on surface migration kinetics. As growth proceeds, the specular beam oscillates in intensity, the period of the oscillations being equal to the time taken to grow a single monolayer of material. Typical oscillations are shown in Fig. 12.4 and a schematic explanation is given in Fig. 12.5. The first few layers are essentially complete before another begins. Since the layer thickness is much larger than the de Broglie wavelength of the electrons (for GaAs 2.83 Å $\gg$ 0.1 Å), the electrons are easily scattered out of the specular beam by the step edges. The step edge concentration is a minimum for a completed layer and a maximum for half a monolayer coverage – hence the

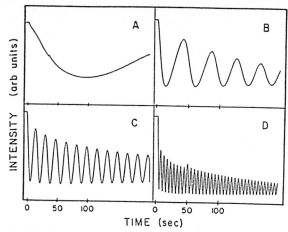

Fig. 12.4. Observed oscillations on a GaAs crystal for increasing flux rates. In *A* the flux is so slow that the substrate steps can scavenge the adatoms before nucleation. The vicinal angle is 1.0 mrad (after [12.10])

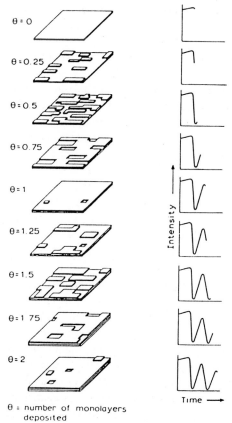

θ = number of monolayers
deposited

Fig. 12.5. A first order growth model of the intensity oscillations (after [12.11])

269

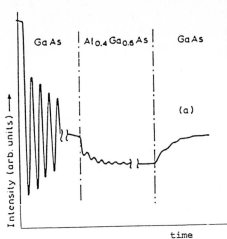

Fig. 12.6. Intensity oscillations across GaAs/-AlGaAs/GaAs interfaces. The short migration length of the Al restarts the oscillations (after [12.12])

oscillations. This simple explanation must be modified to explain the damping of the oscillations as the growth continues. If adatoms fall in a 'ravine' between two steps of width less than the migration length, they will be gettered by the steps edges and nucleation and creation of more step edges will not be possible. As the coverage increases this becomes more likely but so does nucleation of a second layer on top of the first. The result is an increase in the concentration of step edges with growth, tending to a limiting equilibrium value. In RHEED this is seen by the damping of the oscillations and their disappearance, corresponding to the steps gettering all arriving adatoms. A good illustration of this process is shown in Fig. 12.6. Al has a smaller migration length than Ga. GaAs is grown until there are no oscillations. When the aluminium is added oscillations are seen again as the smaller migration length initiates nucleation. When the Al is removed, nothing happens as the growth continues to be by step propagation. An estimate of the migration length can be found by growing at a fixed flux but for differing temperatures on a vicinal surface. On such a surface, cut at a slight angle to a (100) plane, there are terraces of known length (Fig. 12.9). The result is shown in Fig. 12.7. The migration length increases with temperature, and at a certain temperature, growth is only by step propagation. From this, values of l_0 and E_l can be derived. One estimate for Ga atoms on a GaAs (100) surface is $l_0 = 4\,\text{Å}$ and $E_l = 0.3\,\text{eV}$.

It is expected that the best quality growths will be obtained if step propagation is the main growth mode as the randomness of island nucleation and dendritic growth is reduced. One method is to adjust the growth conditions for minimum damping, endeavouring to keep the rate of step nucleation to a minimum. Claims have also been made that vicinal surfaces produce better quality materials. An extreme application of this principle is migration enhanced epitaxy (MEE). If growing GaAs by this technique, first a layer of gallium is grown, then separately a layer of arsenic and so on. The method assumes that the migration length of Ga on a bound As layer or As on a bound Ga layer to be much greater than if

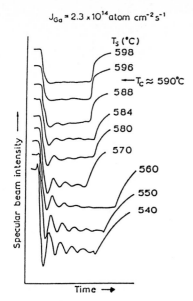

$J_{Ga} = 2.3 \times 10^{14}$ atom cm^{-2} s^{-1}

T_s (°C)
598
596
$T_c \approx 590°C$
588
584
580
570
560
550
540

Specular beam intensity

Time →

Fig. 12.7. Growth of GaAs on a vicinal surface as a function of temperature. At T_c, the migration length is of order the step separation (after [12.13]

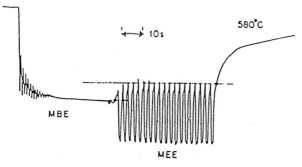

10 s

580°C

MBE

MEE

Fig. 12.8. RHEED oscillations by MBE followed by MEE. The same 2 × 4 reconstruction remained through the growth (after [12.14])

free Ga and As coexist in the same plane. This is borne out by the RHEED data of Fig. 12.8. Here the MEE oscillations are caused by specular reflection changes between Ga and As terminated surfaces, not scattering by step edges. There are smooth transitions from As to Ga terminated surfaces and the oscillations do not damp out with growth. With this technique, high quality materials can be grown at much lower temperatures than normally used, 300° C instead of 600° C.

As previously mentioned, RHEED is highly dynamic; multiple scattering dominates the diffraction and simple kinematic models can not in general be used. One exception is the application of RHEED to measure the terrace widths on vicinal surfaces. The direct and reciprocal lattice for a terraced vicinal surface (not to scale) is shown in Fig. 12.9a. The diffracted intensity is the product of the diffraction due to a terrace of atoms times the diffraction due to a grating of step edges. The rod shown by the dashed lines has the reciprocal width of

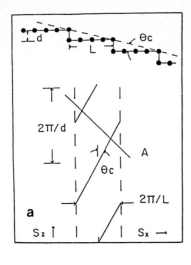 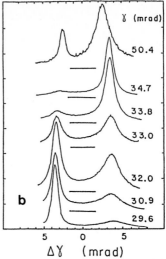

Fig. 12.9. (a) Schematic of a regular staircase and its reciprocal lattice. Curve A is part of the Ewald sphere for a beam directed down the staircase with a RHEED streak split into two. (b) Intensity profiles along a (00) streak as a function of glancing angle γ (after [12.9])

the terrace; the slashes have the reciprocal length of the risers and the angle of the vicinal surface. Depending on the angle of incidence, the Ewald sphere will intersect one or more of the slashes. Figure 12.9 shows intersection with two slashes, producing splitting in the observed RHEED streak. The vicinal angle can be straightforwardly estimated from this data using

$$\beta = (2\pi/kd)\theta_c \cos \phi / (\theta_c \cos \phi + \gamma) \tag{12.2}$$

where β is the measured splitting angle, θ_c is the angle between the vicinal surface and the low index bulk plane, $\gamma = \pi/2 - \theta$ is the glancing angle and ϕ is the azimuthal angle with $\phi = 0$ down the staircase.

So far we have ignored multiple scattering, but one very important multiple scattering phenomenon which is observed in RHEED and LEED is Kikuchi lines. They appear as sharp dark or light lines crossing the phosphor screen (Fig. 12.2a) and they move rigidly with the crystal. Because of this they are used to check the crystal orientation and, with experience, their sharpness provides a qualitative guide to surface cleanliness. Their origin is shown in Fig. 12.10. Diffusely scattered electrons are diffracted inside the bulk crystal. These diffracted electrons lie in cones which intersect the phosphor screen as arcs of large radii. Whether the line is light or dark depends on whether the diffraction is in or out of the diffuse background. This diffraction is from the bulk; the Kikuchi electrons carry underlying information of the solid and not the surface. Their effect is often noticed in oscillation growth experiments. If a Kikuchi line intersects that part of the specular streak which is being monitored, a π out of phase component is observed in the intensity oscillations. This is caused by electrons, which have been

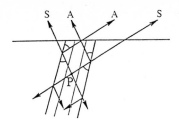

Fig. 12.10. The formation of Kikuchi lines. If diffuse scattering is modeled by emission from a point *P* in the bulk, then diffracted electrons *A* and *S* add and subtract from the diffuse background. Dynamical effects which includes unequal absorbtion for differing paths, gives *A* and *S* different intensities: cancellation does not occur and cones of high and low intensity electrons are emitted from the crystal

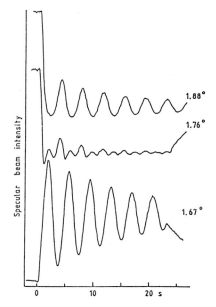

Fig. 12.11. The appearance of harmonics in the oscillations of the specular beam as the glancing angle is varied. At 1.67° the main contribution is from a Kikuchi line. There is a π phase shift compared with the surface pattern at 1.88° (after [12.13])

scattered from the specular beam by surface steps, contributing to the population of diffuse electrons and being diffracted by the Kikuchi process back into the specular streak. At half a monolayer coverage when the specular beam intensity is a minimum, the scattering in from the Kikuchi line is a maximum. An example is shown in Fig. 12.11. Care must be taken in oscillation experiments to filter this harmonic out.

References

12.1   S. Nishikawa, S. Kikuchi: Nature **122**, 726 (1928)
12.2   L.H. Germer: Phys. Rev. **50**, 659 (1936)
12.3   S. Miyake: Nature **139**, 457 (1936)
12.4   R. Uyeda: Proc. Phys. Math. Soc. Japan **22**, 1023 (1940); *ibid* **24**, 809 (1942); Y. Kainuma, R. Uyeda: J. Phys. Soc. Japan **5**, 199 (1950)
12.5   E. Bauer: In *Techniques of Metals Research*, Vol. 2, ed. by R.F. Bunshah (Interscience, New York 1969) p. 501
12.6   E.H.C. Parker (ed.): *The Technology and Physics of Molecular Beam Epitaxy* (Plenum, New York 1985)

12.7   J.H. Neave, B.A. Joyce: J. Cryst. Growth **44**, 387 (1978)
12.8   D.P. Woodruff: In *The Chemical Physics of Solid Surfaces and Heterogeneous Catalysis*, Vol. 1, ed. by D.A. King, D.P. Woodruff (Elsevier, Amsterdam 1981) p. 108
12.9   P.R. Pukite, J.M. Van Hove, P.I. Cohen: J. Vac. Sci. Technol. B2, 243 (1984)
12.10   J.M. Van Hove, P.I. Cohen: J. Cryst. Growth **81**, 13 (1987)
12.11   B.A. Joyce, P.J. Dobson, J.H. Neave, K. Woodbridge, J. Zhang, P.K. Larsen, B. Bolger: Surf. Sci. **168**, 423 (1986)
12.12   B.A. Joyce, J. Zhang, J.H. Neave, P.J. Dobson: Appl. Phys. A45, 255 (1988)
12.13   P.J. Dobson, B.A. Joyce, J.H. Neave, J. Zhang: J. Cryst. Growth **81**, 1 (1987)
12.14   Y. Horikoshi, M. Kawashima, H. Yamaguchi: Jap. J. Appl. Phys. **27**, 169 (1988)

13. Low Energy Electron Diffraction

P. J. Jennings

With 11 Figures

One of the most powerful techniques available for surface structural analysis is low energy electron diffraction (LEED). It is widely used in materials science research to study surface structure and bonding and the effects of structure on surface processes. However because it requires single crystals and ultrahigh vacuum conditions it has limited value for applied surface analysis, which is often concerned with polycrystalline or amorphous materials. LEED has many similarities to X-ray and neutron diffraction but it is preferred for surface studies because of the short mean free path of low energy electrons in solids.

13.1 The Development of LEED

LEED was discovered accidentally in 1924 by *Davisson* and *Kunsman* in the course of their studies of the secondary emission of electrons from a nickel crystal when it was bombarded by a beam of monoenergetic electrons [13.1]. In 1926 *Elsasser* suggested that the observed anisotropy in the elastic scattering of electrons from nickel surfaces was due to diffraction [13.1]. He used the de Broglie relationship to derive an electron wavelength from the momentum p and then related the electron's wavelength to the accelerating voltage V in the electron gun

$$\lambda = \frac{h}{p} \approx \sqrt{\frac{150}{V}} \, \text{Å} \tag{13.1}$$

where V is in electron volts and $V \lesssim 1 \, \text{keV}$. Thus for an electron with a kinetic energy of $150 \, \text{eV}$ the de Broglie wavelength is $\approx 1 \, \text{Å}$, which is similar to the spacing between rows of atoms in a crystal. In 1927 *Davisson* and *Germer* [13.2] carried out a systematic study of the scattering of electrons from Ni(111) and found that the maximum in the reflected intensity of the elastically scattered electrons at any angle α satisfied the plane grating formula

$$n\lambda = a \sin \theta \tag{13.2}$$

where a is the spacing between adjacent rows of atoms, λ is given by de Broglie relationship in (13.1) and n is an integer. At about the same time *Thomson* found diffraction rings when high energy electrons were transmitted through thin metal films. He related the diameter of these rings to de Broglie's wavelength and therefore helped to demonstrate the wave nature of the electron. Davisson and Thomson shared the 1931 Nobel Prize for Physics for the discovery of matter waves. Thus LEED had a momentous birth in the confirmation of de Broglie's hypothesis about the wave nature of matter. *Germer* realized that LEED had great potential for surface studies when he wrote in 1928 [13.4] that, "information of the nature of that which we have obtained may turn out to be of great scientific importance. The whole problem of chemical catalysis concerns itself with what occurs on the surface of a solid body. We believe that this electron diffraction offers the only means which has been suggested for a direct study of questions of this nature".

In the succeeding years high energy electron diffraction, based on Thomson's work, developed rapidly into the field of electron microscopy. However LEED stagnated for many years because researchers found the technique too difficult and too irreproducible. The problem was that it was tedious and time consuming to collect the data and the vacuum technology of the day was inadequate to prepare and maintain clean metal surfaces. During the 1930s most researchers abandoned LEED because of these problems. However in 1934 *Ehrenburg* [13.5] developed a fluorescent screen which enabled him to display a complete LEED diffraction pattern instantaneously. He showed that a typical LEED pattern is a two dimensional array of spots which arise from diffraction by a two dimensional layer or mesh of scattering centres. The LEED pattern therefore reflects the symmetry and structure of the two dimensional reciprocal lattice of the crystal surface. However, despite this advance the LEED experiments were still plagued by a lack of reproducibility due to contaminated surfaces. *Farnsworth* [13.6] in the USA continued to work with LEED through the period from 1930 to 1960 and gradually he improved the apparatus and developed a technique for data collection and analysis. By 1960 ultrahigh vacuum technology was available and LEED had come of age. During the 1960s the field developed rapidly and great advances were made in technique and theory. By 1975 the field was mature with a well established technique and a reliable theory for simulating and analyzing the experimental data. LEED researchers have achieved many substantial successes in the years since then. Some of the most successful applications are described in Sect. 13.6 of this chapter. These days LEED is recognized as the major technique for surface structural studies on single crystals of metals and semiconductors.

13.2 The LEED Experiment

In principle the LEED experiment is very simple [13.7]. A narrow beam of monoenergetic electrons with an energy between 0 and 500 eV is directed onto a planar single-crystal surface at a given angle as shown in Fig. 13.1. A number of diffracted beams of electrons with the same energy as the incident beam are

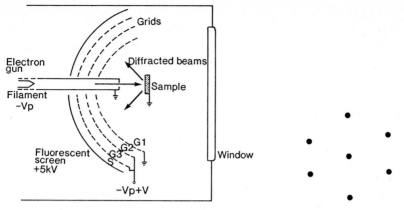

Fig. 13.1. Typical LEED system showing the fluorescent screen S and the hemispherical grids $G1$, $G2$ and $G3$

Fig. 13.2. LEED spot pattern for normal incidence on the (111)surface of Cu with a primary beam energy of 80 eV

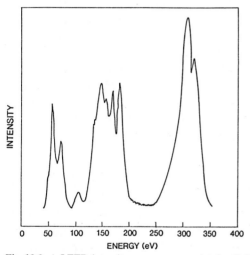

Fig. 13.3. A LEED intensity versus energy plot for Cu(001) at an angle of incidence of 42° along the ⟨11⟩ azimuth

produced in the backward direction. The spatial distribution of these beams and their intensities as a function of angle and energy of the incident beam provides information which can be used to analyze surface structure. Thus, in contrast to X-ray diffraction in which the wavelength is usually held fixed, in LEED there is this extra degree of freedom of changing wavelength via changing energy.

It is customary in LEED to photograph spot patterns produced by the surface layers as shown in Fig. 13.2 and to measure the intensity of one or more beams as a function of the incident beam energy for a fixed angle and azimuth of incidence as shown in Fig. 13.3.

In practice the LEED experiment, like all surface experiments, is difficult to carry out and difficult to analyze for the following reasons:

- The sample surface must be well-oriented, planar and either clean or contaminated in a reproducible manner with some adsorbate.
- The experiment must be carried out in ultra high vacuum to minimize surface contamination by surrounding gases. The background pressure in the LEED chamber should be less than 10^{-10} Torr (10^{-8} Pa).
- Sample manipulation requires sophisticated devices.
- Balancing out of stray magnetic fields is essential.
- The accurate measurement of diffraction angles is difficult, [13.8].
- The collection of intensity data is time consuming although computer based data logging does facilitate this, [13.9].

The LEED apparatus contains four essential components as shown in Fig. 13.1:

1) The *electron gun*. This need not be very sophisticated. The electrons are generated either by on-axis, indirectly heated cathodes or by off-axis tungsten filaments, with energies in the range 0–1000 eV. The electron current is usually a monotonically increasing function of the gun voltage reaching 1–2 μA above about 150 V. The effective diameter of the electron beam is about 1 mm and the energy spread is typically about 0.5°. This implies a coherence width of the incident beam at the sample of 200-550 Å. Thus LEED is only sensitive to small regions of the surface of the order of 200–500 Å over which the periodicity of the lattice is maintained. Steps, kinks and surface imperfections are generally not visible but manifest themselves as background noise.

2) The *goniometer* which holds and orientates the sample. This is usually complicated and costly. Most goniometers allow rotation about an axis in the plane of the surface and about an axis perpendicular to the surface. The sample holder must also allow for temperature control as the samples will need to be heated and cooled (often to liquid nitrogen temperatures).

3) The *detector* to monitor the diffracted beams. The most common is the post-diffraction accelerator of *Ehrenburg* [13.5]. Electrons accelerated through a potential difference V impinge on the surface and are diffracted. Less than 5% of the backscattered electrons are elastically scattered and the remainder are inelastically scattered. These travel through the field free region to grid $G1$. A small negative voltage between $G1$ and $G2$ serves to reject most of the inelastically scattered electrons while the elastically scattered electrons reach $G3$ and a large positive voltage between $G3$ and the fluorescent screen accelerates them to F where they produce fluorescence on impact. The observer looking through the window will see bright spots at the main points of impact. This spot pattern is called a LEED pattern. The brightness of the spots is proportional to the intensities of the corresponding diffracted beams which can therefore be monitored with a spot photometer – this function is usually digitized and stored on a magnetic disk for subsequent plotting and processing.

The screen is hemispherical in profile and centred on the position of the sample.

4) The whole apparatus is enclosed in an UHV *chamber* capable of reaching and maintaining pressures as low as 10^{-11} Torr. At these low pressures a surface will be covered by a monolayer of gas in about 10 h (if the sticking probability is one).

13.2.1 Sample Preparation

Preparation of samples for surface studies proceeds in two stages:

1) *Outside the vacuum chamber:* this includes selection and orientation of the sample, usually an ultrapure single crystal ingot; followed by cutting, lapping and polishing of the sample to produce a flat, smooth surface oriented within 0.5°.

2) *Inside the vacuum chamber* at pressures $\approx 10^{-10}$ Torr (10^{-8} Pa) a series of cleaning procedures aimed at eliminating all foreign atoms from the surface. Common heat treatments are O_2 (to remove carbon) or H_2 (to remove oxygen) followed by scattering or ion bombardment in cycles with annealing (e.g. Ni, Cu). Some materials must be heated to high temperatures to evaporate surface impurities (e.g. W, Mo). Cleavage along favourable crystallographic planes is useful for Si, Ge, Be and Zn.

13.2.2 Data Collection

For purposes of surface structure determination we must photograph the LEED pattern and collect the intensities of a number of diffracted beams as a function of incident beam energy. The angle of incidence θ and the azimuthal angle ϕ must also be specified. The angle is defined relative to the surface normal and the azimuth is defined relative to a specified direction in the surface – (generally one of the lattice directions). The intensities are normally collected by means of a spot photometer or by means of a microprocessor-controlled video camera. The data must be processed before it can be used for structure analysis. This involves:

a) normalisation to crystal incident current (since gun current varies with V)
b) contact potential difference correction (between sample surface and crystal surface)
c) background subtraction
d) corrections for grid transparency as the spot moves across the surface.

13.3 Diffraction from a Surface

If an electron wave with wavelength λ falls upon a two-dimensional net of scattering centres a number of scattered waves are produced. The plane grating condition applies for normal incidence and defines the diffraction maxima or spot pattern

$$n\lambda = a\sin\theta. \tag{13.3}$$

The intensity of the spots varies with energy because the atomic scattering factors are functions of energy and the scattering also occurs from successive layers in the crystal. Thus the Bragg condition also applies for scattering from subsequent layers.

The spot pattern contracts as the energy increases according to (13.3). For the two dimensional scattering the appropriate diffraction conditions are

$$\boldsymbol{k}_\parallel - \boldsymbol{k}'_\parallel = \boldsymbol{g} \tag{13.4}$$

where $\boldsymbol{g}$ is a reciprocal lattice vector

$$\boldsymbol{g} = l\boldsymbol{b}_1 + m\boldsymbol{b}_2 \tag{13.5}$$

and $\boldsymbol{k}_\parallel$ and $\boldsymbol{k}'_\parallel$ are the surface components of the incident and diffracted wave vectors.

Thus the diffracted beams can be labelled by vectors of the reciprocal net $\boldsymbol{g} = (l, m)$.

The Ewald sphere construction provides a useful graphical representation of these two-dimensional diffraction conditions. The construction proceeds as follows:

a) each of the points of the 2-D reciprocal lattice is the origin of a rod,
b) the incident $\boldsymbol{k}$ vector is oriented so that its origin is at the 00 point and its length is $1/\lambda = k/2\pi$,
c) a sphere is drawn with the end point A of the incident vector as origin,
d) the possible diffraction conditions are given by (13.4) or by the intersection of sphere with the reciprocal lattice rods. The direction of the beams is also given by this construction as shown in Fig. 13.4.

Thus the LEED pattern which consists of the intersections of all of the diffracted beams $\boldsymbol{k}'$ with the Ewald sphere is an image of the reciprocal net when it is viewed along the normal direction to the surface at infinite distance.

reciprocal lattice rods

Fig. 13.4. Ewald sphere construction for LEED

As the energy of the incident electrons increases the radius of the sphere will increase and the spot pattern will shrink. Actually this form of the Ewald sphere construction is only valid for scattering by a single layer. For 3-D scattering, as in X-ray diffraction, the rods become points in the 3-D reciprocal lattice. LEED also has some 3-D features as the electrons do penetrate some distance into the crystal and the rods can therefore be thought of as having 3-D lattice points embedded in them. When the Ewald sphere intersects the rod near a 3-D lattice point there will be a maximum in the intensity of that spot. Such maxima are called Bragg peaks.

13.3.1 Bragg Peaks in LEED Spectra

Consider an incoming beam with wavevector k_0 being scattered by a succession of identical, equally spaced layers of the crystal as in Fig. 13.5. The amplitude of the scattered wave is given by

$$\psi_{sc} = \left(\varrho_{0\nu} + \tau_{00} \omega_0 \tau_{\nu\nu} \right.$$
$$+ \tau_{00} \omega_0 \tau_{00} \omega_0 \varrho_{0\nu} \omega_\nu \tau_{\nu\nu} \omega_\nu \tau_{\nu\nu}$$
$$\left. + \ldots \ldots \right) e^{ik_\nu \cdot r}$$
$$+ \text{higher order terms} \qquad (13.6)$$

where ϱ is the reflection coefficient for a layer, τ the transmission coefficient and ω_ν is the phase factor $\exp(ik_\nu^\perp d)$, where k_ν is the wave vector of the diffracted beam.

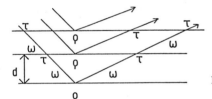

Fig. 13.5. Formation of Bragg peaks in LEED

These contributions may be summed to infinity to give

$$\psi_{sc} \approx \frac{\varrho_{0\nu}}{1 - \tau_{00} \tau_{00} \omega_0 \omega_\nu} e^{ik_\nu \cdot r} \qquad (13.7)$$

The maximum of this function with respect to energy can be determined by assuming that to a first approximation $\tau_{00} \approx \tau_{\nu\nu} \leq 1$. Thus the maximum occurs for $\omega_0 \omega_\nu \approx 1$, i.e. for

$$\left(k_0^\perp + k_\nu^\perp \right) d = 2\pi n . \qquad (13.8)$$

This is the condition for a Bragg peak in the diffracted beam ν. For the 00 or specular beam it becomes

$$\left(k_0^\perp + k_\nu^\perp \right) d = 2\pi n . \qquad (13.9)$$

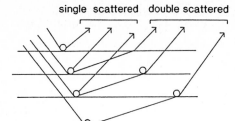

single scattered   double scattered

Fig. 13.6. Mechanism of production of secondary Bragg features in LEED

Actual LEED intensity spectra also exhibit a number of secondary Bragg features which are due to multiple scattering.

The mechanism of production of secondary Bragg features is illustrated in Fig. 13.6. A non-specular beam at the Bragg angle may be scattered back into the specular beam as it leaves the crystal thus echoing its Bragg feature in the specular beam.

The presence of these secondary Bragg features in LEED intensity curves demonstrates that multiple scattering is important. The reason for this situation is that LEED occurs via coulombic scattering rather than via induced electric dipoles as in the case of X-rays. Slow electrons are strongly scattered by a crystal surface and thus penetrate only a short distance. This is useful for surface work but it also implies that a multiple scattering or dynamical theory is required to simulate the observed intensity curves.

13.4 LEED Intensity Analysis

The Bragg peak locations are dependent on the surface structure of the crystal and thus the aim of LEED intensity analysis is to deduce surface structure from the measured intensity curves [13.10, 11]. The usual approach is to postulate a structural model of the surface and to use this as the basis for a simulation of the LEED intensity spectra. The simulated and measured spectra are compared visually or qualitatively and the model is refined to obtain the best possible fit. The simulations are complex and the analysis is difficult and indirect. The procedure involved is basically:

a) the potential in the solid is calculated from the superposition of Hartree Fock atomic potentials. It is spherically averaged using the muffin-tin approximation to obtain the potential of an average atom in the solid;

b) the coulomb scattering from this muffin-tin atom is calculated using the method of partial waves to give a set of scattering phase shifts (as a function of electron energy);

c) the reflection and transmission coefficients for each layer are calculated from the atomic data and the assumed two-dimensional structure using a multiple scattering theory such as the Green's function or KKR method;

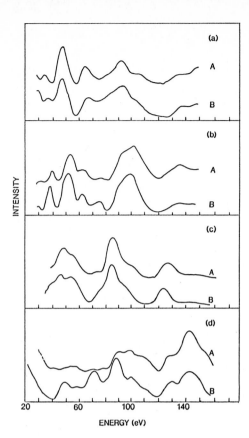

Fig. 13.7. Illustration of the simulation of LEED intensity curves for various surfaces (*A*: experimental, *B*: theory

(a)

A

B

(b)

A

B

(c)

A

B

(d)

A

B

INTENSITY

20   60   100   140

ENERGY (eV)

d) the single layers are combined to form a multilayer crystal using the transfer matrix which takes account of the scattering of each incident beam into the other beams at each layer.

This produces a set of intensities for each of the diffracted beams as a function of the energy and diffraction angles and for the assumed surface structure. The analysis proceeds using simulation techniques in which the crystal model is optimized to obtain the best fit of the computed and measured LEED spectra as shown in Fig. 13.7. The process involved in the calculation is similar to that used in band structure calculations for solids. In fact the LEED intensity curves are dependent on the band structure for the solid along a particular direction through the Brillouin zone corresponding to the direction of the primary beam k_0. The Bragg peaks occur at gaps in the band structure (well above the Fermi level) while the secondary Bragg peaks occur at partial band gaps. If an electron is incident on the crystal surface at an energy where a band gap is present there are no propagating states in the solid for it to enter and thus it is strongly reflected. Thus Bragg peaks correspond to absolute gaps in the band structure. Secondary Bragg peaks occur where the density of propagating states in the solid is low.

13.5 LEED Fine Structure

The LEED intensity curves also show a number of narrow, complex features called LEED fine structure at very low primary energies. These features arise from the scattering of diffracted beams by the surface barrier. They are found only at low energies and they usually form a series of peaks converging on the threshold energy for a new diffracted beam as shown in Fig. 13.8 [13.12].

The mechanism of production of these features is shown in Fig. 13.9. It involves a new beam which has begun to propagate in the crystal but still does not have sufficient energy to surmount the surface barrier. This beam is totally internally reflected by the surface barrier and after subsequent diffraction by the crystal it may be diffracted back into the incident beam where it can produce interference effects in that beam. The spacing of the interference fringes is an indication of the structure of the surface barrier. The analysis of the LEED fine structure enables us to deduce the structure of the barrier. This is an important component in the theory of many types of electron emission phenomena from solids such as photoemission, thermionic emission and field emission.

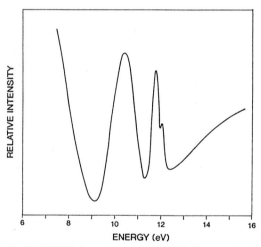

Fig. 13.8. LEED fine structure for Cu(001) for an angle of incidence of 60° along the ⟨11⟩ azimuth

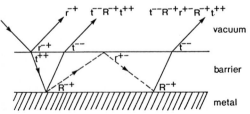

Fig. 13.9. Mechanism for the production of LEED fine structure

13.6 Applications of LEED

LEED is the basic structural technique for solid surface analysis for crystalline materials. It has been applied to many different types of surface but usually it is best suited to structural studies on clean, well-ordered surfaces of simple metals and semiconductors.

13.6.1 Determination of the Symmetry and Size of the Unit Mesh

The LEED pattern provides a map of the reciprocal lattice of the surface. The (00) spot or specular beam can be identified as the spot towards which all of the others gravitate as the energy of the electron beam is increased. At normal incidence we can determine the rotational symmetry of the pattern by inspection. We can also determine the size of the unit mesh from the pattern. To do this we photograph the LEED pattern and measure b_1 and b_2 on the photograph. From a knowledge of the geometry of the LEED chamber we can calculate the diffraction angle θ given by

$$n\lambda = a \sin \theta .\tag{13.10}$$

Hence from a knowledge of θ and λ we can calculate the direct mesh lengths a_1 and a_2, and compare them with the predicted values. Thus we can determine whether the surface and substrate nets are identical. This is usually the case for most simple transition metal surfaces. However most semiconductor surfaces show reconstructed unit meshes which are formed because of the covalent bonds which are broken during the fracture of the surface.

13.6.2 Unit Meshes for Chemisorbed Systems

If a small amount of gas is allowed to adsorb on a metal surface it often forms an ordered structure. Such structures are usually different from those of the clean surface as shown in Fig. 13.10. Studies of LEED patterns make it possible to determine the symmetry and relative size of the unit mesh of the adsorbed layer. If the pattern shows new spots which do not belong to the clean surface net then we must conclude that the surface net is different from the substrate net.

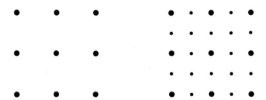

Fig. 13.10. LEED patterns for normal incidence on clean Ni(001) and for $c(2 \times 2)$ O/Ni(001) with a primary beam energy of 80 eV

This pattern analysis is very important and LEED is now established as a major tool for analyzing the structure of chemisorbed overlayers and reconstructed clean surfaces.

Such pattern analysis forms the basis for models of the chemisorption process and is vital to an understanding of catalysis and corrosion. However pattern analysis only tells us the relative size, symmetry and orientation of the overlayer. it tells us nothing about the location of the overlayer relative to the substrate. This is the goal of LEED intensity analysis.

13.6.3 LEED Intensity Analysis

The analysis of LEED intensities generally proceeds from the measured intensity versus energy curves for the specular and several non-specular beams. Because of the strong multiple scattering in LEED a dynamical theory is required. The simple kinematical analysis used in X-ray diffraction is unsuccessful in LEED. Dynamical calculations involve heavy computational effort and so far only the simplest systems have been studied. For example the position of adsorbed selenium atoms on the $\sqrt{2} \times \sqrt{2} R45°$ Se/Ag(001) has been studied by several authors. They all find that the adsorbed selenium atoms are located at alternate fourfold silver sites [13.13].

Similar studies were done for O on Ni but at first they disagreed about the Ni-O bond length, [13.12]. Finally after considerable dispute it was found that the O layer is located about 0.9 Å above the topmost nickel layer [13.14]. This result has recently been disputed by researchers working with ion scattering spectroscopy. In general the field is still in an immature state. LEED intensity analysis is the most widely accepted surface structural technique but it has so far only been applied successfully to simple, clean surfaces and to a few surfaces with relatively simple overlayers. Generally the adsorption site can be found readily and clearly but the accuracy of the bond lengths and surface layer spacings are unsatisfactory. Very little progress has been made so far with the analysis of the structure of surfaces with large unit meshes or surfaces with adsorbed molecules. The problem is ultimately one of immense computational labour and thus researchers tend to use other approaches such as electron or scanning tunnelling microscopy for surfaces with large and complex unit meshes. On the basis of work done to date we can make some useful observations about surface structure:

a) the most closely packed surfaces fcc(111), bcc(110) of clean transition metals exhibit very little or no contraction of the top layer;

b) surfaces with intermediate packing density [e.g. fcc(001), bcc(001)] have a contracted first interlayer spacing. The contractions are typically 2% to 10% depending on the material;

c) the more open surfaces such as fcc(110) and bcc(111) are strongly contracted (of the order of 10–15%);

d) clean semiconductor surfaces are mostly reconstructed – although some such as Si(111) can be stabilized with minute amounts of impurity. Such reconstruction is usually obvious in the patterns [e.g. Si(001) 2×1, Si(111) 7×7]. The structures are often very complicated and may involve four or five layers;

e) the study of adsorbate structures has shown that surface bond lengths are similar to gas phase covalent bond lengths and the surface is often distorted in the process of adsorption. However many simple systems such as $c(2 \times 2)$ O/Cu(001) are still not fully understood because of the complexity of the reconstruction.

13.6.4 Surface Barrier Analysis

The analysis of LEED fine structure at low energies provides information about the structure of the surface barrier (Fig. 13.11). From classical theory we expect that the image force law will apply to an electron at large distances from the surface. This predicts that the potential energy $V(z)$ of an electron at a distance z from the surface is given by

$$V(z) = \frac{-q}{16\pi\varepsilon_0(z - z_0)} . \tag{13.11}$$

Quantum theory predicts a saturation of this potential at short distances leading to a constant average inner potential U_0 which is approximately equal to $\phi + E_F$ at low energies (where ϕ is the work function of the surface, and E_F is the Fermi energy). LEED fine structure analysis provides the best available experimental test of barrier models and also enables us to determine the barrier origin z_0. Studies of this type have so far been carried out only on Ni, Cu and W surfaces [13.12]. A high resolution electron spectrometer is required to resolve the fine structure [13.8].

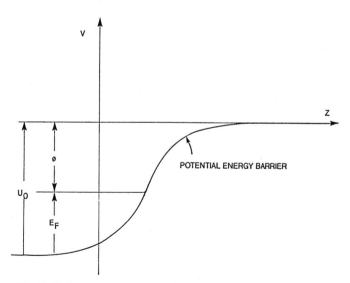

Fig. 13.11. One-dimensional model of the surface potential energy barrier

Recently there have been encouraging advances in this area and it is possible that future LEED studies may be carried out at low energies ($E < 50\,\text{eV}$) where the analysis is simpler but the experimental demands are greater.

13.7 Conclusion

To a large extent LEED has fulfilled the expectations of its early advocates such as *Germer* [13.4]. However it is now clear that it has certain strengths and weaknesses. Its strengths include:

a) the ability to provide direct and accurate information about surface contamination;
b) its suitability for studies of processes on ordered surfaces including chemisorption, phase transitions and epitaxy;
c) its sensitivity to the surface barrier structure.

LEED has now become a standard technique for investigating structure in surface science experiments and it has provided a wealth of valuable data about structure and binding on metal and semiconductor surfaces.

However, it also has some well known limitations which include:

a) it is only suited to studies in ultra-high vacuum on single crystal surfaces. This excludes most industrially and technologically important surfaces. LEED is best suited to fundamental studies of surface processes such as bonding and reconstruction;
b) the analysis of LEED patterns does not give an unambiguous result for complex surface structures. To choose between alternative structural models a LEED intensity analysis is required;
c) LEED intensity analysis requires sophisticated simulation procedures and the computations involved are beyond the power of most computers for complex adsorption systems. However, there are many interesting problems which involve simple surfaces and adsorbates and these are the current focus of attention.

For these reasons LEED is one of the most popular research techniques in surface science. However because of its limitations it is advantageous to use it in conjunction with other compatible techniques such as Auger electron spectroscopy and work function measurements.

References

13.1   C.J. Calbick: The Physics Teacher, May 1963, pp. 1–8
13.2   C.J. Davisson, L.H. Germer: Phys. Rev. **30**, 705 (1927)
13.3   G.P. Thomson: Engineering **126**, 79 (1928)
13.4   L.H. Germer: J. Chem. Ed. **5**, 1041 (1928)
13.5   W. Ehrenberg: Phil. Mag. **18**, 878 (1934)
13.6   H.E. Farnsworth: *Surface Chemistry of Metals and Semiconductors* (Wiley, New York 1959)
13.7   F. Jona, J.A. Strozier, Jr., W.S. Young: Rep. Prog. Phys. **45**, 527 (1982)
13.8   G. Hitchen, S.M. Thurgate: Surf. Sci. **24**, 202 (1985)
13.9   S. Thurgate, G. Hitchen: Appl. Surf. Sci. **17**, 1 (1983)
13.10 J.B. Pendry: *Low Energy Electron Diffraction; The Theory and Its Application to the Determination of Surface Structure* (Academic, London 1974)
13.11 M.A. Van Hove, S.Y. Tong: *Surface Crystallography by LEED*, Springer Ser. Chem. Phys. Vol. 2 (Springer, Berlin, Heidelberg 1979); M.A. Van Hove, W.H. Weinberg, C.-M. Chan: *Low Energy Electron Diffraction*, Springer Ser. Surf. Sci. Vol. 6 (Springer, Berlin, Heidelberg 1986)
13.12 R.O. Jones, P.J. Jennings: Surf. Sci. Rep. **9**, 165 (1988)
13.13 A. Ignetiev, F. Jona, D.W. Jepsen, P.M. Marcus: Surf. Sci. **40**, 439 (1973)
13.14 J.E. Demuth, D.W. Jepsen, P.M. Marcus: Phys. Rev. Lett. **31**, 540 (1973)

14. Ultraviolet Photoelectron Spectroscopy of Solids

R. Leckey

With 8 Figures

Compared with X-ray photoelectron spectroscopy (XPS) or Auger electron spectroscopy (AES), ultraviolet photoelectron spectroscopy (UPS) is not generally considered to be an analytic technique for the surface characterization of materials. It is, however, an extremely surface sensitive technique where even a monolayer coverage of an adsorbate or contaminant is sufficient to grossly alter the signal from a given surface. As we shall see, its main strength lies in its unique ability to explore the electronic structure in the conduction/valence band region of a wide variety of solids. As a technique, it can readily be added to other surface science instrumentation and is indeed often offered as an option by manufacturers of XPS/AES equipment. This chapter is consequently included as a brief introduction to the capabilities of UPS to alert practitioners of other surface science techniques to the information contained in UPS spectra.

UPS relies on the photoelectric effect whereby a beam of monochromatic photons is used to eject electrons from the valence/conduction band region of a material. Traditionally, the photon source used has been a hollow cathode discharge lamp [14.1] running in an inert gas, the most commonly used resonance lines being He 1 at 21.21 eV and Ne 1 at 16.86 eV. Since the bandwidth of the conduction/valence band of materials is in the range 5–10 eV, these photon energies are sufficient to probe the entire bandstructure region of most materials (Fig. 14.1). The photoemitted electrons consequently have energies typically less than 17 eV using He 1 and consist of two main groups.

The first group of electrons are those which, having been excited within the uppermost few atomic layers, escape into vacuum having suffered no inelastic collisions. The energy, and to some extent the direction of emission of such electrons, may be readily related to their original binding energy within the solid and to the momentum associated with their original state. The binding energy E_b of each such electron is simply related to the observed kinetic energy E_k via the Einstein formula $E_k + E_b = h\nu$ as a first (and usually adequate) approximation, as indicated in Fig. 14.1.

The second group of electrons observed in a UPS spectrum consist of those electrons which have either made one or more inelastic collisions (with other bound valence band electrons) or are in fact the secondary electrons which have gained sufficient energy to escape from the material from such a collision. This group of electrons constitutes a largely featureless low energy peak in all UPS spectra with a shape which resembles somewhat a Maxwellian distribution. There is, regretably, no established procedure for subtracting this inelastic background

291

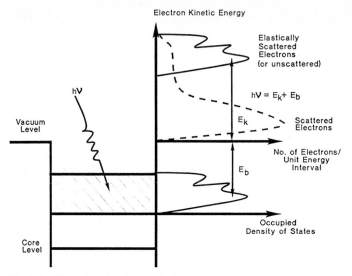

Fig. 14.1. Illustrating the photoexcitation of electrons from a valence band due to monochromatic photons of energy $h\nu$. An idealized energy distribution is also shown with primary emission distinguished from secondary electrons

from a UPS spectrum at the present time; the effect of the background is clearly most severe for low photon energies. For this reason, among others, the helium ion resonance line (He 11 at 40.81 eV) is often used in laboratory UPS experiments.

The ability to select the incident photon energy at will is now available at synchrotron radiation sources. This facility is essential for most serious work involving the determination of electronic band structure as we will see below. Because of the development of synchrotron radiation facilities world-wide, the previous distinction between UPS and XPS has largely disappeared. One can still, however, retain some distinction between the two regimes due to the need to use different types of monochromators for the different photon energies involved. A convenient, if arbitrary, definition of the UPS area would include the photon energy range $10\,\text{eV} < h\nu < 150\,\text{eV}$.

The main aim of many UPS experiments is to gain information about the distribution of electrons in the outermost valence or conduction band region of the material. It is these electrons which are responsible for the chemical, magnetic, optical and mechanical properties of each material. With sufficiently detailed information about the electronic bandstructure, we may begin to tailor-make materials with specific properties – such bandstructure engineering is already occurring in the microelectronics industry with increasing regularity [14.2]. The type of information we need to accomplish such material modification is extremely detailed but can be examined progressively. We can start with the information available from the least sophisticated UPS experiment and can then refine our knowledge of particular materials by examining the interpretation of more exotic experiments involving variable photon energy, angle resolved energy distributions and the effects of the polarization of both incident photons and emitted electrons. In the present chapter, we can clearly only hope to get a brief flavour of the full complexity of modern UPS.

14.1 Experimental Considerations

The minimum requirements for performing a simple UPS experiment are (a) a resonance radiation lamp, (b) an electron energy analyzer capable of operation with 5–50 eV electrons with an energy resolution of ~ 0.1 eV, (c) an atomically clean polycrystalline sample in a UHV environment. The analyzer need not be capable of angular resolution but the energy resolution requirement implies magnetic shielding in the 5×10^{-7} T region. Sample surface cleanliness must be maintained on an atomic scale since electrons of ~ 20 eV energy have an inelastic scattering mean-free path of ~ 5 Å; this requirement in turn leads to the need for a UHV environment (typically 10^{-10} Torr).

With such equipment, the conduction/valence band width may be readily determined for many materials and the distribution of electrons within the band measured (occupied density of electron states). While such information is available for the elements, considerable scope remains for the determination of the density of states of alloys of variable composition, for example. To illustrate this point, UPS spectra from a range of Ag/Pd alloys are shown in Fig. 14.2 [14.3]. This data was taken at both He 1 and He 11 photon energies and we use it here to illustrate a number of points.

Firstly, the presence of the inelastically scattered electrons is evident in the 21.21 eV data but it is clear that, even at this energy, it is reasonably simple to distinguish the primary (unscattered) electrons of interest. Despite this, it is also evident that background subtraction is less of a problem at 40.81 eV photon energy.

Secondly, the primary UPS spectra at the two photon energies are quite different. This is due to two considerations:

1) The probability of a UPS event depends jointly on the number of electrons in the initial state – the occupied density of states and on the number of empty states available at the photoexcited energy – the unoccupied density of states. The larger the photon energy the less structure will be present in the unoccupied density of states. At XPS energies, the photoexcited (final) density of states is assumed to be featureless and this assumption holds quite well also at 40.81 eV. Thus both XPS and He 11 spectra may be expected to mirror variations in the initial (occupied) density of states whereas a He 1 spectrum will represent a mixture of these initial and final state functions which is in general too difficult to interpret.

2) The probability of a UPS event is also controlled by a dipole matrix element term of the form

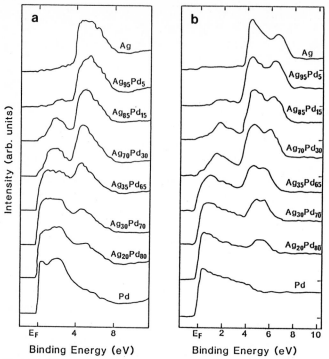

Fig. 14.2. UPS spectra of a series of Ag/Pd alloys taken with (a) 21.21 eV and (b) 40.81 eV photons under angle integrated conditions. From [14.3]

$$|\langle f|r|i\rangle|^2 .$$

Evaluation of such a term involves knowledge of both initial $|i\rangle$ and final $\langle f|$ wavefunctions – neither of which are generally known with certainty. We can, however, say that the strength of a transition will depend on the orbital nature of both initial and final states and experience tells us that emission from d band states is much stronger relative to emission from s or p derived states at 40.81 eV than at 21.21 eV.

Armed with the above information, we can now assert that the He 11 Ag/Pd alloy spectra of Fig. 14.2 are likely to be a good representation of the width and of the density of states of the conduction band and that most prominent structure comes from bandstructure states of predominately $'d'$ nature. As a trivial example of the type of information which follows from these results, we can note the manner in which the onset of strong $'d'$-band emission moves relative to the Fermi edge as the alloy composition is varied. This in turn controls the optical absorption of these alloys and can readily explain the alteration in the colour of these materials with composition.

The main limitations to an experiment such as that described above are the requirement for atomically clean surfaces and that the sample either be conducting or have its surface potential stabilized by use of a low energy (~ 2 eV) flood

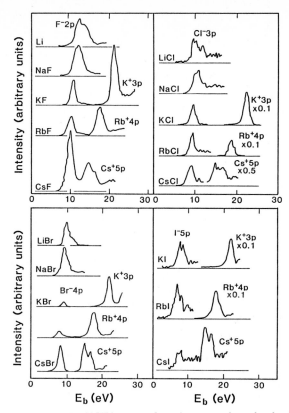

Fig. 14.3. 40.81 eV UPS spectra from the outer valence bands of the alkali halides. From [14.4]

gun. Figure 14.3 includes data from some alkali halides, which, although highly insulating in the bulk, exhibit sufficient conductivity in evaporated thin film form to make the experiment feasible [14.4]. Valence bands originating from both anion and cation in these ionic materials are clearly visible in most of the spectra of Fig. 14.3, this data was originally used to obtain data on band gaps and electron affinities and to assess the merits of various theoretical band calculations.

14.2 Angle Resolved UPS

To proceed beyond the type of experiment outlined above, it is necessary to work with single crystal samples and to acquire spectra with good angular resolution ($\pm 1°$). Under these conditions, we can expect to be able to determine the details of the way crystalline energy levels disperse with momentum in the $E(k)$ bandstructure of the material. This is the fundamental information controlling the electronic, magnetic and optical properties of the solid and is of extreme impor-

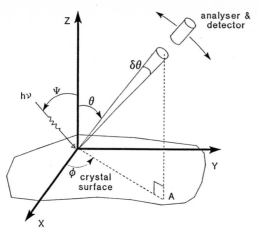

Fig. 14.4. Geometrical arrangement for an angle resolved UPS experiment

tance for solid state electronics, lasers, etc. The geometry of a typical experiment is shown in Fig. 14.4.

Monochromatic photons are incident on a clean single crystal surface in UHV at some angle Ψ to the sample normal. The crystal is orientated so that electrons emitted in a selected plane (AZ in Fig. 14.4) containing a symmetry direction of interest can enter an analyzer capable of detecting the polar angle of emission θ and of measuring the kinetic energy of each electron. In the past this has been accomplished by stepwise moving an energy analyzer possessing a narrow acceptance angle $d\theta$ in small increments of θ, but more recently, display type analyzers have been developed which simplify the process and reduce the data collection time required. Such analyzers have recently been reviewed [14.5]. As an example of one such device, our toroidal electron energy analyzer is shown in cross-section in Fig. 14.5.

In this analyzer, [14.6] electrons emitted in a plane (horizontal in Fig. 14.5) are brought to a ring focus at a position sensitive detector if their energy corresponds with the voltages applied to the various electrodes. The arrival position on the detector is directly related to the polar angle of emission; the detected kinetic energy may be varied by altering the electrode potentials. The device consequently records $N(E, \theta)$ as shown in Fig. 14.6. Typically, the energy resolution is set to 0.05 eV whereas the angular resolution is close to $\pm 1°$ for the analyzer described in [14.6].

A number of features are immediately evident from Fig. 14.6. Perhaps the most striking is the wealth of detail shown here compared to an angle integrated spectrum such as one of those shown in Fig. 14.3. The structure in Fig. 14.6 is dominated by the $3d$ electrons of the conduction band of copper. They are somewhat atomic like in the manner of core levels, i.e. these electrons are rather closely associated with individual ion cores and are consequently partially localized in space, unlike the s-type electrons closer to the Fermi energy. These

296

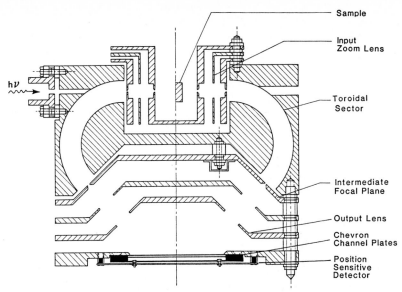

Fig. 14.5. Cross section of a toroidal electron energy analyzer specifically designed for angle resolved UPS. From [14.6]

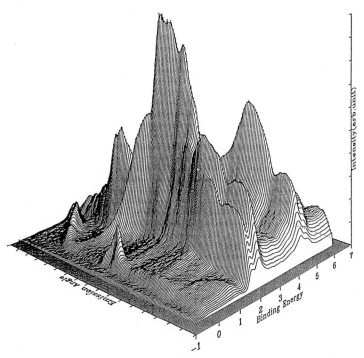

Fig. 14.6. Angle resolved UPS data from Cu(110) in the [112] azimuth. Structures due to *d*- and *s*-derived bands are clearly visible together with a surface state as Γ as described in the text

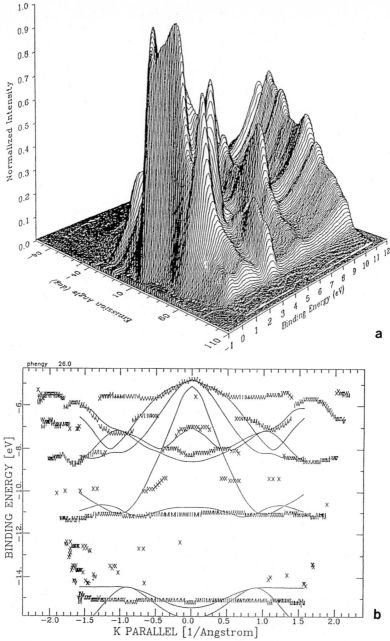

Fig. 14.7. (a) Angle resolved UPS data from GaAs(110) taken at 21.21 eV photon energy. (b) Band-structure diagram $E(k_{11})$ derived from this data (symbols) and as predicted from a LMTO calculation (lines)

298

s-electrons are completely delocalized conduction band electrons and feel the full effects of the periodic potential of the crystal lattice. Emission from such states can readily be identified as the strongly dispersing feature which breaks away from the relatively flat d bands at $\sim \theta = -49°$. The small feature which shows some dispersion and is seen close to the Fermi energy around $\theta = 0°$ is due to a surface state. Such states are not a feature of the bulk band structure but exist as separate solutions of the Schrödinger equation specifically associated with the surface itself. Such states have an important role to play in all surface sensitive situations e.g. catalysis, adsorption and the energy alignment of metal semi-conductor interfaces.

The variation with angle of emission of the binding energy of each identifiable peak in a data set such as Fig. 14.6 may be converted to a graph illustrating the major features of the $E(k)$ bandstructure of the material. To some extent, this is trivial since the existence of periodicity parallel to the crystal surface implies that the parallel component of the momentum of the electron before excitation k_{11} is related to the measured kinetic energy (E) and angle of emission (θ) via

$$k_{11} = 0.511\sqrt{E}\sin\theta .$$

Applying this formula to each peak in an experimental data set enables us to plot a projection of the bandstructure i.e. $E(k_{11})$ which may then be compared with a prediction based on a calculated bandstructure. Such a comparison is shown for GaAs together with the original data set in Fig. 14.7.

Before the full $E(k)$ bandstructure can be experimentally determined, access to the variable photon source of a synchrotron radiation facility is required. The details of the processing of such data are beyond the scope of this chapter, but an example of a set of energy distribution curves obtained over a wide range of photon energies and acquired at normal emission is shown in Fig. 14.8 together with the experimental valence band structure derived from this data. Details of the analysis may be found in [14.7].

In this brief overview, there has not been space to consider UPS experiments on adsorbates nor from quasi-two dimensional materials (layer compounds). The interpretation of UPS data from such systems is quite straightforward due to the fact that the experiment provides $E(k_{11})$ directly and little or no dispersion is expected in the $k_\perp$ direction. Similarly, we have not considered the advances in understanding the magnetic properties of materials which have been the subject of recent experiments using polarized photon sources and polarized electron detectors. More complete descriptions of the capabilities of UPS may be found in the review articles [14.8, 9].

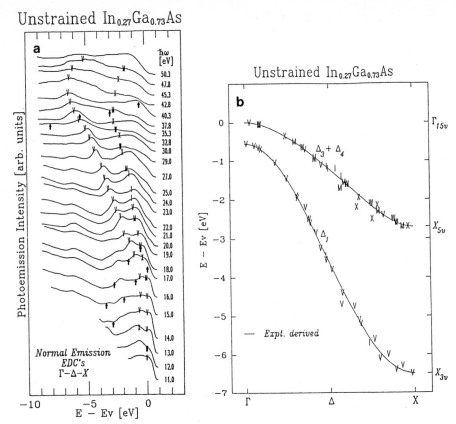

Fig. 14.8. (a) Normal emission energy distribution curves from InGaAs(100) for a range of photon energies. (b) Valence band dispersion curves derived from this data

References

14.1   R.T. Poole, J. Liesegang, R.C.G. Leckey, J.G. Jenkin: J. Electron. Spectrosc. Relat. Phenom. **5**, 773 (1974)

14.2   See, for example, the series of articles in Physics Today, Oct. 86, pp. 24–82

14.3   A.D. McLachlan, J.G. Jenkin, R.C.G. Leckey, J. Liesegang: J. Phys. F **5**, 2415 (1975)

14.4   R.T. Poole, J.G. Jenkin, J. Liesegang, R.C.G. Leckey: Phys. Rev. B **11**, 5179 (1975)

14.5   R.C.G. Leckey: J. Electron. Spectrosc. Relat. Phenom. **43**, 183 (1987)

14.6   R.C.G. Leckey, J.D. Riley: Appl. Surf. Sci. **22/23**, 196 (1985)

14.7   A. Stampfl, G. Kemister, R.C.G. Leckey, J.D. Riley, F.U. Hillebrecht, J. Fraxedas, L. Ley, P.J. Orders: J. Vac. Sci. Technol. A **7**, 2525 (1989)

14.8   F.J. Himpsel: Adv. Phys. **32**, 1 (1983)

14.9   R. Courths, S. Hufner: Phys. Rep. **112**, 53 (1984)

15. Spin Polarized Electron Techniques

J.L. Robins

With 10 Figures

Many of the analytical techniques described in this book involve electron beams. In most cases the parameters measured are the intensity (number of electrons) and the momentum (energy and direction of the electrons). However, electrons have another intrinsic property called spin and if this is also measured additional information can be obtained from the various analytical techniques. The most obvious application is in studying magnetic phenomena although a wide spectrum of analysis of solid state and surface properties is possible. Device engineers will find many valuable applications such as being able to image directly the vector magnetization of the surface of ferromagnets, whereby magnetic domain wall movements can be investigated. Knowledge of such processes is vital in the field of high density magnetic-based information recording, storage and retrieval. Both the pure and device-oriented scientists will see spin polarized electron techniques as a way of measuring exchange and spin-orbit phenomena, giving them the tools to study magnetization in ultrathin films and even on an element-specific basis. Such studies form the basis on which the latest techniques in atomic engineering can be used to fabricate compound and multilayer devices with properties, both magnetic and otherwise, that utilize the subtle effects arising from induced strains, atomic environments, long and short range order, etc. To the theorist, spin related experiments offer a means for very specific testing of models of various types of magnetic interaction and solid state and surface theories.

The widespread application of spin polarized electron techniques is still in its infancy. To date the emphasis has been on developing methods for applying the technique in conjunction with other analytical procedures, in showing what phenomena and properties can be studied and in studying a limited range of well defined systems. Much has already been achieved, but now that the threshold has been crossed the future will see widespread exploitation of these new techniques to answer an ever increasing range of questions. Typical of the questions being addressed are: Can the Curie temperature of the surface of a ferromagnet be different to the Curie temperature of the bulk; can a nonmagnetic material become magnetic when in the form of an ultrathin layer; if chemisorption reduces magnetization does it do so by destroying long range order or by reducing individual atomic magnetic moments; what is the structure of magnetic domain walls on a nanometer scale; what is the role of defects, grain boundaries and additional phases in determining the coercivity of the new neodynium-based permanent magnets? Spin analyzed electron techniques are already supplying answers to such questions [15.1, 2].

In this chapter, a brief description will be given of electron spin and the basic interactions which are influenced by electron spin. The type of additional information which can be obtained from the various techniques through monitoring the spin property will then be given with an emphasis on studying surface magnetic properties of materials. A brief description of the equipment used in producing spin polarized electron beams and for measuring the spin polarization of scattered and emitted beams will be included. By way of example, the chapter will be concluded with a more detailed illustration of the study of magnetic domains using scanning electron microscopy with polarization analysis (SEMPA). Whilst the treatment here will of necessity be superficial the reader is referred for rigorous and detailed treatment to Kessler's book on "Polarized Electrons" [15.3] and, in relation to the interaction of polarized electrons with surfaces, to two other monographs currently available [15.4, 5].

15.1 Electron Spin

Spin is a fundamental quantum mechanical property of an electron [15.3]. For each electron there are two possible spin states which are commonly distinguished by referring to them as "spin up" or "spin down" ($\uparrow, \downarrow$) or, relative to a prescribed direction (such as that of an imposed magnetic field in the z direction), as being parallel or antiparallel to that direction. An electron beam, or other ensemble of electrons, is said to be polarized if there is a preferential orientation of the electron spins and the degree of polarization can then be defined as follows. The polarization in the z direction is

$$P_z = \frac{N_\uparrow - N_\downarrow}{N_\uparrow + N_\downarrow} \tag{15.1}$$

where $N_\uparrow (N_\downarrow)$ represents the number of electron spins parallel (antiparallel) to the z direction. A completely polarized beam has a polarization of unity and an "ordinary" or unpolarized beam has a polarization of zero whilst for a 50% polarized beam the ratio of parallel to antiparallel spins is 3:1.

In most solids there are equal numbers of electrons with spin up and spin down. However in ferromagnetic materials the magnetization results from there being unequal numbers of electrons in the two spin states [15.6]. Figure 15.1a shows a standard schematic representation of the density of states curve for the $3d$ band of a non-magnetic transition metal (or a ferromagnetic material above the Curie temperature). The cross-hatching shows the filled states extending up to the Fermi level E_F on the energy scale. Figure 15.1b shows a similar representation but distinguishes the $e_\uparrow$ and $e_\downarrow$ states and represents a ferromagnetic transition metal below the Curie temperature. In this case an exchange interaction is responsible for shifting the energy of the $3d$ states, the shifts being in different directions for the $e_\uparrow$ and $e_\downarrow$ states. Filling to the Fermi level now results in an excess of $e_\uparrow$ electrons relative to $e_\downarrow$. By way of example, in nickel there are 5

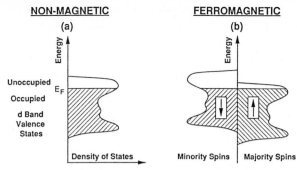

NON-MAGNETIC   **FERROMAGNETIC**

(a)   (b)

Fig. 15.1. (a) Standard schematic density of states representation of the partly filled $3d$ band of a non-magnetic transition metal, or a ferromagnetic material above the Curie temperature. (b) The same for a ferromagnetic material below the Curie temperature but with the spin up $e_\uparrow$ and spin down $e_\downarrow$ states shown separately, illustrating the energy shifts caused by exchange interactions. The cross-hatching shows the filling of the bands to the Fermi level E_F

electrons per atom in the $3d_\uparrow$ band and 4.46 in the $3d_\downarrow$ band. This results in a magnetic moment of about 0.54 Bohr magnetons per atom.

15.2 Interactions Involving Spin

There are two principal types of interaction involving spin. These are spin-orbit and exchange interactions. Whilst they will be described briefly here, the reader is referred elsewhere [15.3] for detailed quantum mechanical descriptions.

15.2.1 Spin-Orbit Interactions

The spin-orbit interaction is most pronounced in electron scattering or emission from heavy materials. It is essentially due to the interaction of the spin of an electron with its own angular momentum in the electric field of a strong scattering potential such as the strongly attractive ion core of an atom. That is, when an electron is scattered by an atom, the relative motion of the incident negative electron and the positive ion core induces an effective magnetic field, the direction of which depends upon whether the electron is passing to the left or right of the atom. The scattering potential is modified by the interaction of the electron's magnetic moment and this magnetic field. Thus the scattering cross section will be different for scattering to the right and left. This asymmetry is important in electron scattering studies and in the development of polarization detectors.

15.2.2 Exchange Interactions

The exchange interaction occurs because of the presence of other electrons which also have spin and is the outcome of the Pauli principle which requires that no two interacting particles have the same set of quantum numbers. The consequence is that the interaction potential between two electrons will be different depending on whether they have parallel spins (same spin quantum number) or antiparallel spins. As one of the interacting electrons can be within an atom or the surface of a solid, exchange effects are always involved when electrons scatter from a surface, although if the surface is non-magnetic or paramagnetic (having equal numbers of spin up and spin down electrons) the effects cancel. In ferromagnetic materials, where the surface has a net spin, exchange effects are dominant. Such magnetic surfaces can thus be studied with spin polarized electrons by measuring the scattering cross-sections as a function of the degree of polarization of the incident beam.

15.2.3 Polarization and Scattering

An unpolarized beam can be considered as being composed of equal numbers of electrons with spins in opposite directions. Thus as a consequence of the asymmetry described in Sect. 15.2.1 above, when an unpolarized beam is scattered through an angle θ, more electrons of one spin state than the other will be scattered into the chosen direction of observation. The result of this is that, due to spin-orbit effects, the scattered beam becomes polarized and this polarization will be in a direction normal to the scattering plane. The polarization P is given by

$$P = \frac{N_\uparrow - N_\downarrow}{N_\uparrow + N_\downarrow} = S(\theta) . \tag{15.2}$$

Here $N_\uparrow$ ($N_\downarrow$) is the number of electrons with spin parallel (antiparallel) to the normal to the scattering plane. The scattering plane is the plane containing both the incident and scattered beam directions and θ is the angle between these directions. $S(\theta)$ is the Sherman function [15.3]. Alternatively, when the incident beam is already polarized normal to the scattering plane, there is an asymmetry of the scattering through an angle θ to the left and right within the scattering plane. This asymmetry A also is governed by the Sherman function $S(\theta)$, as well as by the polarization P of the incident beam, in the case of elastic scattering. This asymmetry is given by

$$A = \frac{N_L - N_R}{N_L + N_R} = PS(\theta) , \tag{15.3}$$

where $N_L(N_R)$ is the number of electrons scattered to the left (right) through an angle θ. Clearly if the incident beam is totally polarized, with $P = 1$, then $A = S(\theta)$.

The above shows that a polarized beam can be formed by scattering and that measurement of asymmetry in scattering can be used to determine the polarization of the incident beam if the Sherman function is known or has been determined for the scattering material and the primary energy being used. In practice, instrumental effects may contribute to measured asymmetries and steps must be taken to recognize these and eliminate their contribution [15.3]. In addition it is possible to study either spin-orbit or exchange interactions independently, by an appropriate choice of the scattering plane and the polarization direction [15.7] or by appropriate manipulation of the observed scattering data [15.8].

15.3 Experimental Techniques

Most analytical techniques involving electrons can be enhanced by using polarized electron primary beams or by measuring the polarization of the scattered or emitted electrons. In general these techniques are discussed at length elsewhere in this book so the treatment here will be restricted to pointing out the type of additional information which can be gained in each case by considering the polarization of the electrons. The necessary physical changes required for the experiments involve at least the replacement of the electron sources or detectors with spin polarized sources or polarization detectors respectively, although this is usually not a trivial exercise. Typical sources and detectors are described in Sect. 15.5.

Polarized Electron Scattering: LEED is the most common technique used in studying the elastic scattering of electrons from crystalline solids. It is usually used for determining the atomic arrangement of atoms, however other features such as the shape of the surface potential barrier can also be studied. When the spin of the incident or diffracted electrons is monitored the technique is termed SPLEED (spin polarized low energy electron diffraction). In general the spin measurements complement the intensity measurements and are sometimes more sensitive to certain features of the target material or its electronic structure [15.7–11]. Alternatively, the study of spin dependence in inelastic scattering events provides an opportunity to study different types of phenomena, such as the production of electron-hole pairs with opposite spin (Stoner excitations) [15.12]. This technique is often described as spin polarized electron energy loss spectroscopy. As with elastic scattering, it is usual to only include either a source or detector of polarized electrons, although some experiments using both have been carried out [15.13, 15].

Polarized Photoemission: Photoemission, in the form of both XPS and UPS, provides information on occupied electron states; the former relating to core levels and the latter to the valence states, as indicated in Fig. 15.2a. To study polarized photoemission, the spin orientation of the emitted electrons is analyzed [15.16–18]. One category of spin polarized photoelectrons is produced when there is magnetic order in the target such as in ferromagnetic materials (see Fig. 15.1b). Another category arises when circularly polarized incident radiation is used, and the dipole selection rules restrict excitation processes such that the emitted electrons are spin polarized. The use of circularly polarized synchrotron radiation from storage-rings will widen the application of these techniques for band structure studies.

Polarized Inverse Photoemission: This process is the "inverse" of photoemission in that electrons of well defined energy, momentum and spin impinge on

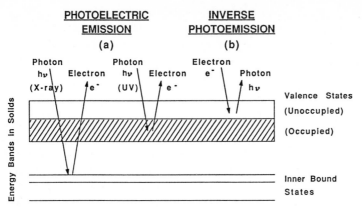

Fig. 15.2a,b. Illustration of photoemission and inverse photoemission processes in terms of the energy bands within a sample, emphasizing the fact that the two processes can be used to probe the occupied and unoccupied states in the valence band, respectively

the target and photons are emitted, as indicated in Fig. 15.2b. Spin-orbit coupling and ferromagnetism lead to a dependence of the emitted photon intensity on the spin of the incident electrons [15.19, 20]. As the final state for the incoming electron will be one of the previously unoccupied states above the Fermi level (see Figs. 15.1b and 15.2b), this technique probes the unoccupied states above the Fermi level. More specifically, it can be seen that with polarized incident electrons the unoccupied majority and minority spin states in ferromagnetic materials can be studied separately [15.21].

Polarized Auger Analysis: Auger analysis is both element specific and non-destructive and is thus a commonly used technique for chemical analysis in surface studies. In combination with other techniques it can also provide information on electron correlations and screening effects which are important in magnetic and near-magnetic solids. If the polarization of the emitted Auger electrons is measured [15.22] it can be used to probe properties such as the local magnetic moment in magnetic alloys and compounds at particular element sites [15.23] and the coupling of partly filled inner shells with the net spin of the magnetic $3d$ electrons [15.24].

Polarized Secondary Emission: True secondary electrons are mainly conduction electrons excited in the cascade process and their spin polarization is proportional to the average magnetization near the surface of the solid [15.22, 25]. They thus act as a surface magnetometer and, depending on the energy and configuration used, varying degrees of spatial and depth resolution can be achieved. To exploit this technique requires an unpolarized exciting beam of electrons together with an energy analyzer and a polarization detector.

Scanning Electron Microscopy with Polarization Analysis (SEMPA): As discussed above, it has now been established [15.24, 26] that during emission true secondary electrons carry with them the spin alignment which they had whilst in the solid. This offers a means of studying the magnetic domain structure of surfaces. This technique requires combining polarization analysis with high resolution scanning electron microscopy and is referred to as SEMPA [15.27, 28]. It is described in detail in Sect. 15.6, with illustrations of the results which can be achieved.

15.4 Study of Magnetic Properties

In a recent review, Pierce [15.2] has listed various magnetic properties about which information is required when studying the surface or a thin film of a magnetic material. His list is reproduced below together with the techniques which would be most suitable for investigating the various properties. Further information and references are given in Pierce's original article [15.2].

Spontaneous Magnetization: Any of the spin polarized electron techniques will show whether or not the sample is magnetic or whether its magnetic moment has been enhanced by a change of lattice constant at the surface or by the lower dimensionality if it is in the form of a thin film. However, polarized Auger analysis [15.22] is element specific which can be an added advantage if the material is associated with another magnetic material, such as in the form of a substrate or in a layered material.

Temperature Dependence of Magnetic Order: Polarized electron scattering offers simplicity and efficiency as only the asymmetry in the intensity of the elastically scattered electrons needs to be measured. Such observations can detect low temperature spin deviations [15.29], critical exponents [15.30] and surface Curie temperatures which are different from the bulk [15.31].

Electronic States: To determine the spin dependence of the electronic structure or the energy distribution of the magnetic states, the most suitable techniques are polarized photoemission [15.16, 17, 32] or polarized inverse photoemission [15.19, 20]. The choice depends upon whether the interest is in the occupied or unoccupied states, respectively.

Magnetic Surface Anisotropy: Either polarized photoemission [15.33] or the SEMPA technique [15.34] as described in Sect. 15.6 are suitable for determining whether the surface anisotropy [15.35] is perpendicular or parallel to the surface.

Elementary Excitations: Polarized electron scattering such as spin polarized electron energy loss spectroscopy [15.32, 36] is the most suitable technique for studying electron induced excitations such as Stoner excitations.

Structure: By making measurements of polarized secondary electron emission at different primary and secondary electron energies, it has been possible to achieve magnetic depth profiling [15.37]. This allows a determination of the spacial variation of the magnetization from the surface into the bulk. In addition, careful analysis of SPLEED intensities gives information on the layer by layer magnetic moment [15.8].

Magnetic Domains: In general, there is a need to be able to monitor the size and shape of magnetic domains, to relate their occurrence to physical features in the surface and to know more about the domain walls. Scanning electron microscopy with polarization analysis can provide images of the domain and domain wall structures with very high spatial resolution, whilst the same measurements, ignoring the polarization effects, can be used to produce secondary electron scanning electron microscopy images of the physical features of the surface for comparison with the domain wall locations [15.27, 28, 32, 38]. This is illustrated in Sect. 15.6.

Magnetization Curves: Polarized electron scattering [15.39] and polarized secondary emission measurements [15.22] have been used to measure surface hysteresis curves. Such observations can show whether the remanent magnetization and coercive field are different at the surface from in the bulk.

Dynamics: Using picosecond laser pulses in conjunction with polarized photoemission permits extremely fast magnetic measurements to be made [15.40]. Such measurements can be used to determine, for example, the time variation of the magnetization when the applied field is changed.

Chemistry: It is to be expected that the magnetic and electronic properties of the surfaces of solids will be influenced by the adsorption of foreign atoms or the segregation of impurities at the surface. Polarized Auger analysis [15.41], polarized photoemission [15.42] and polarized inverse photoemission [15.19, 43] are highly suited to studying such chemical effects.

15.5 Sources and Detectors

Unfortunately there is no simple method of producing a 100% spin polarized beam of electrons [15.3]. Scattering from atomic vapour produces polarization but as well as presenting practical difficulties the yield is very low. Various other principles have been used [15.3] and each has its own advantages. However the technique based on photoemission from GaAs [15.44, 45] is now the most commonly used due to its relative simplicity and high electron yield, despite the relatively low polarization of about 25–40% which it produces. When freshly pretreated with cesium and oxygen, a GaAs surface becomes a highly efficient negative electron affinity photoemitter [15.45]. GaAs is a direct band gap semiconductor and the transitions take place at the Γ point from states that can be described as atomic P states at the top of the valence band to S states at the bottom edge of the conduction band. Due to spin-orbit coupling the P states are split as shown in Fig. 15.3, with the $S_{1/2}$ and $P_{1/2}$ states doubly degenerate and the $P_{3/2}$ state fourfold degenerate [15.46–48]. A photon energy equal to the gap energy E_g (1.52 eV at 77 K for GaAs), such as from a suitably chosen GaAlAs diode laser, will produce transitions from the $P_{3/2}$ level but not from the $P_{1/2}$ level (which is 0.34 eV lower in GaAs). With these parameters the two spin states in the conduction band will still be equally populated. However, if the laser light is circularly polarized with positive helicity σ^+, the dipole selection rules only permit the transitions shown by full lines in Fig. 15.3. The relative transition

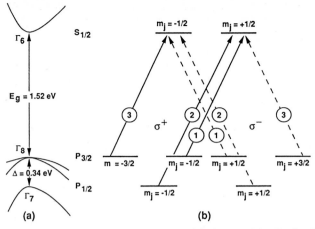

Fig. 15.3. (a) Section of the bandstructure of GaAs around the Γ point. **(b)** Transitions induced across the band-gap by right-circular (σ^+: full lines) and left-circular (σ^-: dashed lines) polarized light. The calculated relative transition rates (shown in circles) lead to a polarization of $\pm 50\%$ for photons with $h\nu = 1.52\,\text{eV}$ [15.46]

probabilities are shown in circles on the figure and it is seen that the two spin states will now be unequally populated. The directions of the spins of the emitted photoelectrons are aligned parallel to the incident photon direction. According to (15.2) the theoretical maximum polarization will be $(3-1)/(3+1) = 50\%$. In practice due to various depolarizing effects [15.4, 45] the polarizations achieved are nearer 25–40%.

It is significant to note here that if the helicity of the circularly polarized light is reversed, the dashed transition lines of Fig. 15.3 become relevant and the direction of the spin polarization is reversed without (in principle) any other parameters being affected. When the incident radiation is normal to the GaAs surface, the photoelectron beam emitted normal to the surface will be longitudinally polarized. If this electron beam is then deflected through 90° by electrostatic fields, which do not alter the spin directions of the electrons at the low energies in use here, a transversely polarized beam is achieved [15.45]. A practical requirement is that ultrahigh vacuum conditions, with pressures in the low 10^{-10} mbar range, are essential in order to reduce contamination effects and thus to prolong the lifetime (typically about 10 hours) of the activated GaAs cathodes.

The most effective method of measuring the polarization of beams of electrons is by Mott scattering [15.49]. This involves measuring the asymmetry of scattering from high atomic number atoms such as gold (see Sect. 15.2.3). If the electron beam is accelerated to about 120 keV the scattering is from the atomic nuclei and the theory is sufficiently well understood that no secondary calibration of the detector is needed. Smaller versions of these so called Mott detectors have now been developed, which operate at lower energies. The design for one of these [15.50] is shown in Fig. 15.4. It operates at 30 to 50 kV, has hemispherical accelerating electrodes, a gold foil scattering target and channeltron electron

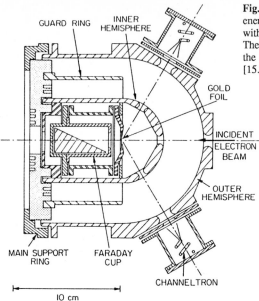

Fig. 15.4. Schematic diagram of a medium energy Mott detector of spin-polarization, with hemispherical accelerating electrodes. There is four-fold rotational symmetry about the direction of the incident electron beam [15.50]

GUARD RING

INNER HEMISPHERE

GOLD FOIL

INCIDENT ELECTRON BEAM

OUTER HEMISPHERE

MAIN SUPPORT RING

FARADAY CUP

CHANNELTRON

10 cm

detectors. Such a configuration measures the component of polarization normal to the scattering plane (see Sect. 15.2.3). Some even smaller Mott detectors and other detectors using different aspects of scattering asymmetry [15.47] have been developed. It is often advantageous for the detector to be UHV compatible. Other advantages of the newer devices are lower working voltages and smaller physical size, partly for convenience and partly to facilitate angular variation measurements. Calibration was usually a problem unless direct comparison with a high energy Mott detector was possible but recently alternative procedures have been proposed [15.51]. Factors which are important in designing, calibrating and operating polarization detectors are documented elsewhere [15.52].

15.6 SEMPA: Scanning Electron Microscopy with Polarization Analysis

The analytical technique described in this section illustrates how polarization analysis can be combined with an existing highly refined technique, namely scanning electron microscopy, to produce a device which can image magnetic domains on a surface with a resolution of 50 nm or better. It also has the capacity to measure all three components of the magnetization vector, both within domains and across domain walls. The instrument [15.27, 28] is based on the underlying principle that when low energy secondary electrons are emitted from a solid they maintain the spin orientation they had whilst in the solid (see Sect. 15.3). This principle is illustrated in Fig. 15.5. If the unpolarized primary electron beam

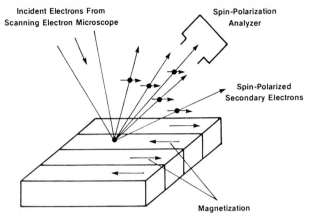

Incident Electrons From
Scanning Electron Microscope

Spin-Polarization
Analyzer

Spin-Polarized
Secondary Electrons

Magnetization

Fig. 15.5. Diagram to illustrate the underlying principle of SEMPA [15.28]. The highly focussed unpolarized electron beam from an SEM causes the emission of secondary electrons which are polarized with the same spin orientation they had in the specimen. As the SEM beam is rastered across the target the secondary electrons need only be collected and their polarization displayed to produce a magnetization image of the specimen surface. Note that the spin and magnetic moment are in opposite directions

can be focussed to a spot which is smaller than a magnetic domain then the magnitude and direction of the secondary electron polarization is directly related to the magnetization of the area probed. Without further focussing the emitted low energy secondary electrons can be transported to the polarization detector via an electrostatic energy analyzer which is used to filter out the higher energy scattered electrons. By rastering the primary beam over the target surface, the output of the polarization detector produces an image of the magnetization of the surface.

A semi-schematic illustration of the instrument [15.28] is shown in Fig. 15.6. Using an SEM with a field emitter source, a 3×10^{-12} A unpolarized incident beam with a diameter of 10 nm or less can be used. The detectors are of a low energy diffuse scattering type [15.28, 53] in which relatively low energy electrons of about 150 eV are scattered from a freshly prepared polycrystalline gold surface and a channel plate electron multiplier is used to measure the scattered intensities. As the channel plate anode is divided into four segments, the two diagonally opposite pairs measure the polarization in two orthogonal directions. Two polarization detectors are used, positioned as shown in the figure. One will measure say the X and Y components within the surface of the target whilst the other, when the beam is electrostatically deflected into it, will measure say the X and Z components. This offers the facility to measure all three components, with a redundancy in one of the components which allows an internal calibration of the sensitivities of the two detectors. The same secondary electrons which are detected for the polarization measurements can also be used simultaneously to produce normal secondary electron SEM images of the surface by ignoring their spin orientation. This results in the production of pairs of images show-

311

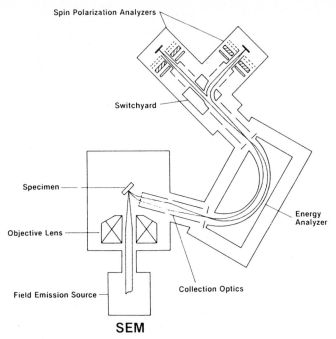

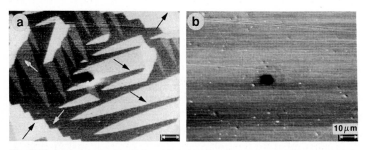

SEM

Fig. 15.6. Schematic diagram of the SEMPA apparatus [15.28]. The highly focussed unpolarized electron beam from the SEM is rastered across the specimen surface and a selection of the emitted secondary electrons are separated from other higher energy scattered electrons in an energy analyzer before being transported to the polarization analyzers. With two orthogonally mounted analyzers, all three components of the magnetization vector in the specimen surface can be detected

Fig. 15.7a, b. A polarization image (**a**) and a secondary electron SEM image (**b**) of the same area of an Fe-3% Si single crystal surface. The grey levels give the four different magnetization directions in the domains, as marked by the arrows [15.28, 54]

ing independent measurements of the physical features and the magnetic domain structure of each chosen section of the surface. Two examples of these are shown in Figs. 15.7 and 15.8.

Figure 15.7 shows a pair of images from an Fe-3% Si single crystal [15.28, 54]. The polarization image in Fig. 15.7a shows one component of the polarization

Fig. 15.8a, b. High resolution polarization (**a**) and secondary electron SEM (**b**) images of an Fe-3% Si single crystal, showing how the walls separating the three magnetic domains in (**a**) are pinned by the three defects (indicated by arrows) in (**b**) [15.28, 54]

with the analyzer polarization axis rotated 28° with respect to the [100] easy axis of magnetization. The four distinct grey levels thus represent domains with magnetizations as indicated by the arrows. The zig-zag domain wall seen in the lower left corner is a common phenomenon which minimizes the energy in such microstructures [15.55]. The disturbance in the domain pattern near the centre of the image can be associated with a defect in the crystal which is clearly shown in the secondary electron SEM image in Fig. 15.7b. Further evidence relating physical defects and magnetic domain structure is shown in Fig. 15.8. The sample here is again an Fe-3% Si single crystal [15.28, 54]. By comparing the polarization image in Fig. 15.8a with the secondary electron SEM image in Fig. 15.8b it can be seen that the walls separating the three domains (indicated by the three levels of grey) in Fig. 15.8a are associated with three defects observed in Fig. 15.8b. It would thus appear that the defects have a role in pinning the domain walls and hence are a major factor in determining the coercivity of magnets, as the coercivity is governed by the ease or otherwise with which domain walls can be moved. Clearly, SEMPA is an ideal technique for studying such phenomena.

In Fig. 15.9 the two images are both magnetization images with the projection of the magnetization along two orthogonal axes, both lying in the specimen surface, being shown in Figs. 15.9a and b, respectively. The linear intensity scale uses white (black) for the maximum value of the magnetization component pointing along the positive (negative) direction. The material is a Co-based ferromagnetic glass [15.56]. In Fig. 15.9a the directions of the magnetization within the sections of the four domains shown are seen to be parallel to the domain boundaries and opposite on each side of each boundary. For a simple Bloch wall an out-of-plane rotation of the spins is to be expected as the wall is crossed. Instead, the magnetization within the domain walls is seen in Fig. 15.9b to be still within the surface and perpendicular to the wall, showing that at the surface the spins rotate in-plane which constitutes a Néel type wall and is in contrast to the Bloch walls which are expected in the interior of the sample. These and similar features are also described in [15.28, 32, 57].

313

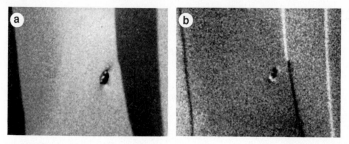

Fig. 15.9a, b. SEMPA images of the magnetization parallel to the line of domain walls (a) and perpendicular to them (b) in the surface of a Co-based ferromagnetic glass. The images are 70 μm across. The magnetization within the walls is clearly observed in (b) indicating a Néel like wall in the surface region [15.56]

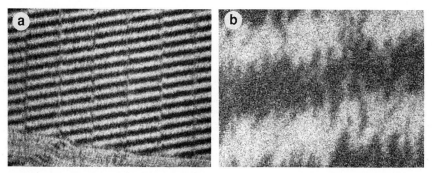

Fig. 15.10a, b. SEMPA magnetization images of written bits in a Co-Ni recording medium shown in (a) at a low magnification where the image is 70 μm across, and in (b) at 10 times higher magnification [15.57]

Figure 15.10 shows magnetization images of a test magnetic pattern written on an information storage disk [15.57] in which sharp well defined boundaries are desirable for a minimum noise signal. The active material is a 70 μm thick layer of Co-Ni (approximately 80%–20%). The images display one component of the magnetization nearly aligned along the domains. It is the jaggedness of the domain boundaries, which is clearly shown here, which sets a limit on the information storage density. The value of the SEMPA technique for investigating such phenomena is obvious.

References

15.1   R.J. Celotta, D.T. Pierce: Science **234**, 333 (1986)
15.2   D.T. Pierce: Surf. Sci. **189/190**, 710 (1987)
15.3   J. Kessler: *Polarized Electrons*, 2nd Ed., Springer Ser. Atom. Plasm. (Springer, Berlin, Heidelberg 1985)

15.4   R. Feder (Ed.): *Polarized Electrons in Surface Physics* (World Scientific, Singapore 1985)
15.5   J. Kirschner: *Polarized Electrons at Surfaces* (Springer, Berlin, Heidelberg 1985)
15.6   See Ref. [15.3] p. 196
15.7   D.T. Pierce, R.J. Celotta: Adv. Electron. Electron Phys. **56**, 219 (1981)
15.8   G.A. Mullhollan, A.R. Köymen, D.M. Lind, F.B. Dunning, G.K. Walters, E. Tamura, R. Feder: Surf. Sci. **204**, 503 (1988)
15.9   R. Feder: J. Phys. C **14**, 2049 (1981)
15.10   F.B. Dunning, G.K. Walters: In *Polarized Electrons in Surface Physics*, ed. by R. Feder (World Scientific, Singapore 1985) Chap. 6
15.11   U. Gradmann, S.F. Alvarado: In *Polarized Electrons in Surface Physics*, ed. by R. Feder (World Scientific, Singapore 1985) Chap. 7
15.12   J. Kirschner: In *Polarized Electrons in Surface Physics*, ed. by R. Feder (World Scientific, Singapore 1985) Chap. 8
15.13   J. Kirschner: Phys. Rev. Lett. **55**, 973 (1985)
15.14   O. Berger, J. Kessler: J. Phys. B **19**, 3539 (1986)
15.15   H. Hopster, D.L. Abraham, D.P. Pappas: J. Appl. Phys. **64**, 5927 (1988)
15.16   H.C. Siegmann, F. Meier, M. Erbudak, M. Landolt: Adv. Electron. Electron Phys. **62**, 1 (1984)
15.17   E. Kisker: In *Polarized Electrons in Surface Physics*, ed. by R. Feder (World Scientific, Singapore 1985) Chap. 12
15.18   See Ref. [15.5], Sect. 2.5, pp. 39–53
15.19   D.T. Pierce, A. Seiler, C.S. Feigerle, J.L. Pena, R.J. Celotta: J. Magn. Magn. Mat. **54–57**, 617 (1986)
15.20   H. Scheidt, M. Glöbl, V. Dose, J. Kirschner: Phys. Rev. Lett. **51**, 1688 (1953)
15.21   See Ref. [15.3] pp. 202–203
15.22   M. Landolt: In *Polarized Electrons in Surface Physics*, ed. by R. Feder (World Scientific, Singapore 1985) Chap. 9
15.23   See Ref. [15.5] pp. 131–132
15.24   M. Landolt, D. Mauri: Phys. Rev. Lett. **49**, 1783 (1982)
15.25   See Ref. [15.3] pp. 207–209. See also Ref. [15.5] pp. 132–134
15.26   J. Unguris, D.T. Pierce, A. Galejs, R.J. Celotta: Phys. Rev. Lett. **49**, 72 (1982)
15.27   J. Unguris, G. Hembree, R.J. Celotta, D.T. Pierce: J. Magn. Magn. Mat. **54–47**, 1629 (1986)
15.28   G. Hembree, J. Unguris, R.J. Celotta, D.T. Pierce: Scanning Microsc. Suppl. **1**, 229 (1987)
15.29   D.T. Pierce, R.J. Celotta, J. Unguris, H.C. Siegmann: Phys. Rev. B **26**, 2566 (1982)
15.30   S.F. Alvarado, M. Campagna, W. Gudat, H. Hopster: J. Appl. Phys. **53**, 7920 (1982)
15.31   D. Weller, S.F. Alvarado, W. Gudat, K. Schröder, M. Campagna: Phys. Rev. Lett. **54**, 1555 (1985)
15.32   J. Kirschner: J. Appl. Phys. **64**, 5915 (1988)
15.33   M. Stampanoni, D. Pescia, G. Zampieri, G.L. Bona, A. Vaterlaus, F. Meier: Surf. Sci. **189/190**, 736 (1987)
15.34   J.L. Robins, R.J. Celotta, J. Unguris, D.T. Pierce, B.T. Jonker, G.A. Prinz: Appl. Phys. Lett. **52**, 1918 (1988)
15.35   U. Gradmann: J. Magn. Magn. Mat. **54–57**, 733 (1986)
15.36   H. Hopster, R. Raue, R. Clauberg: Phys. Rev. Lett. **53**, 695 (1984)
15.37   R. Allenspach, M. Taborelli, M. Landolt, H.C. Siegmann: Phys. Rev. Lett. **56**, 953 (1986)
15.38   K. Koike, H. Matsuyama, K. Hayakawa: Scanning Micros. Suppl. **1**, 241 (1987)
15.39   J. Unguris, D.T. Pierce, R.J. Celotta: Phys. Rev. B **29**, 1381 (1984)
15.40   G.L. Bona, F. Meier, G. Schonhense, M. Aeschlimann, M. Stampanoni, G. Zampiere, H.C. Siegmann: Phys. Rev. B **34**, 7784 (1986)
15.41   R. Allenspach, M. Taborelli, M. Landolt: Phys. Rev. Lett. **55**, 2599 (1985)
15.42   W. Schmitt, K.P. Kämper, G. Güntherodt: Phys. Rev. B **36**, 3763 (1987)
15.43   C.S. Feigerle, A. Seiler, J.L. Pena, R.J. Celotta, D.T. Pierce: Phys. Rev. Lett. **56**, 2209 (1986)
15.44   E. Garwin, D.T. Pierce, H.C. Seigmann: Helv. Phys. Acta **47**, 393 (1974)
15.45   D.T. Pierce, R.J. Celotta, G.-C. Wang, W.N. Unertl, A. Galejs, G.E. Kuyatt, S.R. Mielczarek: Rev. Sci. Instrum. **51**, 478 (1980)
15.46   D.T. Pierce, F. Meier: Phys. Rev. B **13**, 5484 (1976)
15.47   J. Kirschner: In *Polarized Electrons in Surface Physics*, ed. by R. Feder (World Scientific, Singapore 1985) Chap. 5. See also Ref. [15.5] pp. 74–79
15.48   See Ref. [15.5] pp. 74–79

15.49 N.F. Mott: Proc. Royal Soc. (Lond.) A **124**, 425 (1929)
15.50 L.G. Gray, M.W. Hart, F.B. Dunning, G.K. Walters: Rev. Sci. Instrum. **55**, 88 (1984)
15.51 M. Uhrig, A. Beck, J. Goeke, F. Eschen, M. Sohn, G.F. Hanne, K. Jost, J. Kessler: Rev. Sci. Instrum. **60**, 872 (1989)
15.52 See Ref. [15.3] pp. 230–244. See also Ref. [15.4] p. 395
15.53 J. Unguris, D.T. Pierce, R.J. Celotta: Rev. Sci. Instrum. **57**, 1314 (1986)
15.54 D.T. Pierce, J. Unguris, R.J. Celotta: Mater. Res. Soc. Bull. **13**, 19 (1988)
15.55 S. Chikazumi, K. Suzuki: J. Phys. Soc. Japan **10**, 523 (1955)
15.56 M.R. Scheinfein, J. Unguris, R.J. Celotta, D.T. Pierce: Phys. Rev. Lett. **63**, 668 (1989)
15.57 D.T. Pierce, M.R. Scheinfein, J. Unguris, R.J. Celotta: In Mater. Res. Soc. Symp. Proc., eds. B.T. Jonker, J.P. Heremans, E.E. Marinero (Materials Research Society, Pittsburgh 1989) Vol. 151, p. 49

Part III

Processes and Applications

16. Materials Technology

R.St.C. Smart

With 9 Figures

In materials technology, surface analysis is used in three principal modes: *problem solving* in quality control for existing processes and materials; *materials characterization* after surface coating, reaction or modification; and *development* of new materials or processes. To illustrate each of these modes with a few examples, in problem solving, difficulties with contamination, coating chemistry, adherence (e.g. delamination), discolouration and changes in surface reactivity are common. Materials characterization includes the rapidly expanding industry of surface engineering for corrosion and wear resistance, alteration of oxide surfaces for composite (e.g. polymer) compatibility and thin film deposition. The last area encompasses long-term projects in processes as diverse as minerals separation and catalyst design, materials as diverse as slow-release fertilizers and superconductors.

Table 16.1 summarizes the applications, materials and industries in which surface analytical techniques have already become an important part of research and development. It is clearly a daunting list from which exemplification of all materials technology applications would make a book in itself.

In this section, a variety of case studies have been chosen to illustrate some of the more important industrial materials and the type of information available from different techniques. It will be necessary for readers to assess the power of these techniques in relation to processes and materials with which they are familiar by extrapolating from the evidence presented in each of these case studies. An attempt has been made to provide a reasonably general coverage of materials not specifically described in other chapters.

16.1 Metals

Electrowinning is by now so well established in metals extraction as to be regarded as "traditional". Despite its venerable pedigree, one of the mysteries in this process concerns conditions (and their control) for effective detachment of the deposited metal from the so-called "starter sheets" (or cathodes). In this case, stainless steel cathode sheets giving unacceptable adherence of copper sheet (necessitating removal with jackhammers) were examined with XPS, SAM and SIMS. They were compared with sheets showing relatively easy detachment of the copper. Both sets of samples were taken directly from commercial operation without any pretreatment other than distilled-water washing.

Table 16.1. Areas in which surface technology has been extensively applied

Application Areas	Materials	Industries	
– corrosion	– metals	– optical (inc. coatings)	– laser tools, components
– passivation	– ceramics	– vehicle manufacturers	– oil, gas, petroleum
– adhesion	– minerals	– whitegoods	– electricity generation
– contamination	– glasses	– fertilizers, agricultural chemicals	– solar appliances
– minerals processing	– polymers	– minerals, mining	– electronics, microelectronics
– advanced materials development	– paints	– cement, concrete	– scientific instruments
	– composites	– metal goods manfacture	– reprographics
– thin film technology	– wood	– refining, beneficiation	– transport
– joining technology	– paper	– water supply, treatment	– aerospace
– quality assurance	– soils	– agricultural equipment, implements	– medical, biomedical
– plating technology	– semiconductors	– brewing	– household chemicals
– tribology	– superconductors	– wood, paper, composite boards	– bulk chemicals
– metallurgy	– tissue	– coal	– plastics, fibreglass, composites
– medical, biomedical	– blood	– paints	– engineering
– surface coatings	– bone	– lubrication	– dental
– materials failure analysis		– building, building materials	– textiles
– catalysis		– chemical analysis	– forensic
		– biotechnology	– pharmaceuticals
		– photographic	– food
		– waste disposal	– glass, containers
		– gems, jewellery	– soaps, detergents

SAM using secondary electron images showed that both surfaces were roughened on the scale of about 1 μm. No variation in composition across the surfaces was found with this technique although the adherent surface was much more resistive (i.e. had a higher electrical resistance). Most surfaces taken from industrial technology are heavily contaminated with saturated carbon either as graphitic adatoms or as hydrocarbons. These were no exception. *Both* types of surfaces gave initial compositions (in atomic %) of roughly 89% C, 10% O and 1% Cu. No Fe or Cr were detected.

The unacceptable sheets, however, after etching with an Ar$^+$ ion beam (4 kV, 10×8 mm raster), were found to have a carbon layer approaching 10 μm thick on their surface. The depth of etch was verified to ± 1 μm using SAM (and SEM). This layer incorporated copper and iron.

SIMS also revealed impurities at concentrations < 0.1 at.% of Li, Na, Mg, Al, Si, K, Ca, Ti, Cr, F and Cl. After 0.75 μm etching, XPS gave a compositional analysis of 73.9% C; 23.1% O; 2.4% Cu; 2.4% Fe. Scans of elemental regions showed that the metals were present in oxidized form as Cu^{2+} (i.e. with strong shake-up satellites) and broad Fe^{2+}/Fe^{3+} species. The O1s oxygen was very broad (as usual after ion etching) but the high concentration and corrected binding energy near 531.8 eV suggested *both* hydroxide species and oxidized carbon (as in a material like polyethylene oxide). This type of oxidized carbon layer is consistent with the more resistive (i.e. highly charged) nature of the surface. It is not uncommon on metal and oxide surfaces after reaction in solution. The commonly used regenerative treatment of pickling in sulphuric acid solution was shown, using the same techniques, to have no effect on the composition or thickness of this layer.

Cathode sheets with relatively easy detachment characteristics did show incipient development of the same type of carbon layer but etching to 1 μm depth entirely removed it, a thickness consistent with surface roughness. The layer did contain Cu^{2+} but at reduced concentrations (i.e. < 0.6 at.%). No oxidized iron was found in the carbon layer – the underlying surface was consistent with stainless steel (i.e. high Cr^{3+}, low Fe as Fe, Fe^{2+} and Fe^{3+}).

The oxidized carbon layer builds up on the surface during cycling of the sheets in cleaning, pickling and pretreatment. The oxidized metal species come from corrosion of the stainless steel down grain boundaries. This last observation was confirmed separately using metallographic techniques.

This combination of techniques can be used to isolate the process variables introducing the contamination and reaction. Verification of the effectiveness of process changes for their control can be examined in the same way.

16.2 Oxides

Reduced nickel/nickel oxide materials are produced by extractive metallurgy from siliceous nickel ores in several mining operations. After processing to remove silica, silicates and other gangue minerals, the final product is formed by calcination followed by reduction in hydrogen at temperatures in excess of 1000°C. Partial reduction to the metal is achieved during the residence time in reducing condition. The product has highly variable kinetics of dissolution in acid for use in electroplating baths, a major market for this material. Slower dissolution is highly undesirable for this technology. "Poor" material is recognized in process control by slow dissolution (or incomplete dissolution in unit time) and is compensated by longer reduction time in the hydrogen furnace – a costly remedy. The problem is illustrated by the dissolution rate results in Fig. 16.1. "Good" material has high surface area and dissolves in acid (at pH2) at a relatively uniform rate. The "poor" material has lower surface area, much higher nickel metal content (i.e. immediately dissolved proportion), due to the compensating longer reduction time, but slower dissolution beyond about 20% of total dissolution.

XPS comparison of the surface of the "good" and "poor" materials found differences in Ni/Ni^{2+} ratios consistent with the dissolution results but similar compositional analyses (in atomic %) of the first four atomic layers, e.g. C, 15%; Ni, 34–36%; O, 47–46%; and Si, 3.4–3.1%. (The "dust" material had much higher C content at 46%). The Si peak after charge correction (using uncharged C as 284.6 eV), has a binding energy of 103.2 eV corresponding more closely to a silicate than to silica (i.e. $\geq$ 103.8 eV).

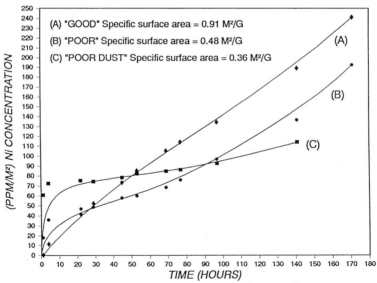

Fig. 16.1. Dissolution kinetics of Ni/NiO product in 10^{-2} mol dm^{-3} nitric acid at 25°C. The dissolution rate (ppm m^{-2}h^{-1}) is given by the gradient of the curves at any particular time. (The "dust" is collected from bag filters in the exhaust gas system)

Fundamental studies of the interaction of NiO with SiO_2 have shown that a surface silicate phase 3–5 atomic layers thick can be formed by vapour phase transport of $SiO:H_2O$ species to the NiO surface at temperatures as low as 550°C [16.1]. The form and thickness of the silicate (as SiO_4^{4-} groups) was confirmed by XPS and infrared spectroscopy. This interaction has the effect of lowering the dissolution rate (per unit surface area) of NiO by factors between 100 and 1000. The dissolution rates of the industrial products in the advanced dissolution phase (i.e. > 50 hours in Fig. 16.1) are similar to those of this passivated (silicated) NiO.

Nevertheless, the surface silicate content of "good" and "poor" samples appears similar from XPS analysis. Analytical TEM (with EDS) revealed the true difference between the products. The "good" sample contained particles which were relatively uniform in size, structure and composition. The "poor" samples had a wide distribution of particle sizes and morphologies but extreme variation of the Si/Ni signal ratios on individual particles from 1.8% to 88%. Particles with high Si/Ni ratios were not found in the "good" sample. Hence, a higher proportion of silica, either as individual particles or as composite particles with the NiO, is found in the "poor" samples leading to more extensive passivation of the oxide and lower dissolution rates.

This combination of techniques, with fundamental and applied research necessary to understand this process problem, is an example of an effective approach to materials technology developed in the last decade.

16.3 Ceramics

In ceramic technology, some of the main surface features requiring definition for control of the properties of the ceramics (see also Chap. 3) are:

- grain size and distribution;
- the location of matrix elements and impurities (e.g. in new phases, lattice substituted in major phases, in intergranular regions);
- the structure and size of intergranular films and pores;
- degradation via leaching/dissolution/recrystallization of the surface.

Properties of concern in optimization of ceramic products include mechanical strength, thermal conductivity, corrosion and wear resistance, all of which depend on the microstructure of the ceramic listed above. Optimization of materials and fabrication conditions to achieve the correct microstructure is the aim of this analysis.

Grain sizes and their distribution can usually be adequately mapped with a combination of SEM/EDS (down to $\sim 1\,\mu m$) and SAM (down to $\sim 0.1\,\mu m$) as illustrated in Chaps. 2 and 3. Comparison of results from each technique applied to the same material is often useful because the different analysis depths, i.e. $\sim 1\,\mu m$ and $\sim 1\,nm$ respectively, can reveal surface segregation of particular elements, reactions of the air-exposed surface, and surface residues from dies or grinding agents. If the grain size is very small, i.e. < 200 nm, it may be necessary to use STEM/EDS to study this distribution. This method has the advantage of allowing structural determination of individual grains by electron diffraction so that compositional inference of phase identification can be confirmed. The disadvantage of STEM/EDS is that it normally requires ion beam etching for thin sections with associated damage to surface layers. Some examples of the use of these techniques applied to a fine-grained ceramic can be found in [16.2, 3].

Locating elements in ceramic microstructures generally relies on the same three techniques, i.e. SEM/EDS, SAM and STEM/EDS, but XPS and SIMS can also be useful in certain cases. New phases can be identified from X-ray diffractometry but, since individual grains can vary widely in composition, STEM/EDS or SAM is needed to locate and analyze a representative set of grains. Substitution of particular elements into grains of the major ceramic matrix phases can also vary considerably. For instance, Cs contents of Ba hollandite (i.e. $Ba_x Cs_y Ti_8 O_{16}$) grains in the Synroc ceramic are found to cover the range $y = 0$ to $y = 1.5$. More extensive examples of determination of lattice substitution and new phase formation can be found in [16.4–6]. Homogeneity in mixing of the precursor materials and any additives is often essential to the performance of the material in practice.

The structure, size and location of elements in integranular films and pores has been discussed and illustrated in Chaps. 1 and 3. The use of XPS and static SIMS in studies of fracture faces of ceramics can be particularly useful in these applications [16.3].

Resistance to surface degradation is now a major area of research and development in ceramics as diverse as Y-Ba-Cu oxide superconductors, perovskites and nuclear-waste disposal materials like Synroc [16.7]. Characterization of loss to solution of particular elements by ion-exchange (leaching) or dissolution (reaction of the matrix) and the formation of protective layers of reprecipitated or recrystallized (in-situ) products form the basis of this research. Depth profiling the reacted surface layer using XPS, AES or SIMS gives important information on these processes but is often difficult to interpret on a multiphase ceramic surface with redeposited crystallites. Figure 16.2 illustrates elemental XPS profiles from such a surface [16.7, 8]. It is clear that Ba, Cs, Mo and Ca are lost from the surface on reaction whilst Ti, Zr and Al are retained. Further study with SEM (Fig. 16.3) revealed, however, that one particular phase, i.e. $CaTiO_3$ perovskite with substituted Sr and rare earth elements, was dissolving preferentially leaving a recrystallized TiO_2 (brookite and anatase) product in the original pervoskite sites. Hence, Ti loss was not significant in the XPS profiles despite the extensive reaction in these sites. Work on the single-phase ceramic perovskite [16.9–1] $CaTiO_3$ has shown that the mechanism for surface degradation is: loss of only 1–2 monolayers of Ca by ion-exchange; base-catalyzed hydrolysis reaction of the perovskite lattice to a depth of ~ 100 nm (ultimately diffusion-limited) releasing the majority of the Ca to solution; and at temperatures above $\sim 90°C$,

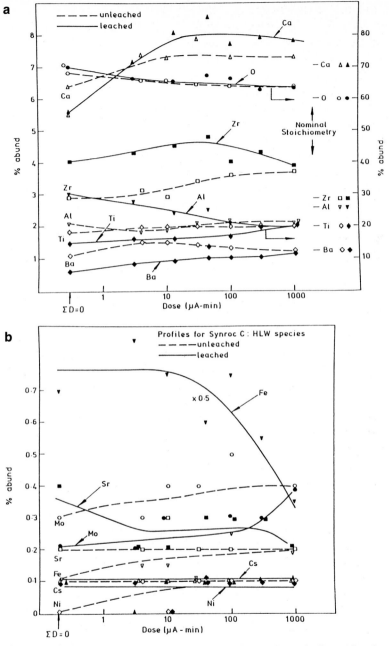

Fig. 16.2). (a) XPS depth profiles from the ceramic Synroc C matrix (i.e. major phase) elements before (*open symbols*) and after (*closed symbols*) hydrothermal attack for 25 days at 350°C. Nominal average bulk stoichiometries are marked as bars on the right with % abundance scales for Ba, Al, Zr, Ca at left and Ti, O at right. $1 \mu A.min$ ion etch is ca. 0.2 nm removal. (b) XPS depth profiles for simulated waste species in the same specimen as Fig. 16.2a recorded under the same conditions. Uncertainties in % abundance are of order ±0.2 at.%. From [16.8]

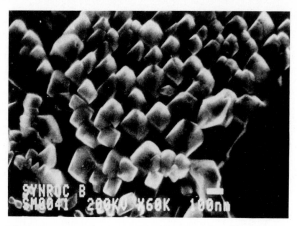

Fig. 16.3. TEM micrograph of TiO_2 (anatase and brookite) crystals formed in-situ from perovkite $(CaTiO_3)$ grains in the surface of a Synroc B ceramic. An ion beam-thinned disc of Synroc B was subjected to hydrothermal attack in water at $190°C$ for 1 day. The original perovskite grain size was ca 1 μm in diameter

recrystallization of the initially-amorphous, reacted layer to crystalline TiO_2. On multiphase ceramic surfaces, a variety of crystalline products can form from solution [e.g. $CaCO_3$, $Ca_3(PO_4)_2$, molybdates etc.] or from recrystallization in situ. Some of the precipitates can be multilayered on the same nucleation site as different products reach saturation in the solution. Mechanisms and their definition using surface analytical techniques are illustrated in [16.7, 12, 13].

16.4 Minerals

16.4.1 Iron Oxides in Mineral Mixtures

Removal of iron oxides is of major concern in many mineral operations with, for instance bauxite, kaolin, quartzite and titanium-based mineral sands. The iron is these systems takes three main form: free particles ($>\sim 1$ μm) of oxides (e.g. hematite, magnetite), hydroxides (usually amorphous) and oxyhydroxides (e.g. goethite); as colloidal ("slimes") particles; and iron substituted into crystal lattices of other minerals (e.g. boehmite, gibbsite, kaolinite). It is usually possible to separate the free particles by flotation. The lattice-substituted iron cannot be separated but is usually < 0.5 wt%, a level acceptable for most processing. The amount and location of the colloidal iron oxides is critical.

For example, after flotation, XPS of a bauxite product (principally gibbsite, boehmite and kaolinite) found a surface composition (atomic %) of: 3.0% Fe; 18.2% Al; 3.6% Si; 57.8% O, 16.1% C. TEM/EDS examination (Fig. 16.4) revealed gibbsite and boehmite particles larger than 200 nm with very small (i.e. 1–10 nm) iron-containing regions dispersed across their surfaces. Colloidal parti-

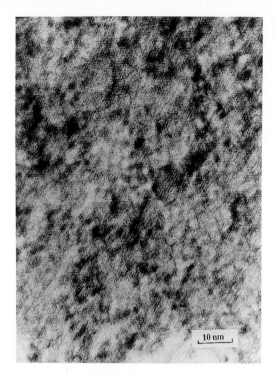

Fig. 16.4. High resolution TEM micrograph of part of a large gibbsite $(Al(OH)_3)$ crystal, from bauxite pisolites, showing lattice fringes (striations). Some fringes are fragmented due to either disorder in the lattice or surface roughness (or both). The surface shows pits and protusions over distances of order 2–3 nm consistent with the high surface area of this material. The dark patches (ca 2 nm diameters) show high Fe/Al ratio consistent with iron oxide particles (as hematite and goethite) attached to the gibbsite surface. From [16.14]

10 nm

cles of hematite and goethite adhere to the surface of the larger platey minerals. They are not removed by ultrasonic agitation but reductive dissolution in dithionite dissolves them (at different rates for hematite and goethite). XPS confirms removal from the surface [16.14].

The efficiency of methods for separation of iron oxides can thus be monitored using XPS with electron microscopy.

16.4.2 Surface Layers on Minerals

In very many cases (perhaps even a majority), the surface of a mineral has neither the same composition nor structure as its bulk. This can be due to reaction with air (e.g. oxidation of sulphides), groundwater (e.g. silicate, carbonate depositions), contamination (e.g. humics, hydrocarbons), intergranular layers (e.g. graphitic) exposed on grinding, and adsorbed species (e.g. sulphates). These alterations to the surface can profoundly influence the behaviour of the mineral in a process.

For instance, a glacial sand deposit to be used for glass-making had unacceptable dewatering and melting properties very different from the dune sand it replaced. XRD and SEM/EDS found only quartz with very little clay and no significant impurities in the quartz after washing. XPS consistently revealed 2.5–5.0 at.% Al, with an aluminosilicate binding energy (not adsorbed Al^{3+}) in the surface. Ion etching suggested that this layer was no more than 200 Å thick. It could not be removed by any practical base-hydrolysis reaction but choice of a surfactant suitable for an aluminosilicate surface reduced dewatering to an acceptable time and adjustment of melting conditions for the presence of the layer allowed acceptable processing [16.15].

16.4.3 Mineral Processing of Sulphide Ores

The complex mixture of minerals and their interactions via dissolved species in most real ores makes the measurement of surface changes difficult to interpret using any technique. Nevertheless, a respectable body of literature has now been developed (e.g. [16.16–22]) in which some useful observations from surface analysis can be found. Studies, particularly using XPS and FTIR, of the effects of conditioning agents (e.g. metabisulphite, sulphide, cyanide) activators (e.g. Cu^+) and collectors (e.g. xanthates, diphosphonic acids) on samples derived from processing plants and from synthetic minerals and ores have elucidated some of the mechanisms of flotation separation.

An example can be seen in the effect of xanthate collector on an ore comprised predominantly of chalcopyrite ($CuFeS_2$), pyrite (FeS_2), quartz, dolomite and iron oxides [16.23]. Table 16.2 shows the composition of the surface at $E_h + 200\,mV$ before (Test 2) and after (Test 4) addition of sodium butyl xanthate ($NaC_4H_9CS_2$) at 0.04 g/tonne. Analysis of the flotation concentrates and tails are compared before and after ion etching to a depth of 25 nm. In the concentrates several changes are evident after xanthate addition:

- the exposure of S (as S^{2-}) is increased roughly six-fold (some of which can be attributed to the adsorbate xanthate);
- exposure of Cu (as Cu^+ only) is markedly increased;
- exposure of Fe is significantly increased and the charge-corrected pyrite Fe ($2P_{3/2}$) signal near 706.5 eV appears whereas only iron hydroxides were found before xanthate addition (Fig. 16.5);

Table 16.2. Quantitative surface analysis of major species (mol%)

Samples	Graphite		CO_3^{2-}		SiO_2		Cu_2S		SO_4^{2-}		FeS_2		$Fe(OH)_3$	
	i[a]	25[b]	i	25	i	25	i	25	i	25	i	25	i	25
Test 2 Conc.	23.1	18.8	15.4	10.7	27.3	33.0	1.0	1.5	0.0	0.0	0.4	2.1	21.6	27.2
Test 2 Tail	6.1	0.8	24.4	15.2	23.1	30.6	0.3	0.9	0.5	1.5	0.0	0.0	21.6	28.4
Test 4 Conc.	24.0	16.4	13.7	7.3	12.3	14.1	12.2[c]	12.2[c]	0.0	0.0	30.6[c]	43.2[c]	4.7	6.8
Test 4 Tail	8.7	2.0	19.9	13.7	30.6	34.8	0.0	0.0	0.0	0.5	0.0	0.0	26.4	32.4

[a] Survey on sample without etch.
[b] Survey on sample with 25 nm etch
[c] Assignment of copper and sulphur is difficult due to presence of Cu xanthate and attenuation of signal

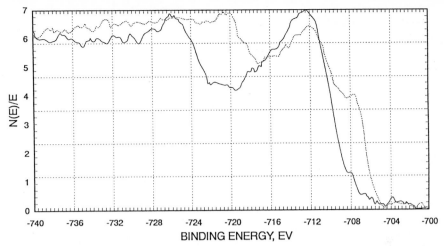

Fig. 16.5. Fe2p ($\sim$ 4 at.% abundance) XPS spectra from two copper ore flotation concentrates from Mt. Isa (Australia) conditioned for 10 min. at E_h − 400 mV with no collector addition (*full line*) and E_h + 200 mV, 0.04 kg/t butyl xanthate collector addition (*dotted line*). The low E_h sample shows only broad iron hydroxide species on the surface. The collector-modified ore reveals new signals near 708 eV and 720 eV from clean chalcopyrite (CaFeS$_2$) and pyrite (FeS$_2$) surfaces

- reduction of the amount of oxygen (as hydroxide) on the surface;
- the presence of carbon (as saturated C and carbonate) on both surfaces.

The tails show little evidence of S (and only as SO_4^{2-} species) or Cu but iron (as hydroxides only) is found in both cases. Much more carbonate is found in the tails samples. The xanthate appears to have a cleaning action partly removing surface hydroxides and exposing the underlying mineral surfaces. Other studies, at different E_h values in solution (i.e. −400, −200, O, +400 mV) show that E_h+200 mV is the most favourable condition for this effect of the xanthate. More detailed examination of the elemental regions in the XPS spectra can identify, for instance, the xanthate contributions, to the C, O and S regions, S^{2-}/SO_4^{2-} ratios, and different forms of Fe.

SAM examination of individual mineral particles in the concentrate after xanthate addition suggests a strong association of Cu with S on some surfaces with Fe and O associated on other surfaces. This result supports the view that the chalcopyrite surfaces are selectively cleaned by xanthate. Results of this kind are still at an exploratory stage but the surface analytical methods show major potential in understanding the chemistry of mineral processing.

16.4.4 Adsorption on Soil Minerals

The adsorption/desorption reactions of anions on soil minerals is fundamental to plant nutrition and the mobility of environmental pollutants. Surface concentrations, chemical states and the structures of adsorbed ions can be determined with a combination of IR and XPS studies (e.g. [16.24, 25]). Goethite (α-FeOOH),

Table 16.3. Atomic concentrations of surface atoms on goethite after adsorption of the anions shown at 5×10^{-2} M concentration, pH3 and pH12

Atomic concentrations [atom %], pH3

Anion[†]	C	O	Fe	X[†]	X/Fe
PO_4^{3-}	19.7	59.9	18.2	2.2	0.12 $\pm$0.008
SeO_3^{2-}	15.8	59.8	21.7	2.6	0.12 $\pm$0.008
SO_4^{2-}	13.1	64.4	21.0	1.5	0.071 $\pm$0.008

Atomic concentrations [atom %], pH12

Anion[†]	C	O	Fe	X[†]	X/Fe
PO_4^{3-}	7.9	68.4	21.5	2.2	0.10 $\pm$0.008
SeO_3^{2-}	14.8	61.3	23.0	0.90	0.039 $\pm$0.008
SO_4^{2-}	16.2	60.4	22.8	0.57	0.025 $\pm$0.008

[†] X = central atom of anion, i.e. P, Se and S respectively

Ratio of atomic % of the central atom X of the adsorbed anion to that of Fe and O at different pH of adsorption on goethite

Anion concentration = 5×10^{-2} M

pH	X	Ratio X/Fe/O
3	PO_4^{2-}	1:8:27
3	SeO_3^{2-}	1:8.3:23
3	SO_4^{2-}	1:14:43
12	PO_4^{2-}	1:9.8:31
12	SeO_3^{2-}	1:24.7:69
12	SO_4^{2-}	1:109.5:310

* For pure geothite Fe/O = 1:2.5

in particular, as the major iron oxide soil mineral, is responsible for retention of phosphate on soils. IR shows that it is bonded as a binuclear HPO_4^{2-} complex, displacing two surface OH groups, across a [001]-directed surface trench. XPS (Table 16.3) shows that it is more strongly coordinated than sulphate (also binuclear) or selenite anions across the pH range from 3 to 12.

Surface analytical methods can monitor these surface species in soil environments in a relatively routine way.

16.5 Polymers

Surface analysis is now routinely used in polymer applications such as: identification of surface functional groups for optimal bonding to other coatings or substrates; monitoring lubricants, accelerators, fillers and other additives excluded to the surface layers during polymerization; contaminations leading to delamination, corrosion or wear. The examples described here will illustrate the type of information available.

16.5.1 Surface Reactions of Polymers Under UV, Plasma and Corona Discharges: Chemical Modification and Reaction Layer Thickness Using Angle Resolved XPS

The surfaces of polymer materials are modified by a variety of chemical reactions. Degradation, i.e. oxidation and crystallization, due to UV radiation is well known. Treatments by plasma discharges and corona discharges are also deliberately used to alter surface properties like wettability, adherence of surface coatings, texture and surface hardness. Information on the chemical changes produced by these processes and the depths of the affected layers is required for control of desirable (e.g. hydrolyzed surface) and undesirable (e.g. corrosion and crystallization) reactions.

A particular difficulty with polymers is that most common carbon-based polymer materials are decomposed by electron and ion beams giving a graphitic residue where the beam has hit the sample. This rapidly obscures the underlying polymer and hence, information is lost. Photon beams, as in XPS and FTIR, are much less damaging; in fact, most polymers are unaffected by these techniques. However, depth profiling (using an ion beam) is not normally available. Angle resolved XPS can be used to profile changes of composition with depth down to $\sim 100\,\text{Å}$, as explained in Chaps. 1 and 7 but, for composition variation beyond this depth, the use of a cryogenic microtome to cut thin sections ($\sim 0.1\,\mu\text{m}$) from the surface is required.

Chapter 7 discusses the relative escape depths of photoelectrons at take-off angles between $10°$ and $90°$. In polymers, with a longer mean free path of electrons, this can vary from $\sim 15\,\text{Å}$ ($10°$) through $\sim 60\,\text{Å}$ ($45°$) to $\sim 90\,\text{Å}$ ($90°$) depending on the polymer and the X-ray source. Figure 16.6 illustrates the different XPS spectra measured at $90°$ and $10°$ angles on a plasma-treated polystyrene sample. The O signal is clearly stronger in the top surface atomic layer. Table 16.4 gives the relative atomic concentrations of C and O, for this and other plasma treatments, illustrating that the top four atomic layers of the polymer are heavily oxidized by this treatment. Figure 16.6 also illustrates changes in the C 1s region where a range of oxidized species are found as shown in the $10°$ angle resolved spectrum. Table 16.4 lists the concentrations of each species, resolved by curve fitting. The H_2O plasma is seen to be more effective at oxidizing the surface than the mixed $H_2O + H_2$ plasma. The oxidized surface is wettable and has strongly altered compatibility with other materials.

This kind of depth information is often sufficient to monitor surface reactions, corrosion, exclusion etc.

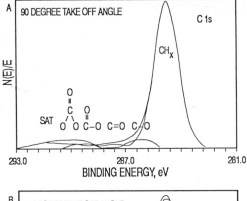

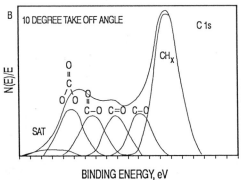

Fig. 16.6. Comparison of high (90°) and low (10°) take-off angle, high resolution C 1s XPS spectra from a linear polystyrene surface treated in a water plasma (14.1 W, 5 min). The much-increased oxidized products are confined to the surface layers, i.e. 10°, ~ 1 nm escape depth, 90°, ~ 9 nm depth. Contributions from the oxidized functional groups have been resolved with curve fitting software. From [16.30]

Table 16.4. Atomic concentrations and relative functional group concentrations from ARXPS of plasma-treated polystyrene

Sample	Exit Angle	%C	%O	CH_x	C-O	C=O	$\overset{\overset{O}{\parallel}}{C\text{-}O}$	$\overset{\overset{O}{\parallel}}{O\text{-}C\text{-}O}$
As received	90°	100.0	0.0	100.0	–	–	–	–
H_2O Plasma	90°	86.0	14.0	87.9	4.8	2.4	2.4	2.5
H_2O Plasma	10°	64.5	35.5	45.2	12.5	13.0	13.6	15.7
H_2O Plasma + H_2 Plasma	90°	93.0	7.0	94.6	5.4	–	–	–
H_2O Plasma + H_2 Plasma	10°	83.8	16.2	82.6	12.1	2.6	1.8	0.9

16.6 Glasses

Surface analysis has been used extensively, particularly XPS and FTIR, to characterize glass surfaces before and after weathering or aqueous attack (e.g. [16.26–28]). Like ceramics, attack by water at glass surfaces proceeds by four processes: ion exchange in the first few atomic layers; the formation of an amorphous, so-called siliceous layer from base-catalyzed hydrolysis disrupting the network

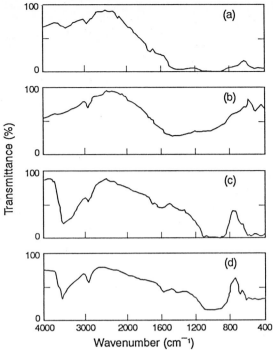

Fig. 16.7a–d. Infrared spectra from the surface of a base glass (a) initially; (b) after evacuation ($\sim 10^{-5}$ Torr) at 170°C for 1 h; (c) after hydrothermal reaction at 200°C for 18 h, drying at 200°C for 1 h in air and cooling in a vacuum dessicator; (d) sample (c) evacuated at 170°C for 1 h. Note the changes in the hydrogen-bonded $Si(OH)_x$ vibrations in the broad absorptions near 3500–3300 cm^{-1}. The sharp shoulder near 3700 cm^{-1} arises from "free" (not H-bonded) OH groups. The broad absorption in (a) and (d) near 1640 cm^{-1} is the bending vibration of trapped H_2O molecules in the reacted layer. From [16.7]

bonding; surface segregation of particular elements into the hydrogen-bonded hydroxylated siliceous layer; and recrystallization in-situ or precipitation from solution [16.7]. These reactions are important in control of glass technology such as surface powdering, mould transfer, opacity, surface hardening and corrosion.

An example of this attack on a zinc-containing glass is presented here to illustrate the type of information leading to elucidation of these mechanisms. A simulated nuclear waste borosilicate glass containing 20.9 wt.% Zn (or 11.7 at.%) with 14 other cations, including Na, Mg, Ca, Fe and Ni, was reacted at 170–200°C in water. Figure 16.7 illustrates IR spectra, from both in-situ (ATR) and detached (transmission) samples of the altered hydrolyzed layer. The initial ν(OH) absorption on the unattacked surface near 3500 cm^{-1} is entirely removed by evacuation at 170°C. After aqueous reaction at 170°C for 60 h, strong, broad, hydrogen-bonded ν(OH) (3200–3500 cm^{-1}) and $\gamma(H_2O)$ (near 1600 cm^{-1}) is found. Evacuation removed some of the $\gamma(H_2O)$ absorption but not the ν(SiO) region around 1000 cm^{-1} [16.29] indicating possible crystalline products within the siliceous layer.

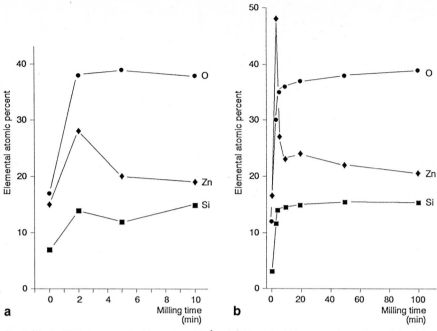

Fig. 16.8a, b. XPS depth profiles ($\sim 1\,nm\,min^{-1}$) of O, Zn and Si in the surface of a polished disc of a zinc-containing base glass (**a**) before and (**b**) after hydrothermal attack in water at 200°C for 1 h, dried 200°C, 1 h. Note change of depth scale from (**a**) to (**b**). From [16.27]

Figure 16.8 shows XPS profiles for the same glass before and after a relatively short duration of aqueous attack at 200°C for 1 hour. Note that the Si and O signals are relatively unaltered in the siliceous layer but there is a strong accumulation of Zn in the first 10 nm of the surface. SIMS profiles of longer duration attack have shown Zn accumulation in the surface to depths greater than 100 nm. Other elements, notably Mg, Fe and Ni, are also found segregated into the siliceous layer in this and other glasses after aqueous attack whilst the alkali metal cations (e.g. Na^+, K^+, Rb^+) are almost always lost from the reacted layer.

SEM studies of the surface after extensive reaction show a relatively uniform (conducting) surface layer with small crystallites embedded in it as illustrated in Fig. 16.9. TEM with electron diffraction, however, established that more than 90% of the layer is amorphous. The crystalline forms were indexed to the mineral zinc hemimorphite $Zn_4(OH)_2Si_2O_7.H_2O$. In other zinc-containing glasses, different crystal forms: e.g. Zn_2SiO_4 have been identified. No crystalline forms containing the segregated elemenmts of Fe, Mg or Ni were found so that it appears that they are at least partly preferentially bound in the siliceous layer instead of dissolving in the solution.

Even more complex behaviour of different elements can be found in other glass systems. This combination of techniques, however, can give a full description of the changes in composition and chemistry of the surface layers.

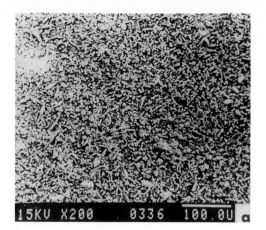

Fig. 16.9. SEM micrographs of the surface of a polished disc of a zinc-containing glass after hydrothermal attack in water at 170°C for 60 h, dried 170°C, 1 h. The white bar in the micrograph is 100 μm

References

16.1   W.R. Pease, R.L. Segall, R.St.C. Smart, P.S. Turner: Evidence for modification of nickel oxide by silica. J. Chem. Soc. Faraday Trans. I **76**, 1510 (1980)

16.2   R.L. Segall, R.St.C. Smart, P.S. Turner, T.J. White: Microstructural characterisation of Synroc C and E by electron microscopy. J. Amer. Ceram. Soc. **68**, 64 (1985)

16.3   J.A. Cooper, D.R. Cousens, J. Hanna, R.A. Lewis, S. Myhra, R.L. Segall, R.St.C. Smart, P.S. Turner, T.J. White: Intergranular films and pore surfaces in Synroc C: structure, composition and dissolution characteristics. J. Amer. Ceram. Soc. **69**, 347 (1986)

16.4   T.J. White, R.L. Segall, P.S. Turner: Radwaste immobilisation by structural modification – the crystallochemical properties of Synroc, a titanate ceramic. Angew. Chemie (Int. Ed. Engl.) **24**, 357 (1985)

16.5   P.E. Fielding, T.J. White: Crystal chemical incorporation of high-level waste species in aluminotitanate-based ceramics: Valence, location, radiation damage and hydrothermal durability. J. Mater. Res. **2**, 387 (1987)

16.6   W.J. Buykx, K. Hawkins, D.M. Levins, H. Mitamura, R.St.C. Smart, G.J. Stevens, K.G. Watson, D. Weedon, T.J. White: Titanate ceramics for the immobilization of sodium-bearing high-level nuclear waste. J. Am. Ceram. Soc. **71**, 678 (1988)

16.7   S. Myhra, R.St.C. Smart, P.S. Turner: The surfaces of titanate minerals, ceramics and silicate glasses: surface analytical and electron microscope studies. Scanning Microsc. **2**, 715 (1988); see also S. Myhra, D.K. Pham, R.St.C. Smart, P.S. Turner: "Dissolution of Ceramic Surfaces" in *The Science of Ceramic Interfaces*", ed. by J. Nowotny (Elsevier, Amsterdam 1991)

16.8   S. Myhra, A. Atkinson, J.C. Riviere, D. Savage: A surface analytical study of Synroc subjected to hydrothermal attack. J. Amer. Ceram. Soc. **67**, 223 (1984)

16.9   T. Kastrissios, M. Stephenson, P.S. Turner, T.J. White: Hydrothermal dissolution of perovskite: implications for Synroc formulation, J. Amer. Ceram. Soc. **70**, C 144 (1987)

16.10 D.K. Pham, S. Myhra, F.B. Neall, R.St.C. Smart, P.S. Turner: The chemical durability of perovskite $CaTiO_3$: surface analytical approach. J. Amer. Ceram. Soc., to be published

16.11 D.K. Pham, S. Myhra, F.B. Neall, R.St.C. Smart, P.S. Turner: The chemical durability of perovskite $CaTi_3$: investigation by scanning and transmission electron microscopy. J. Amer. Ceram. Soc., to be published

16.12 S. Myhra, D. Savage, A. Atkinson, J.C. Riviere: Surface modification of some titanate minerals subjected to hydrothermal attack. Amer. Mineral. **69**, 902 (1984)

16.13 P.E.D. Morgan, D.R. Clarke, C.M. Jantzen, A.B. Harker: High-alumina tailored nuclear waste ceramics. J. Amer. Ceram. Soc. **64**, 249 (1981)

16.14 K. Kuys, J. Ralston, R.St.C. Smart, S. Sobieraj, R. Wood, P.S. Turner: Surface characterisation, iron removal and enrichment of bauxite ultrafines, Min. Eng. **3**, 421 (1990)

16.15 R.St.C. Smart, B.G. Baker, P.S. Turner: Aspects of surface science in Australian industry. Aust. J. Chem. **43**, 241 (1990)

16.16 A.N. Buckley, I.C. Hamilton, R. Woods: 'Investigation of the surface oxidation of sulphide minerals by linear potential sweep voltammetry and X-ray photoelectron spectroscopy', in *Flotation of Sulphide Minerals*, ed. by K.S. Forssberg (Elsevier, Amsterdam 1985) pp. 41–60

16.17 A.N. Buckley, R. Woods: The surface oxidation of pyrite. Appl. Surf. Sci. **27**, 437 (1987)

16.18 L. Ranta, E. Minni, E. Suoninen, S. Heimala, V. Hintakka, M. Saari, J. Rastas: XPS Studies of adsorption of xanthate on sulphide surfaces. Appl. Surf. Sci. **7**, 393 (1985)

16.19 J.W. Strojek, J. Mielczarski: Spectroscopic investigations of the solid-liquid interface by the ATR technique. Adv. Colloid Interf. Sci. **19**, 309 (1983)

16.20 S.C. Termes, P.E. Richardson: Application of FTIR spectroscopy for in situ studies of sphalerite with aqueous solutions of potassium ethylxanthate and with diethyldixanthogen. Int. J. Min. Proc. **18**, 167 (1986)

16.21 E.W. Giesekke: A review of spectroscopic techniques applied to the study of interactions between minerals and reagents in flotation systems. Int. J. Min. Proc. **11**, 19 (1983)

16.22 J.G. Dillard, M.H. Koppelman: XPS in the study of mineral materials. Am. Geophys. Union Trans. **56**, 12 (1975)

16.23 S. Grano, J. Ralston, R.St.C. Smart: Influence of electrochemical environment in the flotation behaviour of sulphide ores. Int. J. Min. Proc. **30**, 69 (1990)

16.24 R.J. Atkinson, R.L. Parfitt, R.St.C. Smart: Infrared study of phosphate adsorption on goethite. J. Chem. Soc., Faraday I **70**, 1472 (1974)

16.25 R.R. Martin, R.St.C. Smart: X-ray photoelectron of anion adsorption on goethite. Soil Sci. Soc. Amer. J. **51**, 1 (1987)

16.26 D.E. Clark, L.L. Hench: An overview of the physical characterisation of leached glass surfaces. Nucl. Chem. Waste Manage. **2**, 93 (1981)

16.27 R.A. Lewis, S. Myhra, R.L. Segall, R.St.C. Smart, P.S. Turner: The surface layer formed on zinc-containing glass during aqueous attack. J. Non-Cryst. Solids **53**, 299 (1982)

16.28 N.S. McIntyre, G.C. Strathdee, D.F. Phillips: SIMS studies of the aqueous leaching of a borosilicate glass. Surf. Sci. **100**, 71 (1980)

16.29 V.C. Farmer: The Infrared Spectra of Minerals. Mineralog. Soc. London (1974) pp. 285–303

16.30 J.H. Evans, J.K. Gibson, J.F. Moulder, J.S. Hammond: PHI Interface (Perkin Elmer Phys. Elec. Dir., MN, USA) **7**, 1 (1984)

17. Characterization of Catalysts by Surface Analysis

B.G. Baker

With 14 Figures

Solid catalysts are the basis of many important industrial processes. Reaction of gases or liquids to form particular products occurs at specific sites on the catalyst surface. The structure and composition of the catalyst surface is critical in determining the reactivity and selectivity of a catalyst. The techniques of surface analysis provide the means of characterizing a catalyst in terms of the actual composition and structure of the surface rather than by its bulk properties. The objective of such studies is to provide a scientific basis for improving catalyst formulations and understanding the processes of activation and deactivation which the catalyst undergoes. Supported catalysts, the type most widely used in industry, consist of an active component dispersed on the internal surface of a porous inorganic oxide. High area solids and light loadings very highly dispersed are often employed to maximize the catalytic activity of expensive components. Metals and oxides may be formed on a support by decomposing or reducing a salt which has been introduced by solution impregnation. The preparation pretreated and actual catalytic reaction conditions may result in reaction between the components including the support. It is the purpose of the surface analysis to reveal these processes.

Specific objectives of the surface analysis of catalysts are:

1) to determine the surface composition of the catalyst in reactive form;
2) to identify the valence state of elements present at the surface of the active catalyst;
3) to monitor the extent of interaction between the catalyst components and the support and possibly explain the effects of varying the support;
4) to determine the effects on surface composition of various catalyst preparation and pretreatment procedures.

The most useful techniques are X-ray photoelectron spectroscopy (XPS), Auger electron spectroscopy (AES) and ion scattering spectroscopy (ISS). Others include SIMS (Chap. 5), FTIR (Chap. 8), RBS and NRA (Chap. 9), LEIS (Chap. 11), LEED (Chap. 13) and UPS (Chap. 14). The application of these surface techniques to the study of catalysts poses a number of experimental problems:

i) The active components of the catalyst are generally present in low concentration and are at the surface of a high area porous solid. Most of this surface is internal and therefore not directly accessible to photon, ion or electron beams.

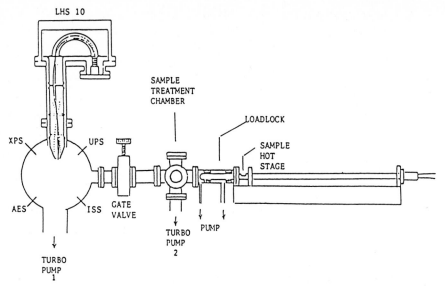

Fig. 17.1. Electron spectrometer (Leybold) with sample treatment chamber

ii) The catalyst support is generally an insulating solid which tends to charge electrostatically during the analysis. This has the effect of displacing peaks and makes the determination of chemical shift effects more difficult.

iii) The active form of the catalyst persists only in a controlled environment and will not stand exposure to the atmosphere.

To overcome these problems the spectrometer for catalyst studies should have a sample preparation chamber which provides for heating, reactant gas exposure and vacuum transfer to the analysis chamber. The method of mounting the sample needs to be compatible with these processes and particular attention must be given to the effects of electrostatic charging in order to quantify chemical shift effects.

A spectrometer, modified for catalyst studies is shown in Fig. 17.1. Based on the Leybold LHS10 electron spectrometer, this system has a rod with a heated stage [17.1]. The sample may be heated in the treatment chamber in either a gas flow or static atmosphere to create conditions comparable to practical service of the catalyst. High vacuum is then achieved before opening the gate valve and transferring to the spectrometer chamber. The sample stage can be heated during analysis if required. Other systems provide for transfer of a sample holder and allow the treatment chamber to be isolated during analysis. Systems combining a microreactor, with facilities for kinetic studies, with an electron spectrometer have also been employed [17.2]. The objective in each case is to examine catalysts in their reactive form. The mounting of powder or granular samples of catalysts for analysis must allow for good thermal and electrical contact with the holder and ensure mechanical stability so that sample is not spilled during evacuation. An effective method is to press a pellet of sample in a small stainless steel cup.

The top of the pellet can be cleaved off to expose fresh surface which has not contacted the plunger of the press. The cup is then mounted on the stage of the spectrometer providing good electrical and thermal contact. In XPS, an insulating sample tends to acquire a positive charge which decreases the observed kinetic energy of the photoelectrons. This shift is in the same direction as a chemical shift where the element has a more positive valence state. Catalysts frequently show both effects and in addition often exhibit differential charging, i.e., some parts of the sample charge while others are uncharged. This can be of use in identifying particular elements as being in combination in a separate phase. Charging is detected by comparison of observed energies with values tabulated for reference peaks in the spectrum. A low energy electron "flood gun" may also be used to directly neutralize charge at a surface. This is particularly useful in cases of differential charging. The case studies that follow contain examples of many of these techniques.

17.1 Examples of Catalytic Systems

17.1.1 Alumina

A variety of forms of alumina are used as catalyst supports. The valence state of aluminium is invariant at +3 and the oxide is insulating. Analysis by XPS provides a good measure of the extent of charging. This analysis also reveals surface impurities which are likely to be incorporated into a catalyst prepared on this support. Gamma alumina prepared by the Bayer process has an area of $\sim 70 \, m^2 \, g^{-1}$ and typically contains about 0.2 percent sodium as impurity. This sodium is readily detected by XPS showing that it must be concentrated at the alumina surface [17.3]. The spectra in Fig. 17.2a show $Al(2p)$ and $Na(1s)$ on separate scales. The binding energy for $Al(2p)$ at 77.6 eV indicates charging of ~ 3.4 eV, typical of Al in alumina. The binding energy for $Na(1s)$ at 1072.3 eV equals the reference value for uncharged sodium. This extraordinary example of differential charging is explained by structural identification of β-alumina ($Na_2O.11Al_2O_3$) by electron diffraction. This exists as a surface phase in this type of γ-alumina. It is an electrolytic conductor and thus no local charging of sodium occurs. The $Al(2p)$ spectrum is dominated by the more abundant bulk Al_2O_3 and so is charged. Heat treatment of this alumina causes a phase change to produce α-alumina and decompose the β-alumina. Spectra after heating, Fig. 17.2b, show both $Al(2p)$ and $Na(1s)$ charged by ~ 3.4 eV. The sodium in the heat-treated (H.T.) alumina is now as islands of soluble sodium oxide on the reconstructed alumina surface.

17.1.2 Tungsten Oxide Catalysts

Catalysts containing tungsten oxide in a partially reduced state have been shown to have activity for the skeletal isomerization of alkenes [17.4]. Conditioning and

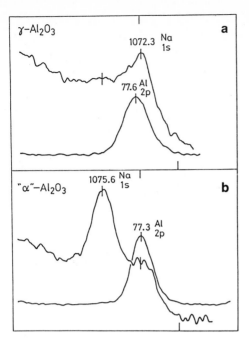

Fig. 17.2. X-ray photoelectron spectra for sodium (1s) and aluminium (2p) in (a) γ-Alumina (Merck) and (b) heat treated (1200°C, 30 min) alumina. The apparent binding energies are displaced ~ 3.4 eV except for the sodium peak in the initial γ-alumina. Sodium in this material is present in a β-alumina surface phase which conducts. Heat treatment destroys the β-alumina leaving sodium compounds on the surface

operation of the catalyst requires a controlled reducing atmosphere achieved by hydrogen and water vapour. The chemistry of tungsten in this catalyst has been studied by following the 4f peaks in XPS [17.1]. Unsupported WO_3 is semiconducting and gives no charging problems. The binding energy of the 4$f_{7/2}$ peak is 35.2 eV compared with 30.2 eV for tungsten metal. The chemical shift of 5.0 eV results from the change from zero to +6 valence state. Reduction of WO_3 in wet hydrogen results in a shift of the 4f peaks to lower binding energies (Fig. 17.3). Intermediate stages of reduction show mixed spectra indicating sample heterogeneity but a number of intermediate oxides of tungsten $W_{20}O_{58}$, $W_{18}O_{49}$ are known and the blue colour of these oxides is observed. The reduction by this method is limited to WO_2.

The supported form of the catalyst consists of 6 percent loading of WO_3 on alumina. The final state of the preparation involves heating in air to decompose tungstic acid. This preparation step has also been followed by successive XPS analyses (Fig. 17.4). Prolonged heating at 400°C results in a shift of the 4f peaks to even higher binding energy (Fig. 17.4d). This chemical shift is not due to a change of valence state but to the change of tungsten from the octahedral sites of WO_3 to the tetrahedral sites of aluminium tungstate $Al_2(WO_4)_3$. The tungsten oxide which has reacted with the support is not readily reduced. Spectra in Fig. 17.5 show that reduction in dry hydrogen at 400°C reduces little of the sample from Fig. 17.4d whereas the sample (Fig. 17.4a) can be reduced to metallic tungsten. Aluminium tungstate is not a catalyst for the reaction and catalysts overheated in preparation had poor activity.

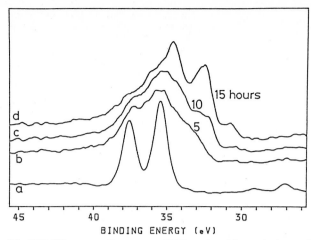

Fig. 17.3. XP spectra of unsupported WO₃ reduced in wet hydrogen (pH₂/pH₂O=40) at 405°C. (*a*) WO₃ before reduction: (*b*) after 5 h reduction: (*c*) after 10 h reduction: (*d*) after 15 h reduction

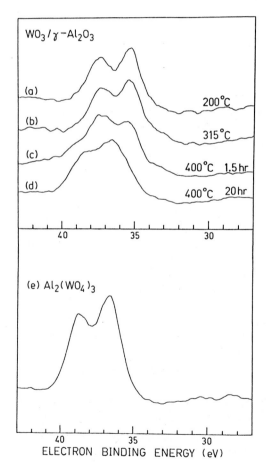

Fig. 17.4. XP spectra of a sample of 6% WO₃/γ-Al₂O₃ heated in air. (*a*) Heating to 200°C; (*b*) heating to 315°C: (*c*) heating at 400°C for 1.5 h: (*d*) heating at 400°C for 20 h: (*e*) XP spectrum of Al₂(WO₄)₃

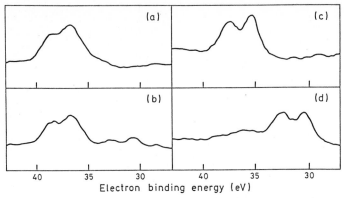

Fig. 17.5. XP spectra (**a**) of sample from Fig. 17.4d after exposure to wet hydrogen at 400°C. (**b**) Subsequent exposure to dry hydrogen at 400°C (15% reduction). (**c**) of sample from Fig. 17.4a heated in air to only 200°C. (**d**) Subsequent exposure to dry hydrogen at 400°C (75% reduction)

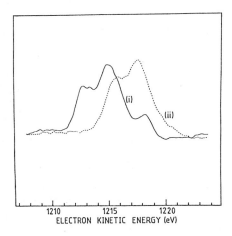

Fig. 17.6. A sample of unreduced WO$_3$/HT-Al$_2$O$_3$ (air, 450°C) serves as an example of how charging by X-rays can complicate a spectrum. (*i*) Spectrum recorded without using the flood gun. (*ii*) Flood gun used to discharge sample: 2.7 eV charge (see text)

Some charging effects were observed in spectra from supported tungsten catalysts. In Fig. 17.6, the spectrum (*i*) from unreduced WO$_3$ on alumina shows three peaks. Operation of an electron flood gun shows that only the two peaks at lower kinetic energies are from charged regions of the sample. The recognition of multiple charging effects is obviously important before interpreting spectra in terms of elemental valence states. However, spectra recorded with the flood gun operating are broadened and, as the setting of the flood gun is arbitrary, peak positions are indefinite.

Ion scattering spectroscopy may also be applied to the study of catalysts. The technique requires the analysis of the energies of positive ions and so requires the analyzer to operate with potentials the reverse of those used for electron spectroscopy. The incident beam from an ion gun is typically 1 keV helium ions. The loss of energy of the ion in a single binary elastic collision depends on the mass of the surface atom. The spectrum recorded as E/E_0 where E is the measured kinetic energy of the scattered ion and E_0 the incident energy.

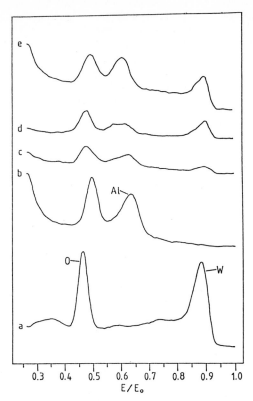

Fig. 17.7. He$^+$(E_0 = 1 kV) ion scattering spectra of (*a*) bulk WO$_3$; (*b*) HT-Al$_2$O$_3$: (*c*) 6% WO$_3$/HT-Al$_2$O$_3$, untreated; (*d*) 6% WO$_3$/HT-Al$_2$O$_3$, heated in air at 400°C for 20 h: (*e*) bulk Al$_2$(WO$_4$)$_3$

Electrostatic charging of insulators is a problem in applying the technique to catalysts and there are difficulties in quantifying the data [17.5]. The technique is however more surface sensitive than XPS and is of particular use in defining the location of a component on a catalyst support.

Analysis of the tungsten oxide catalyst system is shown in Fig. 17.7. Oxygen, aluminium and tungsten are identified. The relative intensities of tungsten and aluminium in the prepared catalyst indicates that tungsten oxide is well dispersed on the support effectively covering the surface.

17.1.3 Palladium on Magnesia

The following experiment to test the reactivity of palladium metal on magnesium oxide shows the nature of the metal-support interaction for a catalyst based on these materials. A thin film of palladium was evaporated on to a single crystal magnesium oxide surface. The sample was heated in vacuum at 400°C and then argon ion milled to the metal-oxide interface. Analyzes by XPS (Mg $K\alpha$) are shown in Fig. 17.8. Oxygen 1s peaks from PdO and MgO differ in binding energy by 1.8 eV for chemical reasons. The observed spectra in Fig. 17.8a differ by 4.5 eV due to electrostatic charging of the MgO only. The complicated palladium 3d spectrum in Fig. 17.8b is interpreted as a superposition of three sets

343

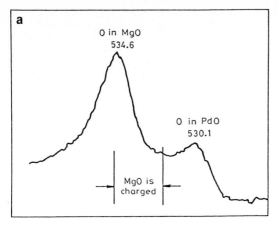

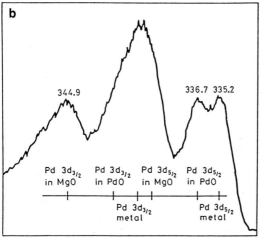

Fig. 17.8a, b. X-ray photoelectron spectra of a palladium film deposited on single crystal magnesium oxide after heating and after argon ion milling to the metal-oxide interface. (a) Oxygen in MgO distinguished from oxygen in PdO by chemical shift and by electrostatic charging. (b) Palladium $3d_{3/2}$ and $3d_{5/2}$ peaks showing chemical shift due to oxide formation and identifying Pd in the charged MgO

of the $3d_{3/2}$ $3d_{5/2}$ doublet; the metal, palladium oxide which is conducting and palladium in the insulating magnesium oxide.

17.1.4 Cobalt on Kieselguhr Catalysts

This type of catalyst is used for the Fischer-Tropsch synthesis of hydrocarbons from carbon monoxide and hydrogen. Under these reducing conditions, cobalt may be reduced to the metallic state. However, promotors such as MgO and ThO_2 are present and may interact with cobalt. The purpose of a surface analysis by XPS was to determine the state of cobalt under various reducing conditions and to define the role of the promotors [17.6]. The unreduced catalyst containing both promotors was found to be electrically conducting and exhibited no charging. The XPS analyses (by Al, $K\alpha$) are shown in Fig. 17.9. All of Mg, Th, Si and Co are detected and the oxygen $1s$ is region shows a complex structure due to the overlapping of the various oxide components. The binding energy and shape of

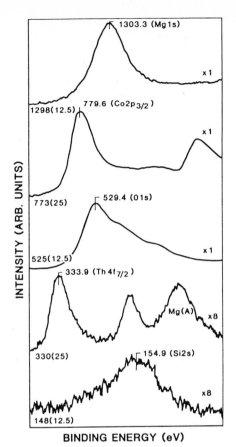

Fig. 17.9. Representative XPS spectra of a Co-ThO$_2$-MgO-Kieselguhr catalyst (Al $K\alpha$) in the unreduced (calcined) state. Data are shown for Mg 1s, Co 2$p_{3/2}$, O 1s, Si 2s, and Th 4$f_{7/2}$ regions. This catalyst was electrically conducting and exhibited no charging

the Co 2$p_{3/2}$ peak identify the presence of Co$_3$O$_4$. The largest component of the O 1s peak is also identified with Co$_3$O$_4$.

After reduction of the catalyst, the Mg 1s and Th 4f lines indicated no reduction belong Mg^{2+} and Th^{4+} whereas the cobalt and oxygen spectra were markedly changed. Four catalysts, reduced at 400°C for 1 hour gave cobalt spectra as in Fig. 17.10. Difference spectra are plotted to reveal that the amount of unreduced cobalt is large in catalysts containing MgO, but very small when MgO is absent. The oxygen spectra showed overlapping peaks and they required a curve fitting procedure to resolve the contributions. An example is shown in Fig. 17.11 for the unpromoted catalyst. The reduced form contains only small contribution of CoO to the O 1s peak, consistent with the observation that cobalt is reduced to metal in this catalyst. The corresponding result for MgO-promoted catalysts shows CoO present after reduction. It is concluded that there is a strong interaction of MgO with cobalt and that CoO possibly forms a solid solution which is resistant to reduction.

345

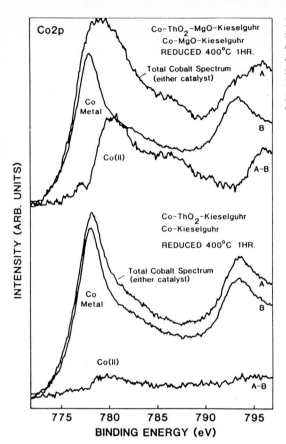

Fig. 17.10. Difference Co $2p_{3/2}$ spectra for magnesia-promoted and other cobalt catalysts after reduction at 400°C for 1 h in H_2. The cobalt metal lineshape (*B*) was deconvoluted from the total spectrum (*A*) to give the residual Co(II) unreduced component (*A*–*B*)

17.1.5 Iron Catalysts

Catalysts containing iron are used in the synthesis of ammonia and for the Fischer-Tropsch synthesis of hydrocarbons from carbon monoxide and hydrogen. Under these reducing conditions it might be expected that metallic iron is the active component. However, structureal stabilizers or supports and a variety of promotors are present and the possibilities of reaction with iron are many. The control of the chain growth process in the Fischer-Tropsch synthesis and selectivity to the formation of olefins can be achieved by a lightly loaded supported iron catalyst [17.7]. In this type of catalyst, promotors and support are present in excess so that the selective reaction may well depend on compounds of iron rather than on the metal. Such a catalyst is, however, very difficult to study by XPS because the active components are highly dispersed and of very low concentration. The alternative is to study model systems designed to test the feasibility of reaction which may occur in the conditioned catalyst.

In the first type of experiment, a catalyst was prepared by precipitating iron, aluminium and praseodymium from solution in the atomic ratio 0.7: 0.2: 0.2. After drying and calcining, a sample was reduced and tested and found to have

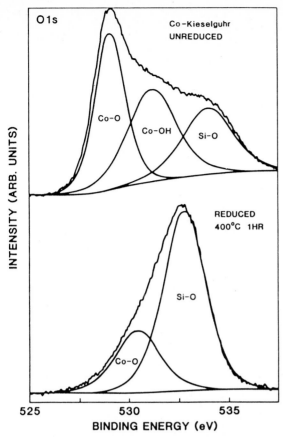

Fig. 17.11. Oxygen $1s$ spectra of unreduced and reduced Co-KG catalyst at $400°$C including curve-fitting analysis (three-peak fit). Both the Co-O and Co-OH components are considerably attenuated after reduction, due to conversion of cobalt to the metallic phase. All peaks are normalized to the peak maximum so the change in Si-O intensity is only apparent. For most of the catalysts the Si-O intensity is a measure of the exposed KG surface and is approximately constant

modest Fischer-Tropsch activity. This material is an unsupported version of the catalyst formed by supporting iron on alumina with a rare earth promoter. The difference, apart from the lower surface area and activity, is that alumina is present as a minor component. Analysis by XPS should then reveal the behaviour of iron without the problems of the dominant insulating support. As a preliminary experiment, a sample of Fe_2O_3 was tested by reducing in hydrogen at $320°$C in the sample preparation chamber of the electron spectrometer. Analysis of the Fe $2p$ peaks showed that complete reduction to zero-valent iron occurs. The precipitated catalyst was then analyzed at various stages of reduction under the same conditions. The spectra in Fig. 17.12 show effects due to both electrostatic charging and chemical shifts. Initially Al $2p$ is charged while Fe $2p$ and Pr $3d$ are not charged. Oxygen $1s$ shows evidence of two states. Evidently at this stage separate phases exist. As the reduction proceeds the multiple charging effect on

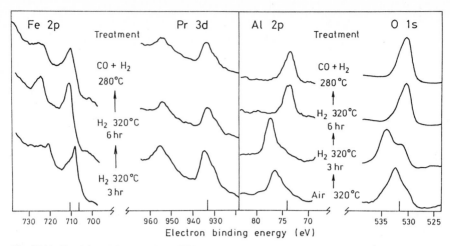

Fig. 17.12. Precipitated iron catalyst. XPS analyses at various stages of reduction. The reference marks on the energy scales indicate the peak position for an uncharged sample. In the final state of reduction the sample is uncharged but iron is not in the metallic state (see text)

the Al $2p$ and O $1s$ spectra at first increases, then disappears. The final result is that both Al and O are in a single uncharged state. The Pr $3d$ signal shows no charging. The Fe $2p$, by position and shape, is present in the +2, +3 valence states. This condition persists after treatment under reaction conditions with CO and H_2. It is concluded that iron has reacted under the reducing conditions to form a conducting compound with Al and O. Reaction could also have involved Pr but since the oxide of this element is also conducting, there is no direct evidence. The possible products of reaction are aluminates within the composition range Fe_3O_4 to $FeAl_2O_4$. Surface phases of this composition would be expected to have considerable conductivity.

A second approach to the investigation of the reactivity of oxide supports towards iron is to directly introduce the metal to the external oxide surface by vacuum evaporation. Experiments of this kind have been conducted on alumina, silica, titania and praseodyminium oxide [17.9]. The objective is to observe any reaction of the metal with the oxide support under the conditions for catalyst service. Very light loading of metal is necessary to ensure that excess metal does not obscure the analysis of the interface. The XPS analyses for iron on alumina are shown in Fig. 17.13. Trace (*b*) after deposition of the iron film shows two charge states; Fe $2p_{3/2}$ at 706.7 eV and 709.6 eV. The former, probably represents a conducting, continuous iron film and the latter represents islands or clusters of discontinuous Fe on the alumina. After reaction with H_2 at 320°C for 3 hours, the spectra are markedly changed [trace (*c*)]. The binding energies of Fe($2p_{3/2}$) XPS spectra can then be attributed to Fe(II) and Fe(III) at 709.5 eV and 711.0 eV with charging ~ 4 eV. The iron has reacted with alumina under conditions known to reduce iron oxide to metal. This result is confirmed by analysis by electron-induced Auger electron spectroscopy. Auger spectra, recorded before and after

348

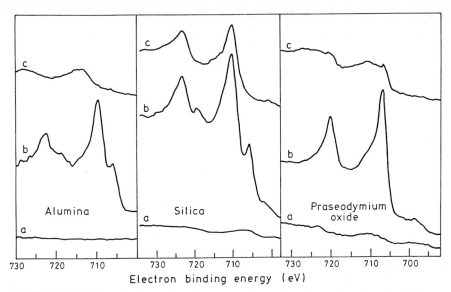

Fig. 17.13. XPS Fe $2p$ peaks from an iron film on an oxide support. Traces (*b*) as deposited; (*c*) after heating in hydrogen at 320°C

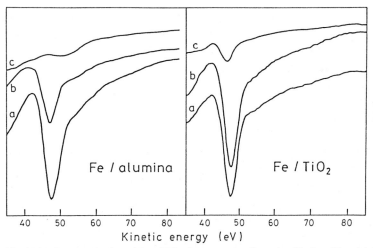

Fig. 17.14. Iron on alumina and on TiO_2 Auger spectra from the thin iron film. (*a*) before, and (*b*) after taking the XP spectra, and (*c*) after heating in H_2 at 320°C

the XPS spectra, (Fig. 17.14a, b) show a strong Fe peak at 47.5 eV corresponding to Fe in the metallic state. After the treatment in hydrogen the Auger spectrum shows complete loss of the 47.5 eV Fe(O) peak and appearance of weak, broad oxide peaks at 51.0 eV and 37.5 eV. This is characteristic of oxidized iron [17.10].

A similar experiment for iron on titania showed that iron remains metallic on TiO_2 after the treatment in hydrogen. The XPS Fe $2p$ peaks by position and shape indicate zero-valent iron and the Auger analyses (Fig. 17.14) show the peak

349

characteristic of metallic iron is present after the hydrogen treatment. The peak has moved to 46.5 eV due to charging and is of decreased intensity but there is no evidence of oxide peaks at 51.0 and 37.0 eV. It is concluded that titania is unreactive towards iron at 320°C in hydrogen and that it may be classed as an inert support in contrast to alumina which was found to be reactive.

An iron on silica experiment was conducted under the same conditions. The XPS analyses (Fig. 17.13) show metallic iron, charged at 710.8 eV and uncharged at 706.3 eV (trace b). After heating in H_2 at 320°C the sharp peak at 710.8 eV is consistent with zero-valent iron but there is a broad oxide emission near 714 eV. This suggests limited reaction of iron with silica has occurred. Auger analysis of the reduced sample was difficult due to electrostatic charging but a weak broad peak at 56 eV and a peak near 48 eV are consistent with the conclusion that part of the iron is reacted.

Reference was made previously to the effect of praseodymium oxide as an additive to iron Fischer-Tropsch catalysts. To test the reactivity of this oxide towards iron, a sample of the oxide (Pr_6O_{11}) was reacted with an iron film under reducing conditions. The XPS results in Fig. 17.13 are free of the effects of electrostatic charging: Pr_6O_{11} is a semiconductor. The iron $2p$ peaks are well defined for the metal film (trace b). After heating in hydrogen at 320°C most of the iron is in a higher valence state. Extensive reaction of iron has occurred. Possibly the compound formed is $PrFeO_3$ however the amount present as a surface phase is too small for structural identification.

This series of results, after the reduction of the deposited iron layers in H_2 at 320°C, can be summarized as follows:

- Fe on Al_2O_3 reacts totally to form a new surface phase in which the Fe is in oxidized form.
- Fe on TiO_2 does not react and remains as separate Fe metal particles in islands on the TiO_2 surface;
- Fe on SiO_2 is partly oxidized suggesting that the role of this oxide as a catalyst promoter is in the formation of a compound with iron.

The extreme differences in reactivity of the metal with different oxide substrates correlates with the behaviour of the supported metal catalysts where alumina is the preferred substrate for selective Fischer-Tropsch activity. Silica is less effective and titania is not useful as a substrate.

17.2 Conclusion

The above examples show that surface techniques, particularly XPS, can reveal much of the complex chemistry of a catalyst surface. The reactions which occur during the conditioning or activation of the catalyst involve only materials contacting at a surface. The surface phases which form are not readily characterized by bulk techniques nor do they necessarily follow the thermodynamics of bulk preparations. Direct analysis of the surface is the only way to reliably characterize the catalyst surface.

References

17.1 P.J. Chappell, M.H. Kibel, B.G. Baker: J. Catalysis **10**, 139 (1988)

17.2 D.J. Dwyer: In *Catalyst Characterization Science*, ed. by M.L. Deviney, J.L. Gland (American Chemical Society 1985) pp. 124–132

17.3 B.G. Baker, N.J. Clark: New Fischer-Tropsch Catalysts. Report NERDDP EG85/470 (Australian Govt. Printing Service 1985)

17.4 B.G. Baker, N.J. Clark, H. McArthur, E. Summerville: Catalysts. U.S. patent 4610975, New Zealand patent 216337, Australian patent application No. 73419/87

17.5 J.C. Carver, S.M. Davis, D.A. Goetsch: In *Catalyst Characterization Science*, ed. by M.L. Deviney, J.L. Gland (American Chemical Society 1985) pp. 133–143

17.6 B.A. Sexton, A.E. Hughes, T.W. Turney: J. Catalysis **97**, 390 (1986)

17.7 B.G. Baker, N.J. Clark, H. McArthur, E. Summerville: U.S. patents 4610975, 4666880, 4767792; New Zealand patent 207355; Australian patent 565954

17.8 B.G. Baker, N.J. Clark, M.N. Tkaczuk: Proc. Aust. X-ray Analytical Association (AXAA-88)393–396 (1988)

17.9 R.St.C. Smart, P.S. Arora, B.G. Baker: Proc. Aust. X-ray Analytical Association (AXAA-88) 219–228 (1988)

17.10 C. Klauber: Ph.D. Thesis, Flinders University (1984) p. 7.42

18. Applications to Devices and Device Materials

J.M. Dell and G.L. Price

With 11 Figures

As the complexity and density of VLSI and LSI circuits has increased, the requirement for a better understanding of device and material parameters, and improved process control have become paramount. Nearly all semiconductor devices are fabricated in the first 1 or 2 μm of the semiconductor surface and some devices such as MOSFETs can be classified as truly semiconductor surface devices. With the recent advent of materials growth technologies such as molecular beam epitaxy (Chap. 12), new semiconductors can be made with internal structures which have atomic layer dimensions. These structures can be positioned within a few atomic layers of the surface. It is not surprising then that many of the surface analysis techniques described in this book have become standard tools for the semiconductor device and material technologist.

The areas of application of surface analysis techniques in device and semiconductor materials technology can be divided into three categories:

- device and circuit failure analysis
- process technology development
- material parameter evaluation.

The problems are of such complexity in the device area that one or two technologies are rarely sufficient. In this chapter we select one topic from each of the three categories and show how a number of techniques can be used together. The treatment of each technique is necessarily brief (there is no space to cover the application of even one technique to the device area in detail) but emphasis is on the complementary nature of the methods. The first of the three categories, device failure analysis, is the most routine application of surface analysis techniques, mainly being concerned with location of defect structures and identification of material contaminants. An outline of the important factors influencing the selection of analysis technique in this area will be given. A brief coverage of the non standard applications of SEM is also given. The other two areas are somewhat different to the first in that the surface analysis techniques are used to obtain some measure of a device or material parameter, or provide an insight into some aspect of device operation. In the second section, GaAs ohmic contact development will be used to describe the role of surface analysis techniques in process technology development. For the third category of material parameter evaluation, the application of surface analytical techniques to the measurement of semiconductor heterostructures and quantum wells will be discussed. Measurement of such basic material parameters is essential to device design and process control.

Some techniques which are not normally thought of as surface oriented and are used to determine material properties will be described.

18.1 Device and Circuit Failure Analysis

Device reliability is an important criteria for component selection in complex systems with some components being more prone to failure than others [18.1–3]. For example, semiconductor laser diodes used in optical communication systems are often individually tested for many hours to ensure that the final packaged device is reliable. With many VLSI and LSI devices, such exhaustive testing is uneconomic and a more limited testing schedule must be used in conjunction with a knowledge of the general reliability performance of a particular process technology. These are combined to develop a set of technology design rules and select materials used in the process. When device failures do occur, as evidenced by a reduced yield in production testing, or worse, as failures in the field, it is essential to be able to quickly determine the location, the type of failure and the mechanism of failure. The process technology can then be changed or materials quality improved to counteract the problem.

An important criteria for analysis techniques used is that, at least initially, they are nondestructive in nature since often one analysis technique is insufficient to determine the failure mechanism or contaminant. There are particular difficulties if decapsulation of the device is required, especially if it is a plastic package. Modern plastic packages are quite resistant to solvents and special techniques are used to achieve decapsulation without affecting the device itself. Detailed surveys of the methods used in device failure physics have been given in the literature [18.1, 4, 5]. Table 18.1 is a summary and ranking of these techniques [18.1, 5]. It is clear from this table that many of the surface analysis techniques described in this book have application in the device failure area. One technique, SEM, however stands out as particularly useful and has some applications peculiar to the semiconductor device field.

SEM is one of the most powerful techniques for device failure analysis [18.6–8]. Simple direct observation can reveal small faults such as dendrite formation in metallizations, cracks in passivation layers, electromigration, pinhole defects, voids and short circuits. In addition, electron beam induced X-ray emission can be analyzed in either a wavelength dispersive or energy dispersive mode to give information on composition of various materials. The SEM accelerating voltages used in device failure analysis are often quite low and voltages down to 4.5 kV are used to help prevent damage to the circuit under test.

An important application of SEM is as an electrical test instrument for both simple dc analysis and time response measurements. When a bias is applied to the integrated circuit under test, those areas which are negatively biased appear brighter than those which are positively biased. The technique is called voltage contrast SEM or SEM-VC [18.9–12]. It allows quick location of faults and is

Table 18.1. Summary of methods and ranking of device failure analysis techniques (after [18.1, 5])

Order of Investigation	Procedure
(1)	Visual/optical inspection
(2)	Radiography
(3)	Simple electrical tests (Electrical probing)
(4)	Package tests, the hermicity testing
(5)	Decapsulation
(6)	First-rank analytical techniques (mainly imaging) Optical/interferometric microscopy SEM SEM-EDAX; SEM-VC; SEM-CL; SEM-EBIC Metallography-sectioning Radiography
(7)	Second-rank analytical techniques (imaging, compositional and constitutional techniques) Auger electron spectroscopy (AES) X-ray photoelectron spectroscopy (XPS or ESCA) Scanning acoustic microscopy (SAM) X-ray diffraction Gas analysis – moisture content Trace analysis – chemistry
(8)	Third-rank analytical techniques Secondary ion mass spectroscopy (SIMS) Rutherford back-scattering (RBS) Thermal imaging (IR microscopy) Liquid crystal display/location Laser induced mass analysis (LIMA)
(9)	Comparison of failure mechanisms with literature/library
(10)	Simulation tests

widely used in analysis of large memory circuits. By using stroboscopic viewing techniques, in a similar manner to a sampling oscilloscope, quite high clock rates can be used allowing measurement of gate propagation delays at internal nodes of the circuit [18.6]. Figure 18.1 shows a typical voltage contrast image of an integrated circuit.

SEM-VC may also be used to overcome one of the emerging problems of digital integrated circuit design that of testability. As the complexity of an integrated circuit is increased, it becomes increasingly difficult to specify a set of device inputs which can be used to verify correct operation of the device. The inputs must be such that a failure at some internal device is propagated through the circuit and appears as an error in the output. If an exhaustive set of tests is used, testing time becomes uneconomic. To overcome this, SEM-VC offers the ability to directly observe each part of the circuit rather than the effect of a failure at the output.

Fig. 18.1. An example of a voltage contrast SEM display. The lighter areas represent regions of most negative bias. (Courtesy of T.P. Rogers, Telecom Australia Research Laboratories)

18.2 The Development of High Quality Ohmic Contacts to GaAs

The most basic of all semiconductor device building blocks is the ohmic contact which is required to provide current injection or apply potentials to the active regions of virtually all semiconductor devices. For many materials, such as the more ionic semiconductors (ZnS, CdS etc.), ohmic contact formation is relatively easily achieved due to the strong dependence of the surface (Schottky) energy barrier height on the difference between the metal and semiconductor work functions as predicted by standard Schottky barrier theory [18.13]. By choosing a metal which gives Schottky barrier height of the order of kT at the temperature which the device is to be operated, thermionic emission across the barrier allows significant current flow without application of large biases. However, for the more technologically significant semiconductors, such as Si and the III-V compound semiconductors (GaAs, AlGaAs, InP, InGaAsP, etc.), the Schottky barrier height is very nearly independent of the difference between the metal and semiconductor work functions. This is due to large densities of surface states which appear in the semiconductor band gap at the surface and effectively pin the Fermi level there. For semiconductors such as InAs and GaSb, the pinning is such that any common metal will form an ohmic contact to the n type semiconductor and it is virtually impossible to form a Schottky (rectifying) diode. For the wider band gap semiconductors such as GaAs and AlGaAs, without special treatment, all metal semiconductor contacts show Schottky diode behavior.

In the case of these wider band gap materials, it is the purpose of the ohmic contact metallization system and processing to modify in some way the potential barrier or provide an alternative conduction mechanism. This may be via doping the layer immediately below the contact sufficiently that the barrier is very narrow and tunnelling may occur or introduction of a damaged layer beneath the contact which will introduce a large number of states in the bandgap of the material. Conduction is then via tunnelling between these states. The actual mechanism by which the metallization and processing effects the barrier and conduction mechanism is very complex to determine. In addition to making an ohmic contact with low specific contact resistivity[1], the material system must also satisfy several other requirements including

- stability under thermal and electrical stress
- ease of deposition
- the ability to be bonded to
- reproduciblity
- compatibility with other device processing stages.

As a result, ohmic contact metallizations for materials such as GaAs are quite complex, containing several components with particular care taken in processing to obtain reproducible results. The processing usually involves some form of high temperature alloying or sintering which allows some interaction between the metallization components and the semiconductor. As device integration densities become larger, the requirement of improved contact performance has become paramount and a concerted effort has been made to better understand the contacting mechanisms, the effects of processing and the behavior of ohmic contacts under various operating conditions. In this, surface analysis techniques have made a major contribution. The various techniques which have been applied to surface characterization studies of ohmic contact are summarized in Table 18.2 [18.14]. Each technique has advantages and disadvantages. While SEM-EDAX and SEM-WDL allow good lateral resolution, the depth resolution is poor. Alternatively, RBS, with a typical probe size[2] of $1\,mm^2$ has poor lateral resolution but allows depth profiling without recourse to sputtering techniques, such as SIMS and scanning AES, with the associated difficulty of ensuring that a uniform etch rate is maintained and knock on effects are minimized. Ensuring that the etch rate is uniform is a particular difficulty in the case of alloyed ohmic contacts which often show large lateral variations in composition over distances of the order

---

[1] Specific contact resistivity, ϱ_c, is a measure of the electrical performance of the contact and is defined as $\varrho_c = (\partial J/\partial V)^{-1}_{V=0}$ where V is the voltage drop across the contact and J is the current density through the contact.

[2] This can be reduced to approximately $11\,\mu m$ using a focused ion beam.

Table 18.2. Surface analysis methods used for structural characterization ohmic contacts (modified from [18.14])

Method	Information
X-ray and electron diffraction	(i) Identification of film structures before and after alloying (ii) Identification of the regrowth layer structure after alloying and removal of metal films (iii) Check crystallographic perfection of surface by LEED (iv) Identification of intermetallic compounds within alloyed contacts (v) Identification of metallic and intermetallic phases by RHEED
Optical microscopy	(i) Examination of macroscopic and structural defects (ii) Examination of cross-sectional view of metal-semiconductor interface
Scanning electron microscopy	(i) Surface morphology before and after alloying (ii) Observation of regrowth layers after alloying and removal of metal films (iii) Examination of ohmic contact-semiconductor interface by exposing cleaved and beveled surfaces
Transmission electron microscopy	(i) Changes in microstructures at interface caused by sintering or alloying (ii) Identification of metallics and intermetallics on cross-sections through the contacts before and after alloying
Electron probe microanalysis	(i) Elemental analysis and lateral distribution (ii) Elemental profiles in the vicinity of interface by exposing cleaved or beveled surfaces (iii) Elemental analysis of thinned cross-sections in combination with TEM
Auger electron spectroscopy	(i) Determining presence of foreign elements on surface (ii) Chemical composition of upper 10–20 Å layer of surface (iii) AES with controlled sputter etching for depth-composition profiles of contact system
He^+ ion-beam backscattering analysis	(i) Low energy (1–2 keV) backscattering useful for surface analysis (e.g., trace impurities, adsorption, gettering) (ii) Low energy backscattering in conjunction with sputter etching for depth profiles (iii) MeV ion backscattering in conjuction with angle lapped samples for deep profiling (iv) MeV ion backscattering for mass sensitive depth microscopy (v) Lateral variation of contact interdiffusion (vi) Depth extent of various intermetallic phases
Proton induced X-ray analysis	(i) Composition of outer most layers of sample (ii) Detection of trace amounts of elements
Secondary ion mass spectrometry	(i) Distribution profiles of various constituents in the vicinity of the metal–semiconductor interface (ii) Compositional characteristics of monolayers near the surface (iii) Investigate slow diffusion in semiconductors (iv) Investigation of elemental distribution after alloying by use of specially thinned samples and sputtering from the back (preventing the "knock on" effect)
X-ray photoelectron spectroscopy	(i) Surface oxidation states and effect of surface preparation prior to deposition of contact metallization
Internal friction technique	(i) Investigate alloying temperatures of metal-semiconductor system
Raman spectroscopy	(i) Investigate oxide formation on semiconductor surfaces

of a micron. The ability of SIMS to detect trace amounts of elements however is particularly useful in determining the concentration profile of slow diffusing components in an ohmic contact system.

The most common ohmic contact to n type III-Vs has been based on a eutectic mixture of Au and Ge [18.15–17,20], often in combination with Ni which is sometimes found to improve contact performance. The Au-Ge eutectic is evaporated (or Au and Ge evaporated separately but in eutectic proportions) followed by a thin Ni layer and usually, more Au. The metallization is then alloyed in a reducing, or at least, inert environment, to actually form the contact. A clear understanding of how this contact system actually modifies the surface potential barrier so as to allow contact formation is still lacking. Indeed, the role of each of the active elements in the contact, particularly when Ni is involved, is only partially understood. What understanding has been achieved is based almost exclusively on surface analysis techniques. The work up to 1979 has been summarized by *Sharma* [18.14].

A combination of techniques which has recently been particularly useful in the structural analysis of ohmic contacts has been TEM in conjunction with EDX and electron or X-ray diffraction studies. *Kuan* et al. [18.18] were the first to investigate the Au-Ge-Ni ohmic contact system using TEM-EDX on alloyed contact cross-sections. The use of EDX on cross-sectional samples thinned for TEM, in combination with the small probe size of TEM, offers several advantages. The typical electron probe size of ~ 0.5 nm is much less than the grain sizes found in alloyed contacts. In addition, the probe spreading in the thinned samples is very small. These factors combine to offer a quantitative analysis technique with very high spacial resolution. Electron diffraction studies, carried out on the same samples by *Kuan* et al., give the crystal structure and associated lattice constants of the various phases formed during the alloying process. The combination of composition and lattice constant/crystal structure information can then be used to identify particular compounds found in the alloyed contact. A TEM bright-field photomicrograph of an alloyed ohmic contact is shown in Fig. 18.2. Included in this figure is a schematic of the image indicating the phases found as deduced from the EDX spectra. Figure 18.3 shows a typical EDX X-ray spectrum from the Au(Ga,As) phase (actually β-AuGa doped with some As). These figures clearly show the extremely high resolution of the TEM-EDX analysis technique. The major limitation of the technique is that only an extremely small area can be observed on any one sample and the technique must be used in combination with information obtained from other surface analysis techniques if the overall contact morphology is to be deduced.

While it is clear that a combination of many surface analysis techniques, in addition to electrical characterization, is required to obtain some understanding of a device fabrication process, care must be exercized in the choice of analytical technique and interpretation of the results obtained from analysis. This has been particularly true of SIMS analysis of the Au-Ge-Ni ohmic contact system. Early SIMS results indicated that considerable spreading of the alloy constituents occurred after alloying which in turn lead to confusion as to the role of Ni in

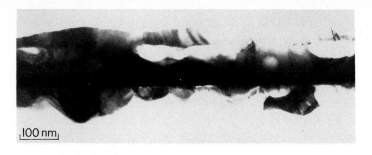

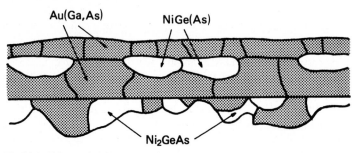

Fig. 18.2. Bright field TEM image of an alloyed Au-Ge-Ni ohmic contact in cross-section. The phases and their chemical compositions are indicated in the diagram below the image (after [18.18])

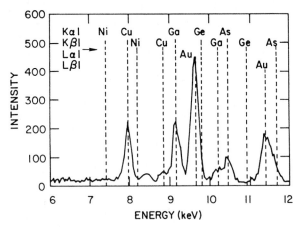

Fig. 18.3. The TEM-energy dispersive X-ray spectra from a gold-rich grain of an alloyed Au-Ge-Ni ohmic contact cross-section (after [18.18])

the contact. However, recent SIMS examination of specially thinned samples where analysis proceeded from the back surface shows no such spreading of the constituents [18.19]. Clearly ion mixing (knock-on) effects are significant in this case.

While the current understanding of the contacting mechanism for the Au-Ge (-Ni) ohmic contact system is not complete, the importance of surface analytical techniques in reaching this stage cannot be overestimated. The areas of application of surface analysis in process development for semiconductor device manufacture is increasing dramatically as the process technology becomes more and more complex [18.21]. It is only the application of these techniques which can lead to a better understanding of the process physics required to obtain improved yield, higher packing densities and integration of novel device structures.

18.3 Semiconductor Interfaces: The Quantum Well

In this section five techniques will be described which together can characterize a fundamental element of modern device technology: the quantum well. The structure is shown in Fig. 18.4a. It consists of a layer of narrow gap semiconductor interposed between two wide gap semiconductors. The whole structure is a single crystal and it is usually grown by one of the modern epitaxial technologies such as molecular beam epitaxy. By choosing the right widths, the conduction and valence band wells can contain one or more levels, e2, e1, hh1 lh1 etc. where hh and lh refer to heavy and light holes respectively. The wells have a great number of applications [18.22]. One which is extremely important in optoelectronics is as the active medium for semiconductor lasers. Quantum well lasers are found to give a more intense output and have a lower threshold current than normal lasers made of bulk material. Part of the reason for this is that the electron-hole pairs e1-h1, e2-h2 etc. are excitons which exist in the lattice in a structure akin to a hydrogen atom; their confinement in the well raises their oscillator strength and enhances their recombination rate. These lasers can also be adjusted to tune

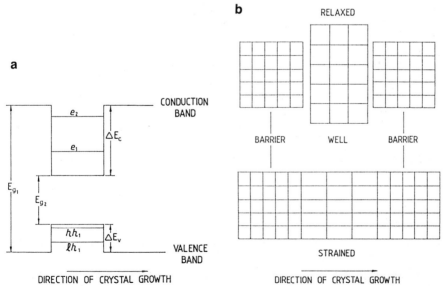

Fig. 18.4. (a) Schematic of the electronic structure of a quantum well defining the terms used in the text. (b) The lattice of a strained quantum well

the wavelength of the light. Light is emitted when electrons and holes combine and the wavelength is determined by the energy gap between them. This gap can be changed by altering the width of the well.

Before being used in a device, the well must be characterized. The bandgaps E_{g1} and E_{g2}, the conduction and valence band offsets ΔE_c and ΔE_v, the positions of the levels and the widths and the crystalline structure of the wells must be known. The small widths of the wells (down to one atomic layer) and their proximity to the surface require either the techniques of surface science or bulk methods refined to approach a surface analysis sensitivity.

The crystalline structure is especially important when dealing with the most general kind of well, the strained layer quantum well. In this case, the well layer is made of a material with a different lattice constant from the barriers. If the well is kept under a certain critical thickness, its structure deforms elastically as shown in Fig. 18.4b. This is an example of where the lattice constant of the well is larger than the barriers, and the lattice is put under compressive strain to match the barriers but elongates normal to the well according to its Poisson's ratio. The whole structure remains a single crystal, essentially defect free. The strain alters the band structure and consequently the device parameters such as the laser wavelength and carrier effective masses and mobility.

This is a complex system, the properties are not generally calculable, and there is only a very small quantity of material. Some of the techniques which will be described are sensitive enough to measure one quantum well. Others require a period array or superlattice of quantum wells which multiplies the amount of material and, in the case of a diffraction technique, provides a grating to enhance the signal from the wells.

18.3.1 Double Crystal X-Ray Diffraction

This is a technique usually associated with bulk materials and thick films but it has been refined to characterize layers as thin as 5 Å.

The geometry is shown in Fig. 18.5 [18.23]. X-rays are first diffracted by a crystal and then are diffracted again by the sample and are detected. To understand

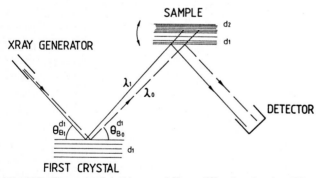

Fig. 18.5. Geometry of double crystal X-ray diffraction showing different wavelengths contributing to the measurement

the purpose of the first crystal, consider a normal diffraction geometry when this crystal is absent and the X-rays from the generator are simply diffracted by the sample into a detector. If the sample consists of two materials, as is the case for a strained quantum well, of differing lattice constants, then the crystal can be rotated between the two Bragg conditions

$$2d_1 \sin \theta_1 = \lambda \tag{18.1}$$

$$2d_2 \sin \theta_2 = \lambda \tag{18.2}$$

and the measurement of $\theta_1 - \theta_2$ would determine $d_1 - d_2$; the lattice constant and strain of the well material could be deduced. However, we are interested in $\Delta d/d \sim 10^{-5}$. Both the spread of wavelengths $\Delta\lambda$ and the divergence $\Delta\theta$ are so great that they prevent such a fine measurement. Filtering and stopping down the beam could improve this, but we require the highest intensity possible because we are dealing with such a small quantity of material.

The purpose of the first crystal is to solve this problem by letting through only those X-rays which will contribute to the diffraction. For InGaAs quantum wells in GaAs, this crystal is chosen to be GaAs. The X-rays leaving the GaAs can have a wide $\Delta\lambda$ spread, but each ray of wavelength λ has a specific θ. As the sample is rotated (or rocked), lattices which satisfy the Bragg condition diffract into the detector.

Figure 18.6 demonstrates the sensitivity of the method. The experimental curve is compared with simulations demonstrating the fit and the sensitivity which can be achieved. For a given superlattice, this technique gives the width, the alloy concentration and the period to within an atomic layer precision.

18.3.2 Reflection High Energy Electron Diffraction

Reflection high energy electron diffraction gives structural information on a quantum well as it is being grown. The distance between the RHEED streaks is inversely proportional to the surface lattice constant. If this distance is recorded during growth, often on video with subsequent computer analysis, the thickness of the well and the state of the strain can be determined.

Figure 18.7 shows the RHEED spacing variation through the growth of a $In_{0.4}GA_{0.6}As$ multiquantum wells [18.24]. They consist of five periods of 24 Å and 60 Å InGaAs wells and 239 Å GaAs barriers. The InGaAs initially keeps close to the GaAs lattice constant. It then passes through a critical thickness at about 5 atomic layers where dislocations accommodate the strain between the two lattices and the streak spacings collapse, corresponding to the InGaAs moving towards its normal unstrained larger lattice constant. When the next GaAs barrier is grown on the InGaAs, the sandwiched well is placed under tension from this cap; and in the case of the 24 Å well, the dislocations are removed from the interfaces and a strained quantum well results. The critical thickness for a quantum well is greater than that for an uncapped layer because there are two forces, top and bottom, making it conform to the substrate. RHEED shows if

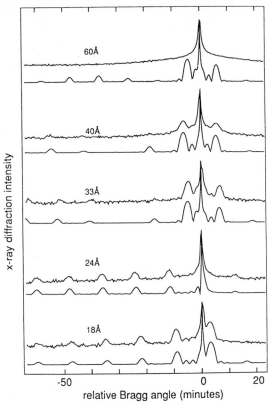

Fig. 18.6. Experimental (*top*) and simulated (*bottom*) X-ray spectra of five, five period $In_{0.4}G_{0.6}As$ multiquantum well structures with the well thicknesses shown on the figure. The loss of features in the experimental spectra above 33 Q well thickness is due to the quantum well critical thickness being exceeded and the consequent misfit dislocations degrading the material (after [18.24])

the strain collapse is fully reversible resulting is a strained quantum well. For the 60 Å well, the critical thickness is exceeded and the growth degrades with each successive well. By measuring the specular oscillations (Sect. 12.2), RHEED also gives a quantitative measure in atomic layers of the quantum well thickness.

18.3.3 Photoluminescence

As quantum wells can be used as the active material in lasers and light emitting diodes, one might expect that optical methods could characterize them. A number of such methods are in fact well established; one of the most useful being photoluminescence [18.25]. The apparatus simply consists of a laser which irradiates a cooled sample (2 to 77 K), and a monochromator which analyzes the emitted light or luminescence. The incident light has energy greater than the bandgap. It ionizes filled levels within the semiconductor, raising the concentration of electrons and holes in the conduction and valence bands, creating higher level excitons in quantum wells and generally populating higher energy levels of

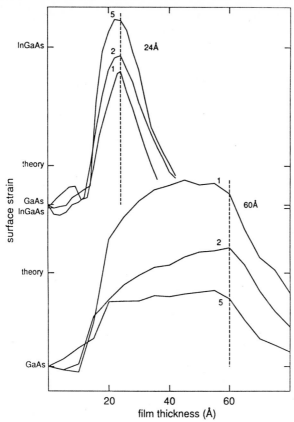

Fig. 18.7. Surface strain found from RHEED streak spacing vs film thickness for the first, second and fifth periods of 24 Å and 60 Å mqw structures. The abscissa zero is the beginning of the InGaAs layer. The dotted line marks the commencement of the GaAs barrier. The surface strain for equilibrium $In_{0.4}Ga_{0.6}As$ and the surface strain change predicted by thoery for the bulk film are marked on the ordinate for both structures (after [18.24])

the band structure. A new equilibrium is set up with the incident beam, pumping the various levels of the band structure and electrons and holes continuously recombining. If they recombine and release their energy as photons, then the monochromator and detector can measure this light intensity and energy. Thus photoluminescence in its simplest form is the pumping of the solid's energy levels with a single laser line and measuring the light emitted. The emissions give the energy level transitions directly from the frequency of the light ($\Delta E = h\nu$). From these energy levels the shape of the quantum well can be deduced.

Examples of photoluminescence spectra are shown in Fig. 18.8 [18.26]. They are for three GaAs/In_xGa$_{1-x}$As/GaAs quantum wells which had been measured by X-ray diffraction to have $x = 0.12$, widths of 144 ± 5 Å and barriers of 450 ± 5 Å. The pump laser was an excimer pumped dye laser at 6600 Å and the temperature of the sample was 78 K. Low temperatures are required in photoluminescence to sharpen the structure which is blurred by the Fermi-Dirac

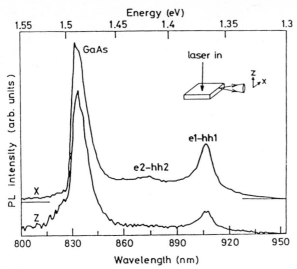

Fig. 18.8. 77 K photoluminescence spectra of $In_{0.12}Ga_{0.88}As$ analyzed with in-plane and perpendicular polarizations. Both the lower energy peaks are x-polarized but the GaAs is unpolarized. Inset shows the experimental geometry (after [18.26])

distribution. One of the spectra is for the emitted light polarized in the plane of the well (x), and one for polarization in the growth direction, perpendicular to the well (z). Both spectra share two peaks, the largest near 1.5 eV and another at ~ 1.37 eV. The ~ 1.5 eV peak comes from GaAs: the incident beam ionizes the GaAs, forming holes in the valence band and electrons in the conduction band. Direct recombination of the electron hole pairs results in light emission with the energy of the bandgap (less the electron-hole binding energy). The peak near 1.37 eV is of lower energy than the GaAs and is attributed to the lowest energy transition possible in the InGaAs well, e1-hh1. Thus we find the most vital piece of information about the semiconductor which is the effective bandgap. For GaAs it is of course well known, but for InGaAs, in a quantum well and strained, it must be determined for each sample. To find which levels contribute to the bandgap and to explain the subsidiary peak at ~ 1.42 eV, we must probe a little further.

A III-V direct bandgap quantum well has a band structure similar to Fig. 18.4. If lattice mismatched, the strain causes the lattice to elongate and the band offsets and energy levels alter markedly. In a quantum well, as explained earlier, the confinement raises the energy of the electrons and holes up to quantized levels $n = 1, 2, 3 \ldots$ Thus given a spectrum we need to assign the levels and determine whether they involve heavy or light holes.

To do this, we compare the two spectra. Conduction band electrons e1, e2 have angular momentum quantum numbers $m_j = \pm 1/2$ as do the light holes. The heavy holes have $m_j = \pm 3/2$. Hence $\Delta m_j = \pm 1$ for an e-hh transition; while $\Delta m_j = 0$ for e-lh; where lh and hh refer to the light and heavy holes respectively. In the former transition this momentum must be conserved by the emission of

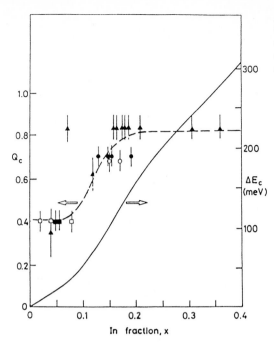

Fig. 18.9. Variation of the conduction band offset as a function of indium concentration for the InGaAs/GaAs heterostructure. Q_c values were taken from a number of sources (after [18.26])

circularly polarized light. Thus for the light hole transition there would be no preferred plane of polarization. The InGaAs peaks are clearly damped compared with the GaAs peak for x polarization. Hence the larger peak is assigned to e1-hh1 and the smaller to e2-hh2.

The shape of the well can now be found. ΔE_c, ΔE_v and ΔE_g can be deduced assuming interpolated values for the hole and electron effective masses and using elementary finite well quantum theory. Results for $0 \leq x \leq 0.4$ are shown in Fig. 18.9. The parameter $Q_c = \Delta E_c / \Delta E_g$ is often used because if constant, it means the offsets scale with bandgap difference.

18.3.4 X-Ray Photoelectron Spectroscopy

ΔE_v can be found with XPS for a semiconductor interface without constructing a quantum well. This account will follow the treatment of *Waldrop* et al. [18.27].

The XPS determination ΔE_v relies on the rather large mean free path of $\lambda \sim 25$ Å possessed by the $\sim 1500\,\text{eV}$ photoelectrons. The heterostructure to be studied is constructed with a thickness less than λ as shown in Fig. 18.10a. The bands are bent owing to surface states but the bending is much greater than λ and can be ignored. The corresponding schematic energy band diagram is given in Fig. 18.10c for two materials A and B. Also shown is the core level E_{cl}. We require

$$\Delta E_v(A - B) = \Delta E_{cl} + (E_{cl} - E_v)^B - (E_{cl} - E_v)^A \ . \tag{18.3}$$

367

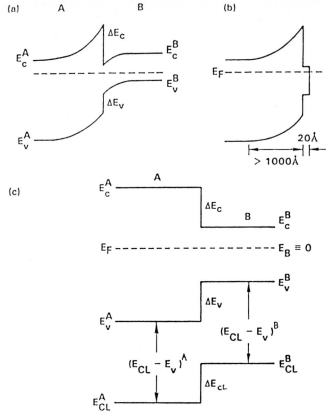

Fig. 18.10. (a) Typical "buried" heterojunction, (b) "exposed" thin heterojunction accessible to XPS analysis and (c) schematic energy band diagram of thin, abrupt heterojunction interface (after [18.27])

In an XPS experiment X-rays of energy $h\nu$ eject photoelectrons of energy E_k, whose intensities are recorded against binding energy $E_b = h\nu - E_k$ with $E_b \equiv 0 = E_F$. The last two terms of (18.3) are properties of the bulk materials. For the InAs/GaAs interface, they are $(E_{GaAs} - E_v)$ and $(E_{In4d} - E_v)$ and are found from bulk experiments. The first term on the right hand side requires the XPS spectra of InAs on GaAs. The spectra of 15 Å of InAs on GaAs is shown in Fig. 18.11 [18.28]. Considerable analysis is required to obtain the position of E_v. It involves curve fitting to a broadened theoretical non-local pseudopotential valence band density of states which simulates the leading edge of the data. The results are shown in Table 18.3. The value of $\Delta E_v = 0.17 \pm 007$ eV gives a $Q_c \sim 0.8$ in agreement with the trend of Fig. 18.9. This technique is experimentally straightforward and has been used to characterize a large number of interfaces.

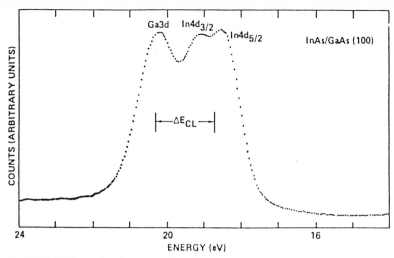

Fig. 18.11. XPS core level spectrum from InAs-GaAs (100) exhibiting a core level from each side of the heterojunction (after [18.28])

Table 18.3. Energies relevant to the determination of InAs-GaAs (100) HJ band discontinuities

E_g^{GaAs}	$= 1.43\,\text{eV}$
E_g^{InAs}	$= 0.36\,\text{eV}$
$(E_{\text{Ga}3d}^{\text{GaAs}} - E_v)$	$= 18.81 \pm 0.02\,\text{eV}$
$(E_{\text{As}3d}^{\text{GaAs}} - E_v)$	$= 40.73 \pm 0.02\,\text{eV}$
$(E_{\text{In}4d}^{\text{InAs}} - E_v)$	$= 17.43 \pm 0.02\,\text{eV}$
$(E_{\text{As}3d}^{\text{InAs}} - E_v)$	$= 40.77 \pm 0.02\,\text{eV}$
$\Delta E_{\text{cl}} = (E_{\text{Ga}3d}^{\text{GaAs}} - E_{\text{In}4d}^{\text{InAs}})$	$= 1.55 \pm 0.06\,\text{eV}$

18.3.5 Summary

In this section we have seen that a number of techniques can characterize a single device structure. X-ray diffraction establishes the sample as a single crystal, gives the lattice constants of the components, their strain, the alloy concentration and the dimensions of the material structure. RHEED gives detail, *in-situ*, of the dimensions and strains of the different components. It works best for large changes over small distances. The optical techniques characterize the electronic structure of the material and are complementary both experimentally and in the information obtained.

References

18.1 E.A. Ameraseka, D.S. Campbell: *Failure Mechanisms in Semiconductor Devices* (Wiley, New York 1987)

18.2 M.J. Howes, D.V. Morgen (Eds.): *Reliability and Degradation* (Wiley, New York 1981)

18.3 J.P. Fillard (Ed.): *Defect Recognition and Image Processing in III-V Compounds* – Materials Science Monographs **31** (Elsevier, Amsterdam 1985)

18.4 C.E. Jowett: Microelectronics J. **9**, 5 (1979)

18.5 B.P. Richards, P.K. Footner: GEC J. Res. **1**, 74 (1983)

18.6 L.E. Murr: *Electron Ion Microscopy and Microanalysis* (Dekker, New York 1982)

18.7 A.G. Cullis, S.M. Davidson, G.R. Border: Proc. Conf. on Microscopy of Semiconducting Materials, Oxford (Inst. of Physics, London 1983)

18.8 P. Burggraaf: Semiconductor International **53**, 53 (1987)

18.9 N.D. Stojadinovic: Microelectron. Rel. **23**, 609 (1983)

18.10 J.J. Bart: 16th Ann. Proc. Rel. Phys. Symp. 108 (1978)

18.11 R. Kossowsky: 16th Ann. Proc. Rel. Phys. Symp. 112 (1978)

18.12 R.B. Marcus, T.T. Sheng: 19th Ann. Proc. Rel. Phys. Symp. 269 (1981)

18.13 A. van der Ziel: *Solid State Electronics* (Prentice-Hall, Englewood Cliffs 1968)

18.14 B.L. Sharma: "Ohmic contacts to III-V compound semiconductors", in *Semiconductors and Semimetals*, Vol. 15 (Academic, New York 1981)

18.15 N. Braslau, J.B. Gunn, J.L. Staples: Solid-State Electron. **10**, 381 (1967)

18.16 J.M. Dell, H.L. Hartnagel, A.G. Nassibian: J. Phys. D **16**, L243 (1983)

18.17 T.S. Kalkur, J.M. Dell, A.G. Nassibian: Int. J. Electron. **57**, 729 (1984)

18.18 T.S. Kuan, P.E. Batson, T.N. Jackson, H. Rupprecht, E.L. Wilkie: J. Appl. Phys. **54**, 6952 (1983)

18.19 J.R. Shappirio, R.T. Lareau, R.A. Lux, J.J. Finnegan, D.D. Smith, L.S. Heath, M. Taysing-Lara: J. Vac. Sci. Technol. A **5**, 1503 (1987)

18.20 N. Braslau: J. Vac. Sci. Technol. **19**, 803 (1981)

18.21 See for example B. Schiflebein: Semiconductor International 62 (Nov. 1988)

18.22 D.A.B. Miller: Opt. Eng. **26**, 372 (1987)

18.23 W.H. Zachariasen: *Theory of X-Ray Diffraction in Crystals* (Dover, New York 1967)

18.24 G.L. Price, B.F. Usher: Appl. Phys. Lett. **55**, 1984 (1989)

18.25 J.I. Pankove: *Optical Processes in Semiconductors* (Dover, New York 1971)

18.26 M.I. Joyce, M.J. Johnson, M. Gal, B.F. Usher: Phys. Rev. B **38**, 10978 (1988)

18.27 J.R. Waldrop, R.W. Grant, S.P. Kowalczyk, E.A. Kraut: J. Vac. Sci. Technol. A **3**, 835 (1985)

18.28 S.P. Kowalczyk, W.J. Schaffer, E.A. Kraut, R.W. Grant: J. Vac. Sci. Technol. **20**, 705 (1982)

19. Characterization of Oxidized Surfaces

J.L. Cocking and G.R. Johnston

With 9 Figures

Metals, in general, critically depend on their surface oxide scales for environmental stability, particularly in aggressive oxidizing atmospheres at high temperatures. The protective capabilities of oxides are dependent on many physical and chemical properties, as well as on their mechanical adherence to the metal surface. In summary, an "ideal" protective oxide would be:

- physically and chemically stable. An ideal oxide would not dissociate nor melt at the temperatures and pressures of interest;
- mechanically stable. The scale would be capable of maintaining intimate contact with the surface of the metal, particularly when sudden temperature changes occur;
- a barrier to diffusion. The function of a protective oxide is to separate the metal from the oxygen in the gas phase. The ideal oxide would, therefore, have a low diffusion rate for both oxygen and metal ions, otherwise the oxidation reaction would proceed at the oxide/metal or oxide/gas interfaces respectively;
- continuous and dense. When pores or cracks are present in the oxide scale the protective capabilities of the oxide are lost.

All of these properties that are essential for oxidation protection are dependent on the chemistry and morphology of the oxide scale, the oxide/metal interface and the near-surface region of the metal.

So that alloys can be developed with improved oxidation resistance, it is essential to know the mechanisms whereby oxidation occurs. Such knowledge can only be obtained through characterization of oxidized surfaces, particularly of surfaces in the initial stages of the oxidation process. The characterization of thin oxide scales, particularly for oxides less than a micron thick, has only been possible in recent years with the development of the array of sophisticated surface analytical equipment described elsewhere in this book. A number of these techniques have been used by the authors to characterize oxidized surfaces of metals [19.1–3]. In this review, application of three techniques, namely scanning Auger microscopy (SAM), microbeam Rutherford backscattering spectrometry (μ-RBS) and extended X-ray absorption fine structure (EXAFS), to the study of the oxidation of selected cobalt- and nickel-based alloys is described. SAM and μ-RBS are powerful tools, particularly when used together, for analyzing multi-phase surfaces where spatial resolution of better than 20 μm is required, whereas EXAFS has great potential for the non-destructive determination of atomic structure in thin surface layers.

19.1 The Oxidation Problem

Nickel- and cobalt-based superalloys are used extensively in demanding high temperature applications such as hot end components in gas turbine engines. The alloy chemistry of the superalloys has been formulated to give maximum strength and not maximum oxidation resistance. Protective coatings are therefore applied to the surfaces of turbine components such as blades, vanes and combustion chamber liners, to increase their oxidation resistance and improve their durability. Optimum oxidation protection of alloys and coatings for high temperature use is provided by oxide scales of alumina (Al_2O_3). Chromia (Cr_2O_3) and silica (SiO_2) are also protective oxides, but Cr_2O_3 has limited use at high temperatures because of its volatility in oxidizing atmospheres, while SiO_2 has a thermal expansion coefficient which is incompatible with most high temperature alloys and hence its adherence is poor. In practice, the protective oxide scales that form are never exclusively a single oxide, but instead are mixtures of oxides and spinels.

The generic formulation for one class of oxidation resistant coatings used both in gas turbine engines (particularly those used in shipboard applications) and in thermal barrier coatings (as the oxidation-resistant layer beneath the outer ceramic layer) is MCrAlX, where M is either Ni, Co or Fe and X is an element such as Hf, Y or Ce, which has a high affinity for oxygen. The disproportionate and beneficial effects produced by minor additions of these oxygen-reactive elements on the adherence of the oxide to the surface was the principal reason that this present work was undertaken. Many mechanisms have been proposed to explain their role in improving the adherence and hence the oxidation resistance of MCrAl alloys, but no consensus has been reached. The mechanisms proposed include:

  i) mechanical "keying" or "pegging" of the scale to the substrate [19.4];
 ii) sulphur "gettering" by the oxygen-reactive element, which is also sulphur-reactive, thus tying up the sulphur impurities in the alloy [19.5]; and
iii) a combination of (i) and (ii).

Almost all previous mechanistic studies of the oxidation of high temperature alloys have been reported for long oxidation times, where the thickness of oxide film is typically much greater than a micron. Similarly where more than one phase is present, little or no attempt has been made to resolve the oxidation behavior of each phase. In the work briefly reviewed here, which was a collaborative study involving Materials Research Laboratory, Melbourne, the US Naval Research Laboratory, Washington DC, the University of Pittsburgh PA, the Royal Melbourne Institute of Technology and the University of Melbourne, the following programme was undertaken:

- the initial stages of oxidation were studied, i.e. oxide films much less than 1 μm in thickness were characterized;
- the oxidation mechanisms for both phases of a two-phase alloy were investigated; and
- the effects of specific elements on the initial oxidation mechanisms of both phases was studied using ion implantation.

The experimental procedures for sample preparation, ion implantation and surface analysis techniques employed have been detailed previously [19.6,7]. With the exception of EXAFS, the analytical techniques from which the results were obtained have been described elsewhere in this book.

19.2 Oxidation of Co-22Cr-11Al

The first example presented in this brief review of the application of microanalytical techniques to oxidation studies is the oxidation of Co-22Cr-11Al (nominal composition, wt%) cast alloys. This alloy has a two-phase structure consisting of a matrix of β-CoAl with α-Co solid solution precipitates ranging in size from 5 to 20 μm. The oxidation experiments were performed, in air, at 700°C for times between 10 min and 96 h. One surface of each sample was metallographically polished to a 0.25 μm diamond finish. Half of this polished surface was implanted with 2×10^{16} Hf (or Y) ions/cm^2 with an energy of 150 keV, while the other half was shielded with a tantalum mask. This technique allowed the in situ comparison of the oxidation of non-implanted and implanted alloys on the same surface.

The effects of specific elements on the oxidation mechanisms are visibly apparent with this experimental arrangement. After oxidation, the two halves of the specimens appear different, even to the naked eye. The most striking effect of ion implantation with Y and Hf is the marked suppression of void formation in the metal at the metal-oxide interface, predominantly at the α/β phase boundaries [19.7]. The scanning electron micrograph in Fig. 19.1 clearly shows the voids in the metal in the non-implanted alloy seen through the thin, transparent oxide film on the β-phase. Across the sharp boundary, it is obvious that Hf implantation has altered the oxidation mechanism and suppressed void formation. The critical contribution of this work is the determination of the chemical composition and thickness of the oxide films, which is now described.

19.2.1 Chemical Characterization

The two most powerful microanalytical tools for analyzing the oxides on both the α and β phases are scanning Auger microscopy (SAM) and microbeam Rutherford backscattering spectrometry (μ-RBS). These two techniques are extremely powerful tools when used in combination. High resolution SAM gives chemical composition data in the top 4 to 10 atomic layers of the surface. For depth in-

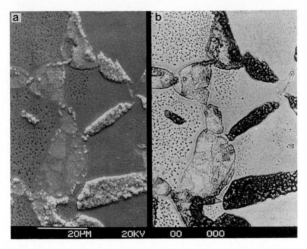

Fig. 19.1. Secondary electron (**a**) and backscattered electron (**b**) micrographs of the boundary region between the non-implanted (left of boundary) and Hf-implanted (right of boundary) areas of a Co-22Cr-11Al surface oxidized for 9.5 h at 700°C. These micrographs show the presence of voids in the non-implanted alloy and their suppression following Hf implantation

formation from SAM, ion beam milling must be used. This technique, however, is at best semi-quantitative with respect to depth because of many experimental problems including:

(i)   preferential sputtering of certain elements with respect to others;
(ii) different milling rates through different phases; and
(iii) redeposition of sputtered material onto the site under analysis.

With the present experimental technique where both the implanted and non-implanted surfaces are on the same alloy face (and in some cases at the boundary, on the same grain), SAM depth profiles give accurate relative depth profiles for the oxides.

RBS and μ-RBS on the other hand are non-destructive techniques that give quantitative depth information from the backscattered spectra. The two techniques (SAM and μ-RBS) therefore provide complementary information for very detailed surface characterization.

19.2.2 Scanning Auger Microscopy

One of the great strengths of SAM is its ability to determine quantitatively the concentration of species present on the surface of a sample in the first 4 to 10 atomic layers. Its relative weakness is in the assigning of depth from the original surface following ion beam milling. To overcome this potential problem, all thickness determinations in this work are reported as relative ion beam milling times rather than as units of thickness. As mentioned above, RBS is used to confirm relative thicknesses. Ion beam milling was performed on areas that included the interface between the implanted and non-implanted regions. Information obtained using the SAM technique is now illustrated with results obtained for oxidized yttrium-implanted alloys.

A scanning electron micrograph of the area of interest is shown in Fig. 19.2a, for a sample which was oxidized for 60 min in air at 700°C. The Y-implanted region is on the right hand side of the micrograph. A number of analysis points were chosen in both the α and the β phases on both the implanted and non-implanted regions. The five elements of interest – Co, Cr, Al, Y and O – were monitored before, during and after ion milling through the oxide into the underlying alloy. Milling was interrupted at selected times to allow elemental area maps to be made. These times were chosen at, or near, the interface between the oxide and the alloy for each phase, and examples are shown in the remainder of Fig. 19.2. The left hand series are the area maps obtained after 5 minutes of milling (near the oxide/alloy interface for Y-implanted β-phase) and the right hand series after 14 minutes milling (near the oxide/alloy interface for the Y-implanted α-phase). Depth profiles of the five elements, at each analysis point, were also constructed, using sensitivity factors to convert the Auger signal to elemental concentrations. Examples of depth profiles for non-implanted β-phase and Y-implanted β-phase are given in Fig. 19.3a and b respectively. The sensitivity factor used to construct Fig. 19.3 pertain to the elements in the oxide and therefore the concentrations in the metal are not accurate. It is possible to construct composite concentration profiles using sensitivity factors for elements in both the oxide and the metal, but this was not done in Fig. 19.3.

The depth profiles are strikingly similar to the concentration profiles generated from RBS spectra (see next section). Preliminary correlations between the two techniques have been published [19.2], with a more detailed publication in preparation. The area maps in Fig. 19.2 show quite distinctly the very uniform thicknesses of the oxides on the same phases. The area maps also show the positions of the implanted layer in each of the α and β phases after oxidation. The Y Auger signal was too weak to monitor during depth profiling – it was only possible to detect it accurately during the compiling of the area maps. In agreement again with RBS results, the Y is in the oxide scale near the oxide/alloy interface. Because of the considerable difference in the thickness of the oxide which formed on the two phases, the Y-rich layer on the α-phase is located at a vastly different depth beneath the outer surface than the corresponding layer on the β-phase.

19.2.3 Rutherford Backscattering Analysis

The μ-RBS technique, which has a spatial resolution of about 10–20 μm and therefore gives information on the individual phases, has been used to characterize both implanted and non-implanted, α and β phases of the alloys. Spectra for the β-phase of the oxidized Co22Cr11Al, both Hf-implanted and non-implanted, are given in Figs. 19.4 and 19.5 respectively. Figures 19.6 and 19.7 give the respective elemental concentration profiles obtained by deconvolution of the spec-

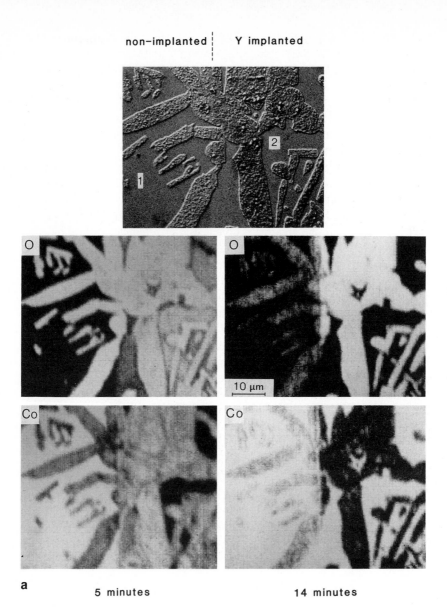

Fig. 19.2a. For caption see the opposite page

non-implanted | Y implanted   non-implanted | Y implanted

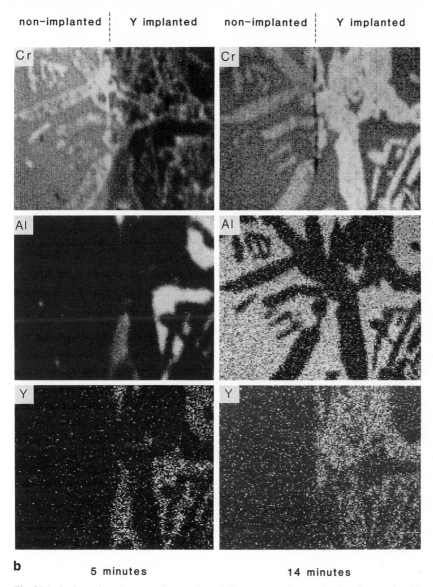

b   5 minutes   14 minutes

Fig. 19.2a,b. Scanning electron micrograph and Auger area maps at the interface region between non-implanted (left of boundary) and Y-implanted (right of boundary) Co-22Cr-11Al oxidized in air for 1 h at 700°. The Auger maps on the left side were recorded after 5 min of ion beam milling, while the maps on the right side were obtained after 14 min of milling. Depth profiles at points 1 and 2 are shown in Fig. 19.3

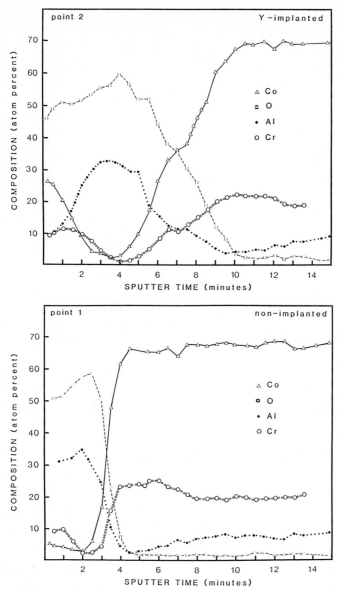

Fig. 19.3. Auger element depth profiles for Co, Cr, Al and O (Y concentration too low to monitor in this mode) at the two points designated in Fig. 19.2, i.e. point 1 – β-phase, non-implanted; point 2 – β-phase, Y-implanted

tra using the RUMP programme [19.3]. The important results obtained from Figs. 19.4 to 19.7 are:

i) The oxide on the Hf-implanted β-phase is more than twice as thick as the oxide on the non-implanted β-phase, in close agreement with the result illustrated above for the Y-implanted alloy;

378

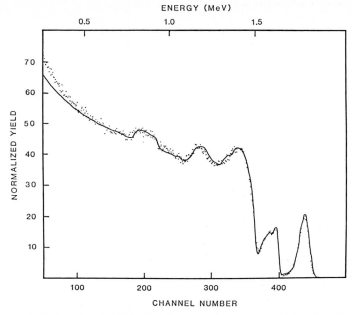

Fig. 19.4. Experimental Rutherford backscattering spectrum (*dots*) of Hf-implanted, β-phase Co-22Cr-11Al oxidized in air for 1 h at 700°C, together with the theoretical spectrum (*solid line*) iteratively generated from the concentration profile shown in Fig. 19.6

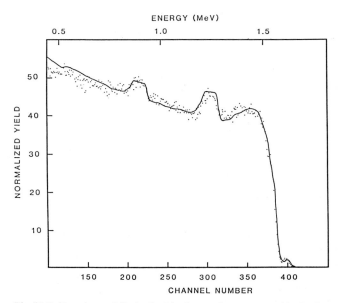

Fig. 19.5. Experimental Rutherford backscattering spectrum (*dots*) of non-implanted, β-phase Co-22Cr-11Al oxidized in air for 1 h at 700°C, together with the theoretical spectum (*solid line*) iteratively generated from the concentration profile shown in Fig. 19.7

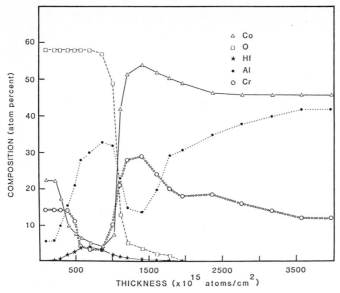

Fig. 19.6. Elemental concentration profiles of the oxide and near-surface region of oxidized, Hf-implanted β-phase Co-22Cr-11Al iteratively generated to produce the spectrum (*solid line*) in Fig. 19.4

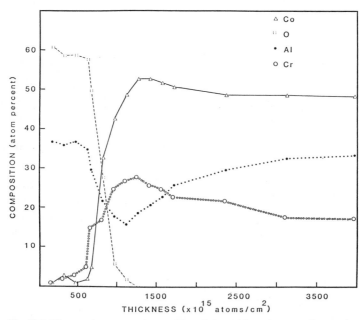

Fig. 19.7. Elemental concentration profiles of the oxide and near-surface region of oxidized, non-implanted β-phase Co-22Cr-11Al iteratively generated to produce the spectrum (*solid line*) in Fig. 19.5

ii) Cobalt is enriched at the outer surface of the Hf-implanted β-phase, where its concentration is at least five times greater than on the corresponding non-implanted phase, again in agreement with the SAM result. Both β-phase oxides are predominantly Al rich (presumably Al_2O_3) as would be expected for the oxidation of CoAl;

iii) the final position of the ion implanted layer of Hf, after oxidation, is near the oxide/metal interface. This is also true for the oxidized α-phase, even though the α-phase oxide is appreciably thicker than the β-phase oxide [19.2], again in agreement with the SAM results.

Perhaps the most important contribution of RBS and μ-RBS to this surface characterization exercise is that the thickness of the various oxide films are quantitatively defined without recourse to milling techniques and without the inaccuracies inherent in the use of sensitivity factors (Sect. 19.2.2). Milling techniques, which are essential for SAM analyses, introduce uncertainties in thickness measurements through physical processes that include preferential sputtering of elements from the surface, activated diffusion on the surface and redeposition of sputtered material. Detailed mechanisms proposed to account for the features will be addressed in a future publication. The aim in the present publication has been to illustrate the great capabilities of SAM and μ-RBS in completely characterizing the oxidized surfaces of individual phases as small as about 20 μm in diameter.

19.3 Oxidation of Ni-18Cr-6Al-0.5Y

The second example presented in this review is the oxidation of Ni-18Cr-6Al-0.5Y (nominal composition, wt%) alloys, ion implanted with V ions. Some samples were produced as castings, which were subsequently polished, while other samples were produced by plasma spraying. The surfaces of the samples were implanted with 2×10^{16} (low dose) or 2×10^{17} V ions/cm^2 (high dose) with an energy of 85 keV. Both non-implanted and ion implanted samples were oxidized in a mixture of 80% Ar and 20% O_2 for 10 min at 900°C. Photoelectron EXAFS (extended X-ray absorption fine structure) spectra were collected from all samples. The EXAFS data were used in conjunction with RBS, SAM and STEM data to monitor the changes induced by the implantation and oxidation processes. Only the EXAFS data will be discussed in this section.

19.3.1 Extended X-Ray Absorption Fine Structure

EXAFS is a technique which utilizes synchrotron radiation. The technique involves measurement of small oscillations which are apparent in the region about 1000 eV above the K- (and L-) shell absorption edges in spectra of X-ray absorption edge coefficients versus X-ray photon energy. These small oscillations arise when an electron leaving the atom from whose shell it has been ejected backscatters from its nearest neighbor atoms, hence causing interference with the photoelectron wave emerging from the atom. Data obtained from the oscillations enable the average distance of a particular type of atom from its nearest neighbors to be determined. In addition, it is theoretically possible to determine the number and the kind of neighbors, as well as the distance to more distant neighbors with the EXAFS technique.

Because each element has different electron binding energies, individual elements in a polyatomic material can be selectively excited, thus enabling the local environment of the atoms in a material to be characterized in a way that is not possible by conventional diffraction techniques. An additional advantage of this technique is that the escape depth of the photoelectrons is of the same order as that of the implantation depth, i.e. around 100 nm.

The EXAFS spectra of the K-edges for Ni, Cr and V in non-implanted, implanted, unoxidized and oxidized samples were collected on the Naval Research Laboratory's beam line at the National Synchrotron Light Source. Fourier transforms of the spectra above the absorption edges were made to obtain information about the coordination of these elements in the NiCrAlY alloy. The Fourier transforms of the Ni edge of the non-implanted, unoxidized samples, such as the one shown in Fig. 19.8a, showed that the Ni had a fcc structure. The large peak near 2 Å is the contribution from the first shell of neighbors. The other peaks correspond to more distant neighbors. The Fourier transform of the Cr edge in the same sample, Fig. 19.8a, showed essentially the same profile. This signifies that the Cr atoms have the same local environment as the Ni atoms, i.e. Cr substitutes directly into the Ni lattice. This result was confirmed by examination and analyses of the same samples using STEM [19.8].

The Fourier transform of the V edge in the low dose, unoxidized sample (illustrated in Fig. 19.8b) is similar to that of Ni and Cr, rather than that of

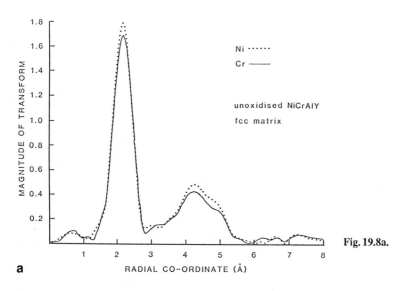

Fig. 19.8a.

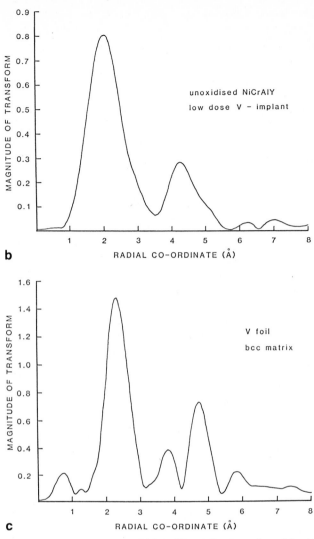

Fig. 19.8. Fourier transforms of (a) the Ni and Cr edges of unoxidized Ni-18Cr-6Al-0.5Y; (b) the V edge of V-implanted, unoxidized NiCrAlY; and (c) the V edge of pure vanadium

pure V, the transform of which is shown in Fig. 19.8c. This means that during implantation V ions substitute into the fcc lattice rather than forming clusters or discrete precipitates. If V had formed clusters it could be expected that the near-neighbor environment would be similar to that of pure V, which has a bcc lattice. Alternatively, if V had formed discrete precipitates with Cr and/or Ni the transforms for these elements would have been modified, which did not occur. This example clearly demonstrates that EXAFS is a powerful nondestructive technique, particularly for the analysis of implanted metals.

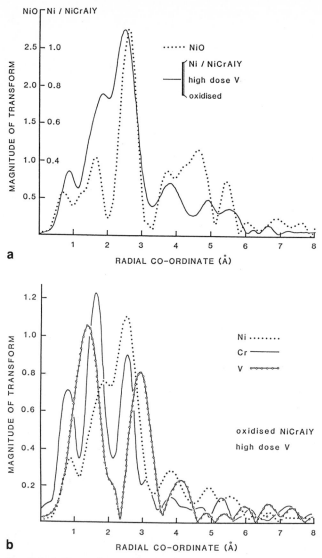

Fig. 19.9a,b. For caption see the opposite page

After 10 min oxidation at 900°C a number of different oxides form on both the non-implanted and implanted alloys. Analyses using STEM showed that the oxides present included NiO, Al_2O_3, $NiAl_2O_4$ and $NiCr_2O_4$ [19.8]. The STEM analyses also showed that the oxides were layered, with NiO richer at the outermost surface and Al_2O_3 richer at the metal/oxide interface. The oxides formed on the V-implanted samples were similar to those formed on the non-implanted samples. No discrete V-containing oxides were resolved in the selected area diffraction patterns, however, EDS analyses of the implanted, oxidized samples

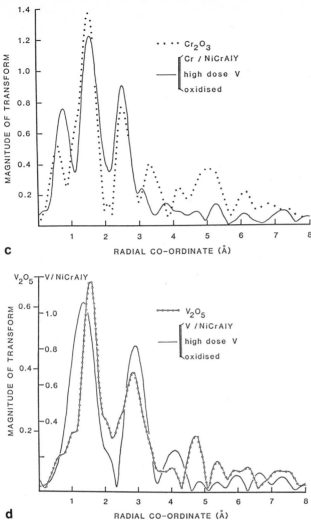

Fig. 19.9. Fourier transforms of (a) the Ni, Cr and V edges of V-implanted Ni-18Cr-6Al-0.5Y oxidized for 10 min at 900°C; (b) the Ni edge of oxidized, V-implanted NiCrAlY (*solid line*) compared with the Ni edge of pure NiO; (c) the Cr edge of oxidized, V-implanted NiCrAlY (*solid line*) compared with the Cr edge of pure Cr₂O₃; (d) the V edge of oxidized, V-implanted NiCrAlY (*solid line*) compared with the V edge of pure V₂O₅

performed in STEM, suggested that V was associated with the Cr-rich oxides. SAM analyses confirmed that oxides tended to be layered. SAM, however, was not able to confirm an association of V with Cr in the oxide scale.

The Fourier transforms obtained from the EXAFS spectra for the Ni, Cr and V edges in an oxidized, high-dose V-implanted NiCrAlY are shown in Fig. 19.9a. An examination of the transforms shows that the local environment of atoms of each of these elements is different. Much microstructural information is contained in these transforms. Comparisons of the transforms from the implanted, oxidized alloy with those of simple, pure oxides (Figs. 19.9b, c and d) confirm the proposal that the oxides formed are complex oxides and not a physical mixture of simple oxides, because of the considerable differences that are readily apparent. Detailed analyses of these results are yet to be performed, but it is possible to characterize the oxides scale on the implanted alloy by following one or both of two paths. Firstly transforms of mixtures of oxide standards may be obtained experimentally and compared with the oxide formed on the oxidized alloys. Alternatively, computer generated transforms may be obtained by modelling the oxides. In order to do this, information provided by other analytical techniques is necessary, particularly at this stage in the development and application of the EXAFS technique. The information provided by STEM will enable comparisons of the transforms obtained experimentally with transforms generated from models of oxides to assist in the characterization of the oxides. The present results highlight the potential of the EXAFS technique for characterizing surfaces.

References

19.1   G.R. Johnston, J.L. Cocking, W.C. Johnson: Oxid. Metals **23**, 237 (1985)
19.2   G.R. Johnston, P.L. Mart, J.L. Cocking, J.W. Butler: Materials Forum **9**, 138 (1986)
19.3   J. Saulitis, G.R. Johnston, J.L. Cocking: Thin Solid Films **166**, 201 (1988)
19.4   C.S. Giggins, B.H. Kear, F.S. Pettit, J.K. Tien: Met. Trans. **5**, 1685 (1974)
19.5   J.G. Smeggil, A.W. Funkenbusch, N.S. Bornstein: Met. Trans **17A**, 923 (1986)
19.6   J.A. Sprague, G.R. Johnston, F.A. Smidt, S.Y. Hwang, G.H. Meier, F.S. Pettit: In *High Temperature Protective Coatings*, ed. by S.C. Singhal (Warrendale, PA, TMS of AIME 1983) pp. 93–103
19.7   F.A. Smidt, G.R. Johnston et al.: In *Surface Engineering*, ed. by R. Kossowsky, S.C. Singhal (NATO ASI Ser., Ser. E 1984), pp. 507–523
19.8   J.L. Cocking, J.A. Sprague, J.R. Reed: Surf. Coatings Technol. **36**, 133 (1988)

20. Coated Steel

R. Payling

With 8 Figures

Metal manufacturing and metal goods figure prominently in the industrial applications of surface analysis [20.1]. This is not surprising when one considers the enormous surface areas being produced each year, for example, in the sheet metal industry. Such sheet metal products undergo complex multistage industrial processing much of which interacts with the particular surface exposed during each process; and the performance and appearance of these products are judged, at least in part, by surface features, such as surface flatness, corrosion resistance, color, gloss, etc.

Coated steel, as the name suggests, is composed of a steel substrate (typically 0.4 to 1.5 mm thick) with various metallic or organic coatings (typically 20 μm thick). The four major areas for the consumption of coated steel are building products, home appliances, office furniture, and the automotive market. The choice of which coating to use in each area depends on the particular application, since coatings differ in their cost, their corrosion resistance, their suitability for forming, welding and painting, in their high temperature stability, and their suitability for containing food stuffs, etc.

The production of coated steel sheet is a multistage process and the changes in surface composition at each stage of processing must be determined if a knowledge of the complete operation is to be obtained [20.2]. Such information is useful when trying to trace a particular product feature, such as surface segregation, to a particular processing stage(s). Since each processing stage will leave telltale chemical traces which are then overlaid by subsequent processing, knowledge of the composition of near surface layers of the product is often more useful than the immediate surface composition. Such information can often separate the manifestation of a problem, such as a corrosion product, from the underlying processing variables which caused the problem.

A typical processing sequence (popularly called a route, or routing) for painted, metallic coated, steel strip, starting with the steel coil from a hot strip mill, is [20.3]:

a) pickling, to remove iron oxide scale in either hydrochloric or sulphuric acid;
b) oiling, immediately after pickling to prevent coil rusting and assist in cold reduction;
c) cold reduction, to reduce the steel thickness to the desired level – here, an emulsion of oil in water is applied to the strip to aid lubrication and heat dissipation;

d) metallic coating by immersion in a liquid metal bath of controlled composition;

e) gas-jet stripping, to control metallic coating thickness;

f) temper rolling and tension levelling, to improve base steel properties and surface flatness;

g) a paint pretreatment process, involving alkaline cleaning, hot and cold water rinsing, a proprietary conversion coating, and a final chromate rinse;

h) a multistage painting process, including the application of primer and top coats, with stoving and water quenching.

Each of these 8 stages acts on the exposed surface, which is changing throughout the sequence from uncoated steel, to a metallic coating, and finally to the surface of the polymer coating. The following series of applications illustrates this journey, and some of the potential hazards along the way, beginning inside the steel and ending with the polymer surface.

20.1 Applications

20.1.1 Grain Boundaries in Steel

Auger analysis of the fracture surface of steel is normally concerned with detecting the presence of segregating elements such as carbon, sulphur, or oxygen [20.4] or aluminum and nitrogen [20.5]. The segregation of these elements is of most general interest when the fracture is inter-granular rather than trans-granular, since the latter may indicate only the existence of inclusions while the former may provide an explanation for the fracture failing at the embrittled grain boundary. Before commencing the Auger analysis, the presence of inter-granular fracture must first be confirmed with scanning electron microscopy, i.e., by the presence of its characteristically smooth secondary electron image.

The detection of segregating elements alone does not always provide the complete answer to fracture analysis, as the mere detection of elements does not, of itself, determine the chemical species of these elements. This information is contained in the position and shape of the individual Auger peaks. Unfortunately, the deciphering of this information is not straightforward theoretically or fully catalogued empirically, hence the need to rely on carefully prepared standards and literature review. In a particular steel which was failing in a through thickness tensile test, it was suspected that segregation of boron nitride was weakening the grain boundaries. Scanning electron microscopy showed that the fracture faces had regions of inter-granular fracture, with the exposed grain faces typically being 20 μm across, large enough to make Auger analysis relatively easy. In the inter-granular regions approximately 25 atomic % boron and nitrogen were detected. The peak positions of both elements were consistent with boron nitride, namely 170 and 382 eV, respectively, [20.6–8] and the peak shape of the nitrogen, in

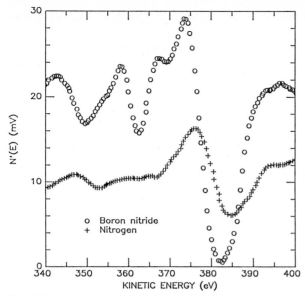

Fig. 20.1. Change in N_{KLL} peak shape between boron nitride and adsorbed nitrogen on steel

particular, was characteristic of boron nitride [20.9] (Fig. 20.1). It was therefore concluded that boron nitride was present on the inter-granular fracture surface.

20.1.2 Steel Surface

Auger analysis of uncoated steel sheet, even after solvent degreasing, shows a complex surface, in which the exposed iron content may typically account only for 5–20 atomic % [20.3]. Other elements detected and their likely sources are: carbon and zinc, from residual oils, particularly extreme pressure additives; manganese, silicon, aluminum, phosphorus, and chlorine from the steel base, by diffusion during annealing; silicon and phosphorus from residual alkaline cleaning solutions; sodium from rolling lubricants; and calcium and magnesium from industrial rinse water. Of these, residual carbon has been of special concern for some years, especially in the automotive industry, because of a strong correlation between residual carbon levels on uncoated steel and the corrosion of painted steel [20.3]. Consequently, sheet steel manufacturers have put considerable effort into determining the amount, type, and sources, of this residual carbon. Scanning Auger mapping shows that the carbon may form a uniform layer or be distributed in patches, and AES depth-profiling provides information on the depth distribution and overall mean thickness of the residual carbon layer. If this thickness is greater than 7–8 nm it can be anticipated that subsequent processing, such as painting, will give problems, providing of course that the sample(s) chosen is representative of the whole surface. Common to any off-line analysis, one is always concerned whether the small area analyzed is representative of the whole

12 tonne steel coil, especially since only the leading or trailing ends of the coil are readily accessible.

Various surface studies have shown the complexity of residual carbon forms on steel. Authors differ in their interpretation of the dominant form [20.10–13] but carbon may be present as organic molecules, graphite, amorphous soot or smut, or iron carbide. One common problem with coil annealing of oiled steel strip is redeposition of carbon near the edge of the strip, so-called "snaky" edge. While surface analysis is useful here, ultimately the solution to such problems is found in the proper control of the production process, in controlling oil levels and furnace conditions, for example.

In XPS studies of oiled steel surfaces, the carbon $1s$ peak is typically found at a binding energy of 286 eV. Degreasing the sample, to remove excess oils, shifts this peak to 284.7 ± 0.4 eV. This peak is present on most real surfaces and corresponds to a form of tenacious hydrocarbon, often called adventitious hydrocarbon, which has a reference value of 284.9 eV [20.14]. Unfortunately, the carbon $1s$ peaks for graphite, amorphous carbon, and iron carbide, all occur around 284.3 eV; so that positive identification of carbon on steel by XPS is uncertain. Since ion-bombardment of all these forms of carbon on steel results in the carbon $1s$ peak shifting to 284.2 ± 0.2 eV [20.3], information on any variation in the type of carbon with depth is therefore also unobtainable.

An alternative XPS method for characterizing oxygen-containing hydrocarbons, such as those found in lubricants used for rolling steel, involves a comparison of the binding energy of the oxygen $1s$ peak with the kinetic energy of the X-ray excited oxygen Auger KLL peak [20.3]. Both peaks appear in the same XPS spectrum and the results are called XPS chemical state plots, or Wagner plots [20.15]. One possible source of residual carbon is the formation of iron soaps between the acid groups in rolling oils and the steel. The Wagner method has been used to demonstrate that organic residues can survive closed coil annealing at 700°C, though energy shifts indicate some structural modification does occur.

Auger analysis is capable of distinguishing carbides from other forms of carbon due to a characteristic peak shape [20.16]. The C_{KLL} (272 eV) peak shape changes markedly from hydrocarbon (or graphite) to carbide, and the sensitivity increases by a factor of about 3 [20.16–18]. When an uncoated steel sample is ion-bombarded to remove most of the oxide film (typically to a depth of 10–15 nm), the surface carbon KLL peak shape, typical of either graphite or hydrocarbon, is replaced by a carbide peak shape [20.11]. But this does not necessarily indicate an underlying iron carbide. The measured carbon level ($\sim 6\%$) is 20 times the bulk value and diffusion of carbon from the surface or bulk and incorporation of oil during iron oxide formation have been proposed to explain this apparent carbide enrichment [20.19]. Steel samples abraded to mid-thickness, however, show exactly the same behavior and the explanation may lie in a combination of other factors. First, in differential spectra a narrow peak gives a greater apparent peak-to-peak height, so that the peak shape change to the narrow carbide shape

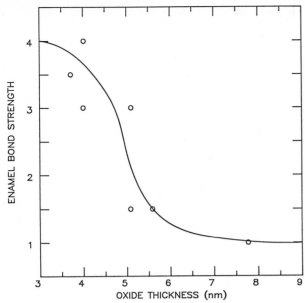

Fig. 20.2. Variation in enamel bond strength on steel in one study as a function of surface oxide thickness

exaggerates the amount of carbide present, by a factor of about 3; secondly, carbon has a very low sputtering rate and the AES technique analyzes what is left on the surface during or after sputtering, which exaggerates the total carbon concentration measured below the surface; and thirdly, carbon mixed with iron – possibly as a result of the sputtering process – may give a carbidic AES peak shape without its being a true carbide, because of short-range Coulomb interactions.

In the enamelling of steel, for the manufacture of enamelled stoves, bath tubs, hot water systems, etc., a frit is applied to the steel and then fired to form the enamel surface. Good adhesion of the enamel requires a sufficient reaction between the frit and the steel surface. In an early study of the factors affecting the adhesion of enamel to steel, it was found that better adhesion was related to a thinner surface oxide [20.3] (Fig. 20.2). A repeat study, however, with a different frit showed variable adhesion even with thin oxides. Since then we have generally found that poor adhesion is associated with a thick oxide ($\geq$ 6 nm), though reformulation of the frit sometimes proves effective in restoring adhesion. Other factors which have been found to be relevant to enamel adhesion are surface carbon (also associated with blistering of the enamel coating), silica, and steel grade. Titanium killed steels, for example, have exceptionally thin oxides (typically $\leq$ 4 nm) but may give enamel adhesion problems, presumably because the surface oxide formed on titanium killed steels inhibits the bonding reaction between the steel and the frit.

20.1.3 Alloy Region

When the steel strip enters the liquid metal bath, alloying begins between the liquid and the steel surface. Control of this alloy growth is important for proper coating properties and for coating adhesion [20.20]. An understanding of the alloying process requires a knowledge of the alloy composition. This is not always straightforward [20.3]. For chemical analysis of the separate alloy phases, the individual layers must be isolated through chemical etching techniques which may alter the composition of the layer of interest. To complicate matters, the layers are generally too thin for definitive electron microprobe analysis and depth-profiling Auger analysis may introduce artefacts through preferential sputtering.

In a particular 54 mass % aluminum, 44 mass % zinc and 1.5 mass % silicon coating studied, the alloy region was a two phase layered structure with a total thickness of 2 μm. It was found that prolonged ion bombardment tended to reduce the relative zinc concentration in the alloy but less so in the bulk metallic coating. To overcome this artefact of prolonged depth-profiling, for AES analysis, the samples were taper-sectioned at 10°, then given only a light argon ion etch, and analyzed in cross-section. For small angles, taper sectioning increases the apparent width of a layer, inversely proportional to the taper angle. Sectioning at lower angles than 10° was not satisfactory because the roughness of the steel surface produced islands of steel, alloy, or coating which then presented the difficulty of determining the precise analysis area. Table 20.1 lists the AES, electron microprobe, and chemical analyses of the two major alloy phases. The Auger sensitivity factors were obtained by setting the AES analysis of the top phase equal to that of the microprobe analysis, where the microprobe and chemical analysis were in good agreement, with the result that the AES analysis matched more closely the chemical analysis of the bottom phase.

20.1.4 Metallic Coatings

Metallic coatings may be broadly divided into three categories: hot-dipped, electrolytic (or electroplated), and vapor deposited coatings. The first two of these broad categories may be further subdivided: hot-dipped coatings into zinc (or galvanized), zinc-aluminum alloy, zinc-iron (or galvanneal), and lead-tin (or terne); and electrolytic coatings into zinc, zinc alloy, and tin (or tinplate). Only hot-dipped coatings and tinplate will be discussed here.

The surfaces of hot-dipped metallic coatings on steel generally differ considerably from the bulk compositions of the coatings [20.21]. The dominant metallic

Table 20.1. Estimates of alloy composition by various methods (in mass %)

Method	Composition							
	Top Phase				Bottom Phase			
	Al	Fe	Zn	Si	Al	Fe	Zn	Si
Electron microprobe	54.2	31.9	7.7	6.2	43.1	48.6	6.8	1.4
Chemical	55.1	32.5	7.2	5.2	52.3	39.3	4.8	3.6
AES	54.2	31.9	7.7	6.2	56.3	36.0	4.4	3.2

Table 20.2. Standard free energy of oxide formation, ΔF, at 400°C (673 K)

Oxide	ΔF_{673} [kJ/mol]
Ce_2O_3	−1066
MgO	−1050
Al_2O_3	−974
ZnO	−563
SnO_2	−444
PbO	−306

species present on the surface of all coatings reported here may be explained simply by consideration of the relative oxygen affinities of the metals in the coatings [20.22]. As the metallic coating solidifies the constituents of the coating compete at the surface for oxygen. Various mechanisms, such as minimum surface energy, charge diffusion, or rejection by the solid coating, may all influence surface composition; but, because of the relatively large energies involved in oxidation, the oxygen reaction would appear to dominate [20.22]. Table 20.2 lists the standard free energy of formation for a number of oxides, referenced to 400°C [20.23]. This temperature was chosen as representative of the high oxidation temperatures of the coatings after exiting the pot. Within the simple theoretical framework considered, the more negative the free energy the more likely the species will be found at the surface. For example, the large negative value for Al_2O_3 compared with ZnO explains the aluminum oxide on galvanized coatings. The large negative value for MgO justifies current interest in magnesium additions. Defining the surface enrichment factor, β^s, by:

$$\beta^s = \frac{X^s/Y^s}{X^c/Y^c} \qquad (20.1)$$

where X denotes the solute concentration and Y the solvent concentration and where superscripts s and c refer to surface and bulk respectively, and assuming the segregation mechanism may be described by the Gibbs adsorption isotherm, then the following equation may be derived:

$$\log_{10} \beta^s = \log_{10} e \, (\Delta F_y - \Delta F_x)/RT \, . \qquad (20.2)$$

Equation (20.2) describes the segregating effect of oxidation, where ΔF is the difference in free energy of oxide formation between the two metallic species, x and y, solute and solvent, respectively and where R is the universal gas constant and T is the absolute temperature. Figure 20.2 shows good agreement of (20.2) with all four categories of hot-dipped metal surfaces considered; except for the slope of the line of best fit which is only 0.07, rather than unity. This discrepancy in slope may indicate the oxidation process is not in thermodynamic equilibrium. Nevertheless, the general energy dependence of (20.2) appears to be supported empirically by the data in Fig. 20.3.

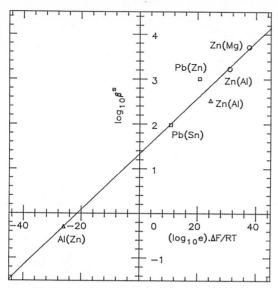

Fig. 20.3. Surface enrichment, β^s, for a range of hot-dipped metallic coatings on steel as a function of the difference in free energy of oxide formation, ΔF, between solute (*inside brackets*) and solvent metals (*outside brackets*)

a) Galvanized Steel

The solid surface formation on hot-dipped metallic coatings is a complex process. A steel strip is passed through a liquid metal bath (or pot). The thickness of the resultant liquid coating, typically 20 μm, is determined by the combined effects of gravity runoff and air jet stripping. The liquid coating, exposed to ambient air, cools rapidly below its solidification temperature. For galvanized coatings this degree of supercooling is about 8°C [20.24]. Nucleating at or near the steel surface, the coating then solidifies rapidly. Inside the galvanized coating, the initial rapid solidification releases sufficient latent heat of fusion that the remaining liquid is brought back to the melting temperature, 418°C [20.24]. Final solidification of the outer coating may then be quite slow, possibly over several seconds, until the remaining latent heat is dispersed, and those characteristic granular structures popularly called spangles are formed.

The chemistry of a galvanized surface is complex even though the galvanizing process only involves passing the steel through a molten bath of zinc and a few minor alloying elements, chiefly aluminum at about 0.2 mass %. During the slow final solidification and while the temperature remains high after solidification until water quenching, significant surface segregation of the metals occurs. Prior to passivation, the surface of the galvanized coating is predominantly aluminum oxide with some zinc oxide and a varying amount of hydrocarbon contamination [20.22] (Fig. 20.4a). Unpassivated galvanized surfaces typically have 20–30 atomic % aluminum and 5–10 atomic % zinc, with the variation largely depending on surface carbon levels which are typically 10–30 atomic % . Though

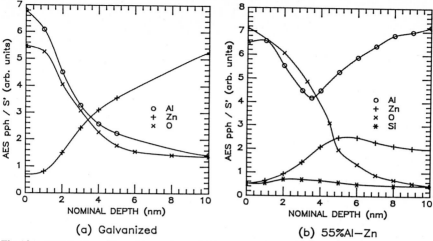

Fig. 20.4. AES depth profiles of the surfaces of **(a)** galvanized steel, and **(b)** 55% Al-Zn coated steel. The AES signals (pph) have been normalized with modified elemental sensitivity factors (S')

significant levels of hydrocarbons are present on all industrial surfaces, the hydrocarbons on fresh metallic coatings are generally very thin, typically less than 1 nm.

The aluminum oxide surface on galvanized steel could be expected to form an excellent barrier to oxygen diffusion and hence a barrier to oxidation of the zinc coating beneath. Unfortunately, if galvanized steel is left unpassivated and stored in a wet environment, the aluminum oxide soon converts to aluminum hydroxide and an unseemly "white" corrosion of the zinc forms [20.22].

The lead content of galvanized coatings varies greatly (typically from 0 to 1.5 mass %), depending on the manufacturing process. Lead has an extremely low solubility in solid zinc and so is rejected during solidification both towards grain boundaries and towards the surface of the coating. This rejection mechanism is separate from the oxidation mechanism presented above, as during solidification the lead coalesces within the remaining liquid to form discrete metallic lead particles at the final solid interfaces rather than a continuous thin oxide layer on the surface. The absence of lead at the immediate surface, however, indicates the lead is under the surface aluminum oxide layer [20.3].

When a galvanized steel sample was prepared by hand-dipping a steel coupon into a laboratory galvanizing pot, to which approximately 0.1 mass % magnesium had been added, the surface was enriched in magnesium to 33 atomic % [20.22]. Clearly, addition of the more oxygen-avid metal magnesium to the bath displaces some of the aluminum oxide at the surface, with magnesium enriching the surface of galvanized steel some three times more readily than aluminum, as determined by the ratio of surface to bulk concentration [20.3].

Since aluminum oxide is a tightly bound, stable oxide in dry conditions, its stability is likely to be affected by defects in the oxide and these in turn be affected by trace impurities. Trace levels are beyond the sensitivity limits of

Auger analysis (typically limited to concentrations above about 0.5% atomic). SIMS does not share this limitation and elements detected by SIMS on the surface of a series of galvanized steel samples but not detected by AES were arsenic, indium, lithium, gallium, and fluorine, though copper, tin and bismuth could not be detected at low levels with the experimental conditions used because of peak overlap. Clearly, knowledge of the surface composition is vital in such work.

b) 55% Al-Zn Coating

Commercial 55%Al-Zn coatings contain 54 mass % (73 atomic %) aluminum, 44 mass % (25 atomic %) zinc, and 1.5 mass % (2 atomic %) silicon [20.3]. A typical depth-profile from a commercial sample is shown in Fig. 20.4b. The commercial coatings are dual phased with aluminum-rich or zinc-rich regions typically 2–8 μm in size. The depth-profile in Fig. 20.4b therefore represents an average of these two phases. The separate phases would be expected to passivate differently and are therefore of separate interest, but the minimum beam diameter of 3 μm in our instrument makes separate analysis quite difficult.

Besides knowledge of the immediate surface, it is essential to know how the composition of the coating varies through its full thickness. Such information not only helps explain the composition of the immediate surface (because of segregation during solidification) but how the coating will perform in long term corrosion. The GDS technique is outstanding in its ability to depth-profile metallic coatings, profiling at typically 1 μm/minute, with high sensitivity (around 10 ppm), for virtually the whole Periodic Table, including hydrogen. Figure 20.5 shows a depth-profile of a 23 μm thick 55%Al-Zn coating on steel. The depth-profile shows the thin oxide covered surface, the nearly uniform centre of the coating and the complex zinc, aluminum, silicon, iron alloy region, described earlier, joining the steel to the coating.

20.1.5 Treated Metallic Coating Surface

Commonly applied passivation treatments for hot-dipped coatings contain either chromic acid (hexavalent chromium, or chromium VI) or mixed hexavalent/trivalent chromium (chromium VI/chromium III) based solutions [20.2]. When chromate solutions were first formulated, that is before surface analysis, it was not known that the surface of hot-dipped galvanized steel was predominantly aluminum oxide. While chromium VI reacts with zinc it does not react with aluminum oxide, so the acid must first disrupt the surface layer to reach the zinc beneath. The chromium III component then forms an inert protective barrier while the chromium VI is reduced to chromium III by reaction with the zinc. Alternatively, a fluoride may be added to the chromate solution to attack the aluminum oxide and thereby increase the reactivity of the surface [20.12].

A valuable method of AES analysis for monitoring total surface chromium levels is integration of the area under the chromium depth-profile [20.25]. This method is of special interest in recording the way chromium concentrations vary over small distances, for example, across grains or between grains, when the stan-

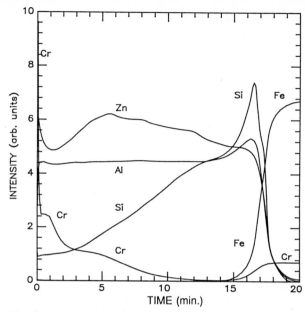

Fig. 20.5. GDS depth profile of 55% Al-Zn coated steel

dard chemical analysis for chromium requires an area of approximately 67 cm².

In a separate investigation, treated tinplate (tin coated steel sheet) was found to have fine scratches on the surface which were detrimental to the product appearance [20.3]. To identify which roll was responsible for the scratching, it was necessary to know whether the scratching occurred before or after the chromate passivation. Auger line scans across the scratches recorded the same chromium levels inside the scratches as outside, indicating scratching occurred before chromating. The faulty roll was subsequently identified and removed from the line.

20.1.6 Metal–Polymer Interface

When studying the adhesion of paint or polymer systems to metallic coatings, it is useful to know the location at which adhesion failure occurs [20.3]. Failure often occurs at the interface between the paint and the metallic coating which would indicate a problem exists either with the paint formulation or with surface cleanliness. But in one investigation, paint was removed from a galvanized surface by adhesive tape following a reverse impact test and XPS showed the underside of the removed paint contained aluminum, zinc, and chromium which had been detached from the metallic coating surface. Auger analysis of the exposed metallic coating showed mostly zinc, with the expected mixed chromium, aluminum oxide surface no longer present. Apparently the paint had adhered so well that separation occurred at the more loosely bound oxide in the metallic coating surface. This result helped focus further work into the nature of the oxide rather than into the paint. Indeed, samples taken before painting indicated the surface oxide was 9 nm thick rather than the usual 4 nm for that coating.

20.1.7 Polymer Surface

The natural weathering of paint or polymer surfaces involves complex processes of chemical and biological attack and thermal and photo degradation, resulting in subtle to gross physical and chemical changes. Specific areas where XPS can provide information on the weathering of paint surfaces are: a) changes in surface composition, for both organic and inorganic paint constituents [20.26]; b) oxidation and hydrolysis, especially from wet storage, by monitoring the oxygen to carbon ratio [20.27–29] and the carbon peak width [20.26]; c) changes in cross-linking, by monitoring the nitrogen peak, when cross-linking agents are

(a) Unexposed   (b) QUV: 50 light hours

(c) QUV: 200 light hours   (d) QUV: 1000 light hours

1 μm

Fig. 20.6a–d. SEM micrographs of a commercial polymer showing the progressive revealing of pigment particles following exposure in a QUV weatherometer

used [20.30]; d) bond saturation, via shake-up satellites [20.26, 28, 31]; and e) variations in pigment concentration with depth, through combined XPS and ion-bombardment. The titanium depth-profile may be particularly useful, because most paints contain titanium (as titanium dioxide, or rutile) and on unexposed paint surfaces the titanium concentration steadily increases with depth, thus providing a possible depth marker for paint film loss during weathering [20.29]. The ion-bombardment of polymers is known to create compositional artefacts under certain conditions, so special care is required with the technique [20.32].

As part of a larger program, described in [20.33], metal panels were coated with a commercial white silicone modified polyester (SMP) and kept unexposed or subjected to 50, 200 or 1000 light hours (2000 total hours) exposure in a cyclic accelerated weathering instrument (QUV). White was chosen because it is a relatively simple pigmented system based on rutile (titanium dioxide). Unexposed and exposed samples were then examined by scanning electron microscopy (SEM) and XPS. The SEM micrographs, in Fig. 20.6, show that the surface morphology changed markedly with exposure, with the smooth unexposed surface becoming quite rough following erosion of the polymer resin and exposure of the pigment particles. XPS analysis combined with argon ion etching indicated the presence of a pigment-free layer (about 20 nm thick) on the unexposed polymer surface which quickly disappeared during accelerated testing. The elements detected by XPS were carbon, nitrogen, oxygen, silicon and titanium. The surface carbon and nitrogen concentrations both dropped during testing, their loss representing the loss of the polymer and of the melamine cross-linking agent in the polymer, respectively. Figure 20.7 shows that the nitrogen/carbon ratio decreased during testing, with rapid loss of nitrogen in the first 200 hours; whereas, the oxygen/carbon ratio increased rapidly in the first 200 hours. An increase in oxygen/carbon ratio could indicate increased oxidation or hydrolysis of the polymer,

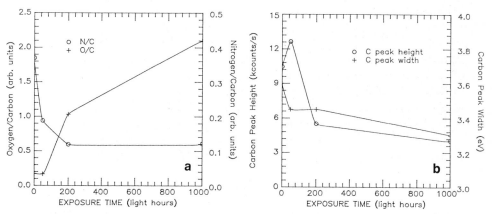

Fig. 20.7a,b. Changes in XPS signals from a commercial polymer following exposure in a QUV weatherometer

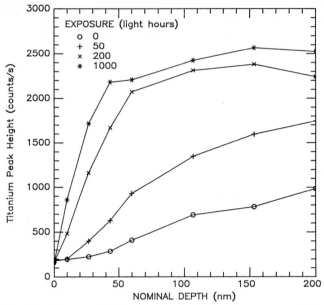

Fig. 20.8. Changes in XPS depth profiles for titanium from a commercial polymer following exposure in a QUV weatherometer

however, the carbon peak width decreased slightly (from 3.6 eV unexposed to 3.3 eV at 1000 horus) which does not support an increase in the number of C-O bonds. The increasing oxygen signal was therefore more likely to be associated with the exposed pigment particles. This exposure of pigment is best illustrated, in the XPS result, by changes in the titanium depth-profiles (Fig. 20.8).

20.2 Conclusion

Surfaces play a vital role in problems of corrosion, adhesion, and appearance of industrial products. A painted, coated stell sheet, for example, contains upwards of nine distinct interfaces, or surfaces, all of which at some stage influence manufacturing and final product performance. The surface composition of such industrial coatings often differ radically from their bulk composition. AES and XPS provide essential information on surface composition and the changes to surface composition during such complex, multistage industrial processing.

Acknowledgements. The author thanks the management of BHP Steel International Group, Coated Products Division, Research and Technology Centre for permission to publish this work, and the many colleagues he has worked with on many of the projects reported here, and particularly Dr. P.D. Mercer, and C.J. Aldrich, D.G. Jones and J.T. Murada who recorded some of the data presented. GDS data were kindly supplied by Mr. H. Hocquaux, UNIREC, France.

References

20.1 M.P. Seah: Surf. Interface Anal. **2**, 222 (1980)
20.2 P.D. Mercer, R. Payling: Nuclear Instrum. Meth. **191**, 283 (1981)
20.3 P.D. Mercer, R. Payling: Proc. 4th Australian Conf. on Nuclear Techniques of Analysis, AINSE, Lucas Heights, NSW, 6–8 Nov. 1985, p. 132
20.4 M.T. Thomas, R.H. Jones, D.R. Baer, S.M. Bruemmer: The PHI Interface **3**(2), 3 (1980)
20.5 R.H. Edwards, F.J. Barbaro, K.W. Gunn: Metals Forum **5**(2), 119 (1982)
20.6 A.P. Coldren, A. Joshi, D.F. Stein: Metall. Trans A **6a**, 2304 (1975)
20.7 M.P. Seah: Surf. Interface Anal. **1**, 86 (1979)
20.8 G. Hanke, K. Muller: Surf. Sci. **152/153**, 902 (1985)
20.9 R.H. Stulen, R. Bastasz: J. Vac. Sci. Technol. **16**(3), 940 (1979)
20.10 J.A. Slane, S.P. Clough, J. Riker-Nappier: Metall. Trans. A **9**, 1839 (1978)
20.11 P.L. Coduti: Metals Finishing **78**, 51 (1980)
20.12 V. Leroy: Mater. Sci. Eng. **42**, 289 (1980)
20.13 R.A. Iezzi: PhD Thesis, Lehigh University (1979)
20.14 L.E. Davis, N.C. MacDonald, P.W. Palmberg, G.E. Riach, R.E. Weber: *Handbook of Auger Electron Spectroscopy*, 2nd ed. (Physical Electronics Div., Perkin-Elmer Corp, Eden Prairie, MN 1978)
20.15 C.D. Wagner, D.A. Zatko, R.H. Raymond: Anal. Chem. **52**, 1445 (1980)
20.16 S. Craig, G.L. Harding, R. Payling: Surf. Sci. **124**, 591 (1983)
20.17 K. Ishikawa, Y. Tomida: J. Vac. Sci. Technol. **15**, 1123 (1978)
20.18 R. Payling, G.L. Harding, S. Craig: Appl. Surf. Sci. **24**, 11 (1985)
20.19 R. Bastasz, G.J. Thomas: J. Nucl. Mater. **76–77**, 183 (1978)
20.20 S.J. Makimattila, E.O. Ristolainen, M. Sulonen, V.K. Lindros: Scripta Metall. **19**, 211 (1985)
20.21 V. Leroy, B. Schmitz: Scand. J. Metall. **17**, 17 (1988)
20.22 R. Payling, P.D. Mercer: Appl. Surf. Sci. **22/23**, 224 (1985)
20.23 C.R. Weast (Ed.): *Handbook of Chemistry and Physics*, 55th ed. (CRC Press, Cleveland, OH 1974) D-45–50
20.24 D.I. Cameron, G.J. Harvey: 8th Int. Conf. on Hot Dip Galvanizing, London (1967)
20.25 R. Payling: Appl. Surf. Sci. **22/23**, 215 (1985)
20.26 D. Briggs: In *Practical Surface Analysis by Auger and X-ray Photoelectron Spectroscopy*, ed. by D. Briggs, M.P. Seah (Wiley, Chichester 1983) p. 359
20.27 D.T. Clark: In *Physicochemical Aspects of Polymer Surfaces*, Vol. 1, ed. by K.L. Mittal (Plenum, New York 1983) p. 3
20.28 A. Dilks: J. Polymer Science: Polymer Chemistry Edition **19**, 2847 (1981)
20.29 S. Skeldar, A. Zalar, R. Zavasnik: 16th FATIPEC Congress, 335 (1982)
20.30 E. Takeshima, T. Kawano, H. Takamura: Trans ISIJ **23**, 652 (1983)
20.31 L.C. Lopez, D.W. Dwight, M.P. Polk: Surf. Interface Anal. **9**, 405 (1986)
20.32 D.E. Williams, L.E. Davis: In *Characterization of Metal and Polymer Surfaces*, Vol. 2, ed. by L.H. Lee (Academic, New York 1977) p. 53
20.33 C.J. Aldrich, E.M. Boge, J.T. Murada, R. Payling, J.R. Bird: *EUREM 88*, Proc. 9th European Congress on Electron Microscopy, York, England, 4–9 September 1988

21. Thin Film Analysis

G.C. Morris

With 9 Figures

The increasing importance of thin films for new technologies has encouraged fundamental and applied research on their physical and chemical structures and on the interfaces made with them [21.1]. Their physical structures (e.g. morphology, topography, crystallite properties, extent and type of defects) are explored by diffraction and microscopic techniques including X-ray and electron diffraction, scanning tunnelling microscopy and ultrasonic microscopy. Their chemical structures (e.g. element type, concentration and spatial distribution) are explored by microanalytical techniques such as Fourier transform infra-red spectroscopy, secondary ion mass spectrometry (SIMS), X-ray photoelectron spectroscopy (XPS), Auger electron spectroscopy (AES), ion scattering spectroscopy (LEIS, HEIS), as well as dispersive X-ray analysis and electron energy loss spectrometry with scanning and transmission electron microscopy. As an ultimate objective, researchers desire a three-dimensional elemental map on an atomic scale for the thin film and its interfaces. Some progress towards that aim has been made, but achievement is some time away.

Thin film analysis may be considered as characterizing materials about one micron thick in which the outer few atomic layers probably have a different structure and composition for a number of reasons especially atmospheric contamination. A wide diversity of application areas, some of which are noted in Table 21.1, need such analysis to solve problems and understand phenomena which occur. The analytical techniques used for the particular application must be carefully selected using criteria such as information depth probed, lateral resolution required, detection limit needed, elemental type and chemical state investigated. Complementary methods are recommended to confirm or deny conclusions made using one technique alone. Thus, a typical analysis might include e.g. X-ray diffraction data, a physical image from electron microscopy, AES and XPS analyses of specific areas exposed by ion sputter removal of successive layers (depth profile analysis). The compositional data from the two techniques, AES and XPS, could be compared. Analytical electron microscopy could also give useful comparative data. Elsewhere in this book, each of the main analytical techniques have been reviewed so that their advantages/disadvantages for a particular application can be assessed.

The aim of this chapter is to illustrate how surface sensitive techniques provide a data base to understand processes and phenomena which occur with particular thin films. Example from several of the different application areas shown in Table 21.1 could be used. However, it is more instructive to use a single sys-

Table 21.1. Some application areas benefiting from thin film analysis

Materials Science/Technology	Film Technology
Corrosion	Adhesion problems
Passivation	Coating thickness
Segregation phenomena	Failure analysis
Failure analysis	Optical problems
Inclusion analysis	Electrical problems
Diffusion analysis	Decorative problems
Alloy composition	Segregation problems
Surface modification	Lubrication phenomena

Metallurgy	Semiconductor/Electronic Materials
Fatigue failures	Dopant distribution
Stress corrosion cracking	Interface widths
Chemical reaction problems	Purity analysis
Diffusion phenomena	Processing phenomena
Segregation phenomena	Diffusion phenomena
Composition studies	Failure diagnosis
Quality assurance	Ageing processes
Temper brittleness	Cleaning processes

Others

Catalyst distribution	Quality assurance
Catalyst deterioration	Medical prosthesis
Mineral composition	Packaging problems
Materials signature	Adhesion phenomena
Tribology studies	Delamination
Surface-bulk differences	Magnetic tape irregularities
Dental materials	Opto-electronic films

tem even though it will have its own complex of problems because the general principles of characterization remain the same, e.g. structure and composition profile are determined, 'good' systems are compared with 'faulty' systems to probe defects, etc. Thin film photovoltaics based on II-VI semiconductors is the system chosen for illustrating how surface analytical methods are an essential probe for examining the materials and their interfaces.

21.1 Thin Film Photovoltaics

21.1.1 Use for Solar Electricity

Photovoltaic devices made from thin films have a future as a viable solar energy alternative if module costs are lowered and device stabilities of 20–30 years can be achieved [21.2]. To establish those aims, efficient, stable photovoltaic systems must be made inexpensively in industrial quantities. The criteria of efficiency and stability require that the effects which arise from contaminations at surfaces, in the bulk and at interfaces, must be minimized, chemical reactions and compositional changes induced by humidity, light and temperature must be inconsequential, and time dependent processes such as impurity segregation and interdiffusion must be insignificant. Surface sensitive analytical methods (SIMS, XPS, AES), combined with structural, morphological (XRD, SEM) and micro-electrical characterization (electron beam induced current – EBIC) play a key role in probing the processes which occur during preparation of the thin films and when fabricating them into photovoltaic devices. This will be illustrated using one of the most promising devices for solar energy conversion, viz. that based on the heterojunction cadmium sulphide/cadmium telluride (nCdS/pCdTe) [21.2].

21.1.2 The Thin Film Solar Cell: Glass/ITO/nCdS/pCdTe/Au

The thin film solar cell glass/ITO/nCdS/pCdTe/Au is fabricated in several steps:

(i) The base material is soda glass coated with a 200 nm layer of indium tin oxide (ITO) , (ii) a layer of about 80 nm CdS is deposited on the glass/ITO out of a chemical solution [21.3] or by an electrodeposition method using a solution containing Cd^{2+} and S_2O_3=ions [21.4], (iii) a layer of about 1.5 μm CdTe is electrodeposited on the CdS from a solution containing Cd^{2+} and $HTeO_2^+$ ions [21.5, 6], (iv) the as-deposited CdS,CdTe films are n-type and the heterojunction nCdS/pCdTe is formed during a conversion step in which the film is heated to about 400°C in air, (v) the desired low resistance contact to the pCdTe is accomplished by a chemical etching of the surface followed by metallization e.g. using a 60 nm thick layer of gold (Au).

Single junction solar cells which convert up to 13% of sunlight into electrical energy have been made [21.4, 6–8]. To fabricate laboratory cells reproducibly and to transfer the technology to industry, variables which affect each process must

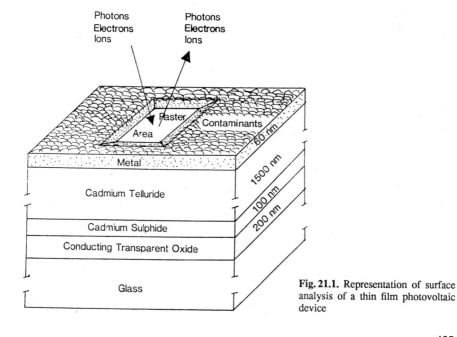

Fig. 21.1. Representation of surface analysis of a thin film photovoltaic device

be recognized and controlled. Development proceeds in an iterative fashion with the knowledge gained at each processing step fed back to define the processes more precisely. Progress in understanding scientifically what is the influence of each variable and what occurs in each process requires surface analysis. This will be illustrated by discussing some typical examples of use in e.g. (i) purity control of the films, (ii) doping profiles in the films, (iii) analysis of layered structures, (iv) analysis of surfaces after processing steps.

Figure 21.1 represents the thin film system and illustrates the main techniques discussed. The primary ion beam is used for SIMS and for sputter profiling away surface contamination and successive layers to expose the underlying surfaces for compositional analysis by XPS and AES. It should be remembered that the uncertainty in such analyses is determined by bombardment induced atomic mixing and by the measured signals being convolutions of species concentrations over the electron escape depth. These points are discussed in Chap. 4.

21.2 Film Purity

21.2.1 Low Level Impurities – Qualitative

Material purity is an important characteristic of photovoltaics. Some desired data include (i) the types, extent and location of impurities in the semiconductor films, (ii) the limitations imposed by those impurities on the final cell's performance, (iii) the minimum purity of starting material to produce adequate cells, (iv) the alterations in various impurity profiles during the fabrication of cells from the films.

Because of the low concentration of impurities and especially because these films (e.g. 1.5 μm CdTe) are on a thick (1 mm) glass substrate, SIMS is an essential rapid and routine analytical tool for multi-element detection. In Fig. 21.2 the positive SIMS spectrum from an electrodeposited CdTe film is compared with that from a single crystal purchased as 5N purity from a commercial source. Spectra were obtained using a Perkin Elmer PHI Model 560 Multitechnique System (XPS, SAM, SIMS). The ion gun was a differentially pumped electron impact Ar^+ source. SIMS data for both the film and crystal were recorded under similar experimental conditions. It is clear that the electrodeposited film had fewer impurities than the commercial 5N crystal used for the data of Fig. 21.2 and, indeed, two other 5N crystals from different commercial suppliers [21.5].

Cross Contamination and Fringe Neutral Problems in SIMS. In such studies, in addition to the obvious precautions of keeping constant the type of primary ion, its energy and current density, secondary lens settings, sample geometry and environment, the 'fidelity' of data for trace element analysis required identification and elimination of transported material from surrounding areas onto the sample analysis area [21.9, 10]. This transported material which was left from a previously sputtered sample contaminated the sample surface producing secondary

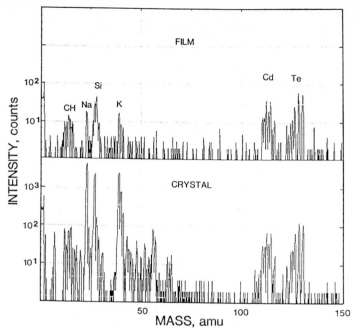

Fig. 21.2. Positive SIMS spectra for an electrodeposited CdTe film and a CdTe single crystal supplied as 5N pure. From [21.9]

ions intermixed with 'true' sample signals. To minimize cross-contamination the sample chamber must be carefully designed to avoid large area metal surfaces near the sample.

Of equal concern for reliable SIMS results are the secondary ions produced by energetic unfocused fringe neutrals bombarding the region around the analysis area. Secondary ions produced by fringe neutrals striking the sample mount and attachments, e.g. spring clips or cover plates, invalidate results of trace element studies [21.9]. The extent of this effect has been quantified as a function of distance from the beam centre and of beam current and beam energy [21.11]. For example, with an Ar^+ current in the range $1 \mu A$–$4 \mu A$ and a beam energy of 4 or 5 keV, the secondary ion signals which arose from neutrals at distances greater than 3 mm from the beam centre was 1.6% of those generated by the primary ion. At distances greater than 6 mm and 9 mm from the beam centre, the values were 0.3% and 0.1% respectively. Hence, to minimize the contribution to the 'true' SIMS signal of fringe neutral signals, two strategies should be used. Firstly, sample mounting hardware such as cover plates, screws, etc., on the front of the sample should be avoided. A small stub projecting about 2 cm from the sample mount covered by the sample has been found useful [21.9]. Secondly, a large crater such as 9 mm × 9 mm should be etched during initial profiling and then the analysis taken over a smaller area such as 3 mm × 3 mm so that data representative of the film's bulk composition could be collected. Thus the

SIMS signal whether from the primary ion or the fringe neutrals would originate initially from an area at a similar depth and with similar composition.

The use of an ion gun with a curved ion path can significantly reduce fringe neutral effects. Analysis of the CdTe films with such a gun (using Cs^+ and O^- primary ions) detected each element observed with the simpler in-line ion gun thus validating the procedures used.

21.2.2 Low Level Impurities – Quantitative

The thin film used to provide the SIMS data of Fig. 21.2 was made using an A.R. grade (2N) source of cadmium sulphate in the deposition solution although this was subsequently electrolytically purified. Purer films can be made with better starting material as illustrated in Fig. 21.3 where SIMS data for films prepared by A.R. grade (2N) and a 4N grade of cadmium sulphate are compared. Although the data of Fig. 21.3 show that the film made from the 2N grade was 'dirtier' over all regions of the mass spectrum, cells as efficient as those made with the purer starting material were obtained. That result suggested that the cleaning procedures [21.5] removed the electrically active impurities.

Although SIMS is primarily a qualitative technique, it is useful for comparative work provided similar matrix samples are used and the precautions stated above are taken. Useful information as illustrated by the data of Figs. 21.2 and 21.3 can be obtained. However, quantitation is often needed and other techniques must be used to complement the SIMS data. Quantitative XPS and AES data can

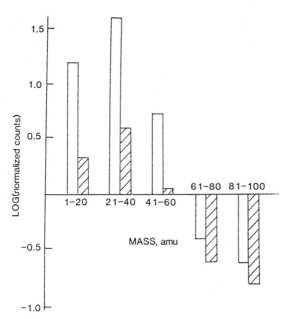

Fig. 21.3. Intensities of positive SIMS signals in various mass regions for electrodeposited CdTe films prepared using a 2N grade (□) and a 4N grade (▨) of cadmium sulphate. The SIMS counts have been normalized to the combined Cd + Te count

be useful, although restricted to elements with concentrations above 0.1%. Inductively coupled plasma-atomic emission and atomic-absorption spectrophotometry have been used to determine the magnitude of impurities in the components of the plating bath solution and six different supplies of cadmium sulphate [21.12]. The origin of an impurity and in some cases an upper limit to its concentration in the films has been determined.

With the type of studies outlined above, the purity of the films produced under varying conditions could be monitored to ensure that the films were purer than 5N single crystals purchased from commercial suppliers.

21.2.3 Doping Profiles in Thin Films

One of the common uses of SIMS is to look at doping profiles e.g. as a consequence of ion implantation of ^{11}B or ^{31}P in Si to alter the electrical character. Provided that the rate of sputter profiling of the host materials was known, quantitative implantation profiles of ^{11}B in Si have been tracked by SIMS [21.13] over six orders of magnitude to levels of $\sim 10^{13} \, cm^{-3}$.

The sputter rate of the host material needs to be determined from other experiments, usually with reference to standard materials such as Ta_2O_5 film on Ta. For electrodeposited CdS and CdTe films, thickness of various films were determined by SEM , by a surface profiler, by near infra-red interference spectrophotometry, and by the charge passed during deposition. The rate of Ar^+ sputtering of these films under defined ion beam conditions was compared with the rate of sputtering standard Ta_2O_5/Ta film [21.14]. The ratios of sputtering rates were $CdS:CdTe:Ta_2O_5$ as 6.3:3.9:1 [12.11].

The data in Fig. 21.4 illustrate a typical doping profile formed in this case by heating n CdTe in an ampoule with phophorus and tellurium. The SIMS signal intensity from the species of $^{31}P^+$ and $^{31}PO^+$ were determined as a function of

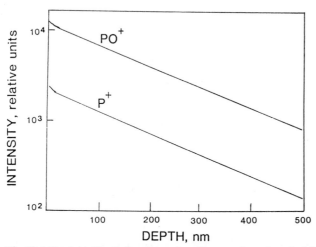

Fig. 21.4. Depth profile of phosphorus doped into an electrodeposited CdTe film measured by SIMS signals from PO^+ and P^+. Rate of depth profiling was $20 \, nm \, min^{-1}$

the depth into the cadmium telluride. Note that the PO$^+$ species had a similarly shaped profile to the P$^+$ species and was considerably more intense, making it the preferred ion species to monitor the phosphorus level.

21.3 Composition and Thickness of Layered Films

A common problem in thin film analysis is determining the composition and thickness of adventitious or deliberately added thin layers on surfaces or at interfaces. When the layers have a varying composition profile, this can be a complex problem. Examples will illustrate how some types of problems can be solved.

21.3.1 Composition Gradation in Films, e.g. $Cd_xHg_{1-x}Te$ Films on Platinum

Films with composition $Cd_xHg_{1-x}Te$ have been cathodically electrodeposited onto platinum [21.6, 15]. There was a possibility because of the deposition method used that the film composition could vary as the film thickness increased. Measurement of the atomic concentrations by XPS with ion sputter profiling was impractical because mercury may preferentially sputter from the film surface. Instead, a series of successively thicker films were made using the same deposition conditions and solutions of similar composition. The surface atomic concentration of mercury expressed as (atom percent Hg)/atom percent (Hg+Cd) was obtained by XPS. These data are shown in Fig. 21.5 which illustrates that a composition gradient would exist in the deposited films.

21.3.2 Thin Overlayers on Films

a) Growth of Oxide/Hydroxide layers on CdTe Films

Electrodeposited CdTe films exposed to the atmosphere grew oxide/hydroxide layers which SIMS depth profiling showed to be about 2 nm thick. Angular resolved XPS which provided the data of Table 21.2, confirmed that the oxygen species were concentrated at the surface layers. The binding energy for the XPS O 1s peak differed for the surface and bulk species. The surface oxygen was probably present as an hydroxyl species [21.16].

When the n-CdTe was type converted by heating the film in air at 350–400°C for about 10–15 minutes, an oxygen-rich film, readily removed by a basic etch, was formed on the surface as shown by the XPS spectrum for the Te 3d region (shown in Fig. 21.6) [21.17]. The oxygen-rich layer, said to be CdTeO$_3$ [21.18]

Table 21.2. Angular resolved XPS data from CdTe films exposed to the atmosphere

	Atomic concentration [Atom %]			O 1s binding energy [eV]
	Cd	Te	O	
Grazing angle	32.0	37.5	30.5	531.9
Normal	45.8	49.3	4.9	530.4

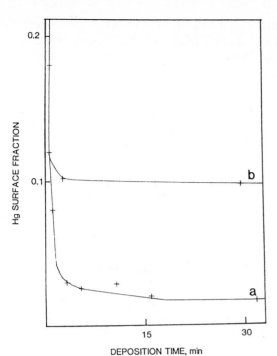

Fig. 21.5. Mercury content expressed as (Hg/Hg+Cd) on the surface of electrodeposited $Cd_x Hg_{1-x}Te$ films of various thicknesses prepared from solutions containing Hg concentrations of (a) 5 ppm (b) 10 ppm

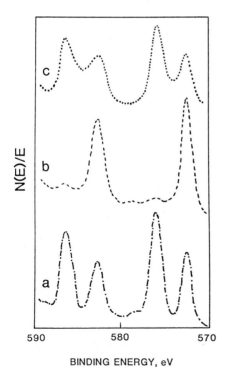

Fig. 21.6. XPS spectrum in the Te $3d$ region of the surface of a CdTe crystal (a) after air anneal 10 min, 350°C, (b) after 20 s etch with 80°C KOH, 30% (w/w), (c) after ion sputtering the oxidized film for 2 min

411

was shown to be a few nm thick by depth profiling until the atomic concentrations of Cd and Te were 50%. The change in the Te $3d$ peak position as a function of ion sputter time was another useful way of displaying the surface nature of the oxygen-rich region as shown in Fig. 21.6c.

b) Surface Modification of Thin Films

Surface modification of materials is finding increasing importance in technology. Probing the chemical changes which surface modifications introduce is essential to suitable control of the modifications. For example, the surface of a p-CdTe film must be modified for a low specific resistance contact with a metal and this has been accomplished with a thin, very heavily doped layer which allowed carrier tunnelling. A favorite recipe to produce the layer has been to use one of several chemical etchants such as bromine in methanol (BM), orthophosphoric acid with nitric acid (OPN) and acidified potassium dichromate (KD). XPS studies have shown the chemistry of what is happening on the surface as will be illustrated by data [21.19] obtained using a BM etch (0.1% Br_2 in methanol). Figure 21.7

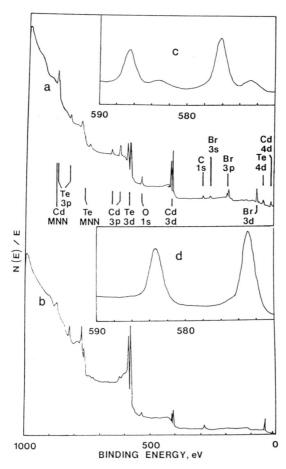

Fig. 21.7. XPS survey spectra of CdTe film after statically etching for 5 s in 0.1% bromine-methanol then (a) blow-dried with nitrogen; (b) washed each in methanol (1 min) and Milli-Q water (1 min) then blow-dried with nitrogen; (c) Te $3d$ region from spectrum (a); (d) Te $3d$ region from spectrum (b)

412

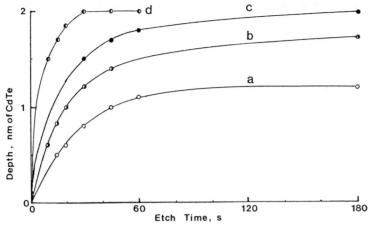

Fig. 21.8. Depth of Cd depleted surface with etch time for CdTe single crystals statically etched by BM of different concentrations: (*a*) 0.1%; (*b*) 5%; (*c*) 10%; (*d*) 50%. From [21.19]

shows the XPS survey spectra of the film after etching and before and after washing with water. The bromine $3d$ peaks were observed near 69 eV when the film was not properly washed. Washing also altered both the intensity and position of the cadmium $3d$ and tellurium $3d$ peaks. The unwashed film showed a strong cadmium $3d$ signal consigned to cadmium bromide on the surface and also showed a split tellurium $3d$ signal assigned to Te^0/Te^{2-} and tellurium bromide. Washing removed the bromide species as evidenced by the disappearance of the Te^{4+} peaks (Fig. 21.7d) and a change in the relative intensity of the Cd:Te $3d$ peaks from 0.7 (unwashed) to 0.5 (washed).

The depth of the cadmium depleted region was determined using XPS with ion profiling as a function of chemical etch time and concentration. A sputter rate of about $0.5 \, \text{nm min}^{-1}$ was used with XPS analyses after successive 1 min sputter times. The atomic concentrations of Cd and Te were plotted as a function of sputter time with the endpoint reached when the Cd:Te ratio was 1.00, the value it was in the crystal bulk. Figure 21.8 shows the result of such a study. The amount of material removed depended on the time and the concentration of the chemical etch but even though the crystal was etched at a rate of $14 \pm 4 \, \text{nm s}^{-1}$ by a static 0.1% BM solution, the depth of the Cd depleted region was limited to about 2 nm. Note that even though the surface material was not CdTe the depth of the Cd depleted region has been expressed as if the surface species sputtered at the same rate as CdTe. This is a convenience, but not accurate e.g. elemental Te sputters about twice as fast as CdTe.

c) Contaminant Modification of Surfaces e. g. Carbon on Gold

As discussed previously, thin layers of adventitious materials can grow on exposed surfaces. For an example, the effects of contaminants on the gold layer evaporated to form a metal contact to the etched p-CdTe films are discussed. The contact resistance increased twenty fold and the cell's efficiency decreased after

413

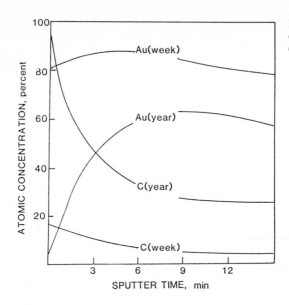

Fig. 21.9. Depth profiles of the atomic concentrations of C and Au in Au deposited on CdTe films

the cell was left exposed to the atmosphere for one year. A freshly evaporated gold surface showed only the presence of gold at and near the surface. However, as shown by the AES data of Fig. 21.9, the gold contact deteriorated with age because of the accumulation with time of carbonaceous materials on the surface. Suitable encapsulation and permanent contacts would normally be used to avoid such problems. Note that the atomic concentrations of carbon and gold in the data of Fig. 21.9 are less than 100% after several minutes of ion beam etching. Tellurium species are observed in this region, but a detailed study of the diffusion of the near surface species has not been published.

21.4 Beam Effects in Thin Film Analysis

Beam induced effects such as preferential sputtering of one element in a compound, preferential diffusion of a species, surface oxidation, adsorbate dissociation, redeposition, etc., are a constant problem in thin film analysis. The use of a multi-technique facility can help to pinpoint the source of a problem. For example, the magnitude of the atomic concentration of oxygen on the surface of either CdS or CdTe was measured by AES to be about ten times larger than the value measured by XPS. This variation was traced to oxidation of the compounds under the electron beam probably by a thermal reaction with residual water vapour in the system [21.9, 20]. Under simultaneous ion beam sputtering the atomic concentration of oxygen measured by AES decreased as the inverse of the ion beam current and at high ion beam currents was the same as measured by XPS. Clearly simultaneous rapid sputtering was needed to provide reliable AES data for oxygen.

414

The experienced researcher generally recognizes when beam effects are altering the 'true' data. Readers are referred to the article by *Kazmerski* [21.21] for further examples.

21.5 Conclusion

Surface analytic methods are a fruitful way of learning some details of thin film structures as shown by examples in this chapter mainly dealing with chemical structures of thin film photovoltaics. The usual methods used to investigate these devices are by current-voltage-time-temperature relationships to probe device parameters and current generation and trapping mechanisms. Various spectroscopies such as time- and energy-resolved luminescence, quantum yield, admittance provide further details, including the type and extent of traps. Ideally, data from these experiments should be correlated with details of the physical and chemical structures available from surface analytic methods. That task is currently too complex for completion for even one type of thin film device made by one preparative method. However, some progress has been made and further development of analytic methods of near atomic resolution will hopefully lead to inexpensive, durable, efficient thin film photovoltaics.

References

21.1   H. Oechsner (ed.): *Thin Films and Depth Profile Analysis*, Topics Curr. Phys. Vol.37 (Springer, Berlin, Heidelberg 1984)
21.2   U.S. Department of Energy, National Photovoltaics Program Five Year Research Plan, 1987–1991, DOE/CH1000093-7, 1987
21.3   W.J. Danaher, L.E. Lyons, G.C. Morris: Solar Energy Materials **12**, 137 (1985)
21.4   G.C. Morris, P.G. Tanner, A. Tottszer: 21st Photovoltaic Specialist Conference, IEEE (1990)
21.5   L.E. Lyons, G.C. Morris, D.H. Horton, J.G. Keyes: J. Electroanal. Chem. **168**, 101 (1984)
21.6   B.M. Basol: Solar Cells **23**, 69 (1988)
21.7   G.C. Morris, L.E. Lyons, P. Tanner, C. Owen: Technical Reports of the 4th International Photovoltaic Science and Engineering Conference, Sydney (1989), p.487
21.8   A 10% efficient solar array covering about 5% of the area of Australia would generate more electrical power than the world's power stations
21.9   G.C. Morris, L.E. Lyons, R.K. Tandon, B.J. Wood: Nucl. Instrum. B **35**, 257 (1988)
21.10 C.W. Magee, R.K. Honig: Surf. Interface Anal. **4**, 35 (1982)
21.11 G.C. Morris, B.J. Wood: Materials Forum **15**, 44 (1991)
21.12 L.E. Lyons, G.C. Morris, R.K. Tandon: Solar Energy Materials **18**, 315 (1989)
21.13 D.S. Simons, P. Chi, R.G. Downing, J.R. Ehrstein, J.F. Knudsen: Proc. Sixth International Conference on Secondary Ion Mass Spectrometry, ed. by A. Bonninghoven, A.M. Huber, H.W. Werner (Wiley, Chichester 1988) p.433
21.14 National Physics Laboratory, Certified Reference Material NPL No. S7B83, BCR No. 261
21.15 G.C. Morris, M. Marychurch: Materials Forum **15**, 143 (1991)
21.16 C.T. Au, M.W. Roberts: Chem. Phys. Lett. **74**, 472 (1980)
21.17 W.J. Danaher, L.E. Lyons, G.C. Morris: Appl. Surf. Sci. **22/23**, 1083 (1985)
21.18 F. Wang, A. Schwartzman, A.L. Fahrenbruch, R. Sinclair, R.H. Bube, C.M. Stahle: J. Appl. Phys. **62**, 1469 (1987)
21.19 W.J. Danaher, L.E. Lyons, M. Marychurch, G.C. Morris: Appl. Surf. Sci. **27**, 338 (1986)
21.20 A. Ebina, K. Asano, Y. Suna, T. Takahashi: J. Vac. Sci. Technol. **17**, 1074 (1980)
21.21 L.L. Kazmerski: Solar Cells **24**, 211 (1988)

22. Identification of Adsorbed Species

B.G. Baker

With 10 Figures

Adsorption at a solid surface is the initial step in many heterogeneous processes. The reactivity of solids, corrosion, inhibition, catalytic reaction and some methods of separation depend on adsorption. The process may involve specific chemical interaction with surface sites and, in many cases, results in dissociation of the adsorbate molecule. The formation of the initial monolayer needs to be understood in detail in order to explain the behavior of adsorbent materials in contact with gas or solution.

The methods of surface analysis are inherently suited to the study of the atomic detail of adsorbates. Measurement of surface concentration, element identity and valence state can be made by X-ray photoelectron and Auger spectroscopy. When these techniques are applied to particulate or polycrystalline materials, the results may represent a combination of the adsorption behavior of a variety of surfaces. Different crystal planes of pure substances provide surfaces which may have markedly different adsorption properties. Techniques which aim to define the spatial arrangement of adsorbates or the details of chemical bonding are applied to prepared single crystal surfaces. The behavior of practical materials must then be interpreted in terms of a series of studies of several different crystal planes.

The following techniques have been employed to study adsorbate layers: low energy electron diffraction (LEED) to detect ordered spatial arrangements; X-ray photoelectron (XPS) and Auger electron spectroscopies (AES) to measure surface concentration and states of chemical combination; ultraviolet photoelectron spectroscopy (UPS) for detecting molecular surface species and changes in electron densities resulting from adsorption; electron energy loss spectroscopy (EELS) and reflection infrared spectroscopy for measuring vibrational states of adsorbed species. The application of these techniques is contained in the examples of adsorption studies that follow. Others are discussed in Chaps. 1, 5, (SIMS), Chap. 10 (STM), Chap. 11 (LEIS) and Chap. 12 (RHEED).

22.1 Examples of Adsorption Studies

22.1.1 Nitric Oxide Adsorption on Metals

The practical objective of most of the scientific interest in nitric oxide adsorption is the control of emissions of this gas into the atmosphere. The control of automotive emissions by catalysts containing noble metals is now established technology. The high cost of Pt, Pd and Rh and the question of reliable supply provides an ongoing incentive to discover alternative materials. Furthermore the exhaust catalyst to control nitric oxide requires the reaction of NO with carbon monoxide. This is only possible when oxygen levels are controlled at a very low level. Emissions from diesel engines and industrial furnaces contain too much oxygen to allow control by existing catalyst technology. Empirical testing of large numbers of materials has failed to solve the problem. Perhaps a more detailed understanding of the atomic and molecular features of adsorbed nitric oxid will reveal new catalytic strategies.

The decomposition of nitric oxide and the reaction of nitric oxide with carbon monoxide have been investigated on polycrystalline iron and nickel [22.1]. The initial reaction, on both metals, at 500–550 K is the decomposition of nitric oxide to nitrogen and nitrous oxide and the incorporation of oxygen into the film to form an oxide layer. The rate of the decomposition reaction decreases as the thickness of the oxide layer increases and becomes immeasurably slow at an oxide thickness > 50 nm. The oxidized surface catalyzes the reduction of NO by CO to form N_2 and CO_2 but with relatively low activity.

The initial stages of the action of nitric oxide on nickel has been studied in a number of experiments on single crystals. The chemisorption of nitric oxide on (110) nickel has been investigated by Auger electron spectroscopy, LEED and thermal desorption [22.2]. It was found that NO adsorbs irreversibly at 300 K forming a faint (2×3) LEED pattern. At 500 K this pattern intensifies. The process of dissociation is revealed by Auger spectroscopy. The data in Fig. 22.1 was

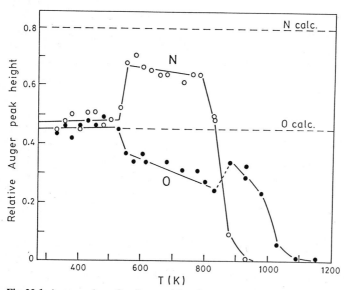

Fig. 22.1. Auger analyses for nitrogen (○) and oxygen (●) from nitric oxide adsorbed on (110) nickel. The analyses were recorded at room temperature after successively heating the crystal to the plotted temperatures at a rate of $10 \, \text{Ks}^{-1}$ and cooling quickly

obtained by successive excursions of the temperature followed by measurement of the Auger peak heights for oxygen and nitrogen. Comparison of the measured peak heights with those calculated by theory for a monolayer of atoms shows agreement only for oxygen not for nitrogen. This is interpreted as being due to NO being bound to the surface via nitrogen so that oxygen is outermost. Above 500 K the nitrogen signal increases and oxygen decreases indicating that the molecule has dissociated leaving the surface covered with atomic N and O. By measuring the rate of the decomposition as a function of temperature the dissociation energy is calculated at $125 \, kJ \, mol^{-1}$. At $\sim 860 \, K$ nitrogen desorbs. The rate of this desorption was measured by AES and by quantitative thermal desorption. It is shown that the desorption of N_2 is first order and that the binding energy is $213 \, kJ \, mol^{-1}$.

The behavior of nitric oxide on the (110) crystal plane is unlike that observed on the close packed (100) and (111) planes [22.3, 4], The initial adsorption on these planes involved dissociation at temperatures $< 250 \, K$. At increased exposures of nitric oxide gas, a molecular form is found to adsorb. This form is stable only to about 400 K when it desorbs as NO gas. There is no dissociation process occurring at $\sim 500 \, K$ as found on the (110) plane.

Comparison of the single crystal work with the results on polycrystalline nickel, previously mentioned, shows that the (110) plane most closely corresponds to the observations on polycrystalline metal. Decomposition of an adsorbed nitric oxide molecule to form adsorbed oxygen and nitrogen at 500 K is the common feature. In the single crystal adsorption experiment, these atomic species persist. In the higher pressure reaction experiment, the oxygen diffuses to form a surface oxide while the nitrogen is displaced to the gas phase by incoming nitric oxide.

The polycrystalline surface is likely to expose various crystal planes including the (100) and (111). The behavior of the (110) plane may well be typical of various rough or stepped planes. The reactive properties of the polycrystalline surface may derive from structural features which constitute only a minor fraction of the total surface.

Ultraviolet photoelectron spectroscopy (UPS) is an important technique in the study of adsorbate-surface bonding. In the case of nitric oxide, where there is obviously interest in both molecular and dissociated forms, the UPS technique is most important. A detailed identification of molecular orbitals is possible, adsorbed atomic oxygen and nitrogen are identified, and the change in electron density at the surface resulting from the adsorbate interaction is recorded.

The photon source for UPS is typically a helium discharge lamp providing HeI radiation at 21.2 eV (Chap. 14). This is introduced to the surface via a windowless port with differential pumping between the lamp and work chamber. The energy range of this form of UPS is $\sim 16 \, eV$ i.e. 21.2 eV less the work function. The Fermi energy is usually detected and is taken as a reference so that spectra have an energy scale relative to E_F. The transition metals important in nitric oxide adsorption have a high density of electrons near the Fermi level. Chemisorption of a gas results in a decrease in intensity in the spectrum near E_F. This is shown in

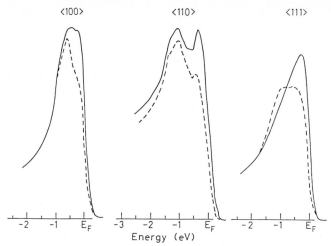

Fig. 22.2. Angle-resolved spectra from clean nickel (100) taken in three crystallographic directions (*full lines*); surfaces with adsorbed nitric oxide (*broken lines*)

Fig. 22.2 for the adsorption of nitric oxide on (100) nickel [22.5]. These spectra are taken in angle-revolved mode in the apparatus. The surface orientation is (100). Three directions of emission from this surface reveal differing d-band features. In fact, the shape for the directions $\langle 110 \rangle$ and $\langle 111 \rangle$ correspond closely to those observed from surfaces prepared with these orientations. The effect of adsorption of nitric oxide is shown in each case to decrease the intensity at close to E_F. This represents electrons from the metal becoming involved in the process of chemisorption.

Nitric oxide may be considered to be intermediate electronically between CO and O_2, two molecules capable of reaction with NO. The series CO, NO and O_2 represent occupancies of the outermost Π^* antibonding orbital of 0, 1 and 2 electrons. Back-donated charge from a metal surface is accommodated in the vacant Π^* orbital of CO without causing dissociation of the chemisorbed molecule. The corresponding process in O_2 results in dissociation which is always observed as the initial result of chemisorbing oxygen on metals. But NO with one electron in the Π^* anti bonding may form molecular or dissociated adsorbate depending on the conditions.

A study of the adsorption of NO on (110) iron by UPS has revealed considerable complexity in the modes of molecular adsorption [22.6]. The initial adsorption of NO at 92 K results in the UPS spectrum shown in Fig. 22.3. The broken lines mark features not present in the spectrum of the clean (110) iron surface (see also Fig. 22.4). They are attributed to a molecular adsorbed state of NO referred to as α_1. Repeated temperature flashes show a series of spectral changes until at temperatures above 330 K a single major feature is centred at -5.5 eV (Fig. 22.3b). This feature indicates photoemission from N $2\tilde{p}$ and O $2\tilde{p}$ and hence atomic adsorbates. The important characteristic spectra are summa-

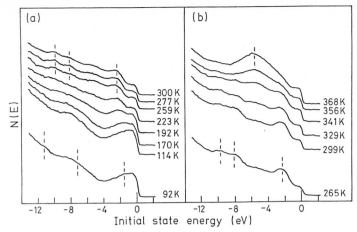

Fig. 22.3. (a) Initial adsorption of NO on Fe(110) at 92 K (α_1 state) followed by a sequence of temperature flashes to form the β state. **(b)** Initial adsorption of NO on Fe(110) at 265 K (β state) followed by a sequence of temperature flashes to form the dissociated (d) state

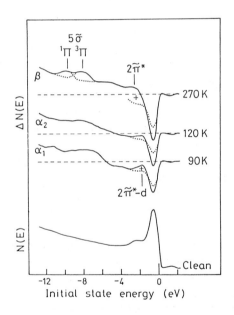

Fig. 22.4. HeI UPS difference spectra, weight averaged and smoothed, comparing the α_1, α_2 and β molecular adsorptions. An estimate of the inelastic scattering contribution is shown by dotted curves

rized in Fig. 22.4 where they are represented as difference spectra by subtracting the spectrum of the clean Fe(110) crystal.

Thus the adsorption of nitric oxide on an Fe(110) single crystal surface at temperatures 90–350 K results in the formation of at least four distinguishable adsorption states, depending on the substrate temperature. The molecular adsorption states, α_1 over 90–110 K and α_2 over 110–170 K, are both thought to involve NO chemisorbed, N end down, possibly at an off-normal angle. Only slight differences exist between the valence electronic structure of the α states. The

β adsorption state exists over a very wide temperature range of 170–290 K, although its concentration can be made to peak at 270 K. A model for the β state is that the initial dissociative adsorption occurs randomly and non-incorporatively, thereby allowing the formation of single vacant sites at which further dissociative adsorption cannot occur. Additional adsorption is thus restricted to molecular adsorption at these single sites. Schematically this can be represented:

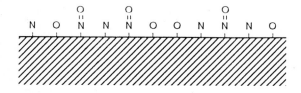

The β state is apparently stabilized at all substrate temperatures at which NO would dissociate given the opportunity, i.e. an adjacent vacant site.

Nitric oxide was found to initially dissociate at all substrate temperatures above 110 K. The extent of dissociation increased with temperature, being complete above 300 K (d state). No measurable molecular NO was found to exist above 330 K. The lower temperature adsorption states are shown to undergo an irreversible conversion to the next highest temperature state upon heating the substrate: $\alpha_1 \rightarrow \alpha_2 \rightarrow \beta \rightarrow d$. A LEED pattern showed an ordered C(2 $\times$ 2) structure for low doses at 350 K but all other adsorption conditions resulted in disorder. Auger analysis of the NO intensity ratio indicated that atomic N is readily incorporated into the sub-surface region.

Of the three molecular states the β state showed the closest correspondence to a state that would be expected for a weak chemisorption of NO. Stabilization of the β state at temperatures which caused NO dissociation on a clean surface may be due to the blocking of adjacent sites by adsorbed atoms or by an electronic stabilization effect in which adjacent adsorbed atoms reduce back-donation into the antibonding $2\tilde{\pi}^*$ orbital.

The complexity of nitric oxide adsorption and its dependence on the specific structure and composition of the surface might at first sight appear discouraging. However it is this diversity which offers hope of finding materials and conditions to hold nitric oxide in a reactive state for catalytic destruction.

UPS studies of molecularly adsorbed NO on various surfaces of Fe, Co, Ni, Cu, Ru, Pd, Pt, Rh, W, Re and Ir have been summarized in [22.7].

22.1.2 Aurocyanide Adsorption on Carbon

The selective adsorption of aurocyanide $Au(CN)_2^-$ onto the surface of activated carbons is of interest due to its widespread industrial use in enriching solutions of dissolved gold as part of the processing of gold ore. Some understanding of the process has been gained by observing the behavior of adsorption from solution under varying conditions but the determination of the nature of the bonding and the structural identity of the adsorbate species requires the application of a spectroscopic technique. The system has been examined by X-ray photoelectron spectroscopy [22.8]. The detailed interpretation of the spectra provides a good example of the potential of the technique for adsorption studies.

An activated carbon sample was treated with a potassium aurocyanide solution buffered at pH 10. The adsorbate concentration which resulted was much greater than is usual in the industrial process. This was necessary to provide sufficient sensitivity in the XPS analysis but investigation over a range of coverages showed that a consistent adsorption state resulted. Excess gold solution was removed by washing and a series of samples prepared by subjecting them to acid washing of increasing severity. The dried samples were crushed and pressed into indium foil to provide a conducting holder for the XPS analysis. The samples in the spectrometer were maintained at 150 K during analysis to minimize possible radiation damage.

The XPS peaks of greatest importance in this study are the Au $4f$ and N $1s$ peaks. These serve to monitor the overall stoichiometry of the adsorbed aurocyanide and also indicate the nature of the bonding. The N $1s$ peak is taken as a monitor of cyanide since carbon is in excess as the adsorbent. Calibration samples showed that there are two distinctly different environments for nitrogen in these cyanides. Structurally $KAu(CN)_2$ consists of nearly linear $Au(CN)_2^-$ anions with both carbons bound to the central gold atom, the nitrogen atoms being in the terminal positions. In AuCN a linear polymeric chain exists with nitrogen bridging from carbon to gold. The observed core level binding energies (BE) for N $1s$ are 398.6 eV in the terminal position and 399.1 eV in the bridging position. This difference in BE provides the basis for distinguishing adsorbed cyanide species on carbon.

The series of N $1s$ spectra for the aurocyanide ion adsorbed on carbon are shown in Fig. 22.5. Adsorption from alkaline solution (trace (a)) shows nitrogen in the terminal position only and a ratio N/Au = 2 consistent with $Au(CN)_2^-$ as the adsorbate. The first acid treatment results in a broadening of the N $1s$ feature (trace (b)). This has been decomposed to reveal contributions from two other nitrogen states with binding energies 399.0 and 397.5 eV. The spectra in traces (b) (c) and (d) are then interpreted in terms of a diminishing coverage of the adsorbed $Au(CN)_2^-$ species and increasing coverage of the two other states. Taking the lower BE to represent the terminal N_t and the higher BE to represent N_b, the spectra are interpreted to determine total N/Au and N_b/N_t ratios. Note that the initial $Au(CN)_2^-$ species is not present in traces (c) and (d).

It is known that heating an acidified solution of $Au(CN)_2^-$ results in the formation of polymeric AuCN and the evolution of HCN. This reaction clearly has not gone to completion on the carbon surface. The limiting values of the measured ratios N/Au = 1.26 and $N_b/N_t = 1.48$ suggest that the polymerization process has resulted in a limited chain length anion. This oligomer would have both terminal and bridged N and a N/Au ratio greater than unity. The species $Au_4(CN)_5^-$ is proposed. This has N/Au = 1.25 and $N_b/N_t = 1.5$ in excellent agreement with the experiment.

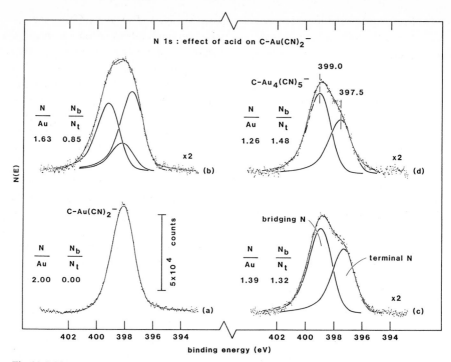

Fig. 22.5. N 1*s* photoelectron peak for (**a**) Au(CN)$_2^-$ adsorbed from alkali then subjected to increasingly severe acid (1M HCl) reaction, (**b**) 15 min at 298 K, (**c**) 180 min at 298 K and (**d**) 15 min at 373 K. Curves normalized to constant gold coverage

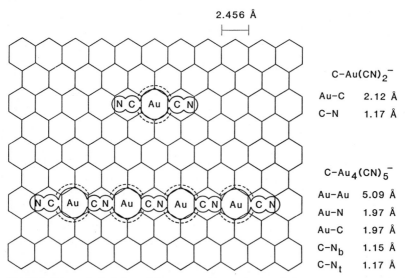

Fig. 22.6. Schematic depiction of the aurocyanide species (Au(CN)$_2^-$ and Au$_4$(CN)$_5^-$ adsorbed in a graphite plane. C and N atoms are scaled according to their covalent radii and Au atoms to the hard sphere ionic radius for Au(I) (*solid line*) and atomic radius (*broken line*)

The monomer and tetramer are represented in Fig. 22.6. The $Au(CN)_2^-$ species adsorbed on activated carbon from alkaline solution is adsorbed on and parallel to the graphitic planes. The bonding mechanism proposed is a π-donor bond from the graphite surface to the central cationic gold atom. The $Au\ 4f_{7/2}$ peak has a binding energy 0.3 to 0.5 eV lower than that observed for the compound $KAu(CN)_2$ suggesting that the surface bond involves a net charge transfer of this magnitude. A shift of the same magnitude exists for the N $1s$ peak from the adsorbate suggesting that there is also a transfer of charge to the terminal nitrogens. This is not an indication of direct bonding of N to the surface but a consequence of the π-donor bond to the gold.

A similar bonding model is applied to the tetramer $Au_4(CN)_2^-$ formed by the acid induced oligomerization of $C-Au(CN)_2^-$. The geometry of the adsorbate on the graphitic carbon lattice is shown in Fig. 22.6. All four gold atoms are bound by π-donor bonds each resulting in the same BE shift for $Au\ 4f_{7/2}$ and hence the same charge transfer. In this case, however, there are four gold atoms and only two terminal nitrogens: the observed BE shift in N $1s$ is larger.

The above study is remarkable in that a single technique, XPS, has been applied to samples prepared in solution to yield results comparable in detail to many in situ, single crystal, multitechnique experiments. The absence of electrostatic charging on activated carbon has facilitated precise measurement and interpretation of binding energy shifts.

22.1.3 Adsorbed Methoxy on Copper and Platinum

Methoxy species (CH_3O) play an important role as intermediates in the oxidation of methanol to formaldehyde. The reaction is catalyzed selectively by copper and silver whereas on platinum and other metals much CO_2, CO and water are formed along with formaldehyde. An explanation is sought in terms of the nature and stability of the reaction intermediate. For polyatomic adsorbates a surface vibrational spectroscopic technique is required to identify the chemical bonds and the mode of interaction with the surface.

Electron energy loss spectroscopy (EELS) has been developed as a technique for the identification of adsorbed species [22.9]. The vibrational modes are observed as energy losses in the range 0–500 meV in electrons reflected from the surface. The primary electron beam must have low energy, typically 1–15 eV, and be monochromatic usually within 5–10 meV. The spectra may be compared directly with infrared spectra by reporting the energy as wave numbers (cm^{-1}). The resolution of EELS, 40–80 cm^{-1}, is inferior to infrared spectroscopy but the range 0–4000 cm^{-1} is impressively broad. The technique requires ultra high vacuum and single crystal conducting samples. It is compatible with other surface techniques, UPS, XPS, AES and LEED, and has been incorporated into multitechnique systems.

The design of the spectrometer used in the study of the methoxy adsorbate is shown in Fig. 22.7 [22.9]. Two electron energy analyzers are located with a fixed scattering angle of 60° from the normal to the crystal. One acts as a

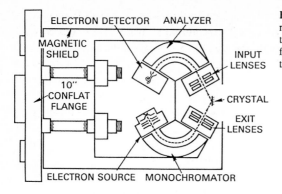

Fig. 22.7. Schematic of a 127° high resolution electron energy loss spectrometer for vibrational studies of surfaces. The incidence angle is 60° from the surface normal

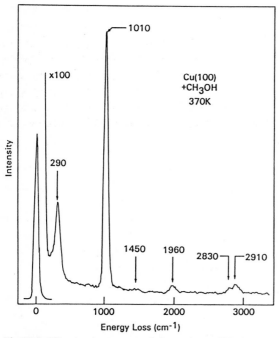

Fig. 22.8. Vibrational spectrum of the methoxy (CH_3O) intermediate chemisorbed on Cu(100) at 100 K. The beam energy was 5 eV. A monolayer of atomic oxygen was reacted with excess CH_3OH and annealed to 370 K to remove water

monochromator running at a low pass energy of < 1 eV and directing a current of $\sim 10^{-9}$ A to the crystal. By adjusting the crystal position and deflectors, the reflected (0,0) beam is directed into the second analyzer. This analyzer is scanned through elastic and inelastic regions to detect the vibrational peaks. Vibrational energies are measured as the difference between elastic and inelastic energies. Identification of peaks is made by reference to infrared and Raman spectra of known molecules and by observing frequency shifts in deuterium-labelled molecules.

426

The spectrum in Fig. 22.8 was obtained by reacting methanol and atomic oxygen on a copper (100) crystal surface [22.10]. Water, a product of the reaction has been desorbed by heating to 370 K. The five fundamental vibrational modes associated with the methoxy species are assigned as follows: ν(Cu-O) 290 cm^{-1}, ν(C-O) 1010 cm^{-1}, δ(CH)$_3$ 1450 cm^{-1}, ν_s(CH) 2830 cm^{-1} and ν_a(CH) 2910 cm^{-1}.

The intense stretching modes (Cu-O) and (C-O) and the absence of bending vibrations of the Cu-O-C chain indicate a perpendicular orientation of the methoxy species relative to the Cu(100) surface, bound through O to Cu. Although the spectrum was measured at 100 K, the preparation involved heating to 370 K. The methoxy layer is stable to this temperature. At higher temperatures, CH$_3$O decomposes to formaldehyde and hydrogen.

An experiment on platinum(111) reveals a similar mechanism for the formation of methoxy from methanol and oxygen [22.11]. The EEL spectra in Fig. 22.9 show atomic oxygen on Pt(111) at 100 K with ν(Pt-O) = 490 cm^{-1} (trace (a)) followed by methanol added at the same temperature (trace (b)). At this stage reaction has not occurred. The methanol spectrum has obscured the Pt-O mode with strong features due to OH bending (680 cm^{-1}), CH stretch (980 cm^{-1}), CH$_3$ deformation (1380 cm^{-1}), CH stretch (2910 cm^{-1}) and OH stretch (3210 cm^{-1}). When the crystal was heated to 170 K the OH bending and stretching modes disappeared and a new Pt-O stretch at 370 cm^{-1} was evident. This is shown in the lower trace in Fig. 22.9 and is the evidence for the formation of the methoxy

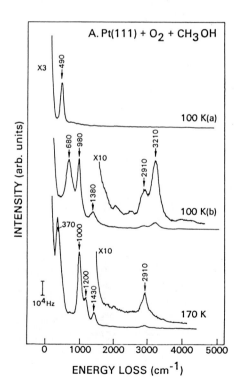

Fig. 22.9. Energy loss spectra of the reaction of methanol (CH$_3$OH) with atomic oxygen on Pt(111) to produce the methoxy intermediate (CH$_3$O). (a) Atomic oxygen $p(2 \times 2)$ structure. (b) Excess methanol condensed on (a). Annealing to 170 K produced the methoxy spectrum (*lower*). The beam energy was near 1 eV

species. The shift to lower energy of the Pt-O as compared to that for atomic oxygen is due to the extra mass of the methyl group on the oxygen.

This methoxy-covered Pt(111) surface is stable only to 170 K. Above this temperature, the methoxy species is not stable on Pt(111). Decomposition in this case is to carbon monoxide and hydrogen; the Pt surface breaks the C-H bonds in the methyl group and the CO inverts to bind carbon to the surface, the usual mode of adsorption of CO.

The breakdown of the methyl group at low temperature clearly precludes the chance to produce formaldehyde. The mechanism may well occur on transition metals generally which strongly chemisorb hydrogen and carbon monoxide. Evidence for the effect of the platinum surface on methanol is shown by the EEL spectra in Fig. 22.10 [22.11]. The upper trace is for bulk methanol, i.e. a multilayer of methanol ice. No oxygen is added. The lower trace is for a monolayer of physically adsorbed methanol at 155 K. In this case vibrational modes are sharper and the OH stretch has moved to 3280 cm^{-1}. This is an indication of decreased hydrogen bonding between molecules in the two-dimensional layer. Also in the two-dimensional layer, new frequencies are observed in the CH region near 2900 cm^{-1}. These occur at lower energies and indicate that there is strong hydrogen bonding of the methyl group hydrogens to the platinum surface. Thus the methyl group in the physically adsorbed state is strongly influenced by the Pt(111) surface. When this monolayer was heated to > 200 K, complete decomposition to CO and hydrogen was observed by EELS and no methoxy groups formed.

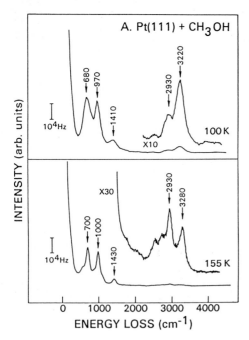

Fig. 22.10. Comparison of the energy loss spectra of multilayer methanol (100 K) and monolayer methanol (155 K) on Pt(111). In the monolayer spectra considerable "softening" of the methyl group CH stretching modes can be seen. The beam was near 1 eV

The distinction between copper and platinum as catalysts for the selective oxidation of methanol to formaldehyde then depends on the stability of the methoxy species. The EEL spectra show that this species is more stable on copper and that, although it can form on platinum at low temperature, it decomposes to adsorbed CO and hydrogen.

The EELS techinque as applied in the above examples has one very important advantage over infrared spectroscopy. The bond of an atom to the surface, Cu-O, Pt-O at energies $< 500 \, cm^{-1}$ are directly measured. These are out of the normal range of surface infrared methods. At higher energies, the superior resolution of IR spectra provides greater detail of the vibrational modes. The two techniques should be seen as complementary in their roles in surface science.

22.2 Conclusion

The examples discussed above have been chosen to illustrate the detailed nature of the information about adsorbates obtainable by surface analytical techniques. The application of LEED, AES, UPS, XPS and EELS to particular adsorption systems has been discussed. These techniques are best combined with other methods of detecting reaction or desorption by monitoring the gas phase by mass spectrometer. It is also generally found that more than one surface technique is needed to obtain a satisfactory description of an adsorbate. The emphasis on single crystal techniques is inherent in this area of surface science. A surface exposing multiple crystal planes would produce mixed spectra from multiple adsorbate states. The behavior on practical surfaces is best deduced by measuring overall properties of adsorption and reactivity of the practical surface and comparing with the detailed studies on various crystal planes of pure substances.

References

22.1   B.G. Baker, R.F. Peterson: Proc. Sixth Int. Congress on Catalysis **2**, 988–996 (1977)
22.2   G.L. Price, B.A. Sexton, B.G. Baker: Surf. Sci. **60**, 506 (1976)
22.3   G.L. Price, B.G. Baker: Surf Sci. **91**, 571 (1980)
22.4   H. Conrad, G. Ertl, J. Küppers, E.E. Latta: Surf. Sci. **50**, 296 (1975)
22.5   G.L. Price, B.G. Baker: Surf. Sci. **68**, 507 (1977)
22.6   C. Klauber, B.G. Baker: Appl. Surf. Sci. **22/23**, 486 (1985)
22.7   C. Klauber, B.G. Baker: Surf. Sci. **121**, L513 (1982)
22.8   C. Klauber: Surf. Sci. **203**, 118 (1988)
22.9   B.A. Sexton: Appl. Phys. A **26**, 1 (1981)
22.10 B.A. Sexton: Surf. Sci. **88**, 299 (1979)
22.11 B.A. Sexton: Surf. Sci. **102**, 271 (1981)

Part IV

Appendix

Acronyms Used in Surface and Thin Film Analysis

C.Klauber

A selection of acronyms encountered in surface and thin film analysis. The majority apply to specific experimental techniques or related aspects. A given technique may have more than one acronym in common use.

ABS	Atomic Beam Scattering
ACAR	Angular Correlation of Annihilation Radiation
AEAPS	Auger Electron Appearance Potential Spectroscopy
AEM	Analytical Electron Microscopy (Chap. 3)
AEM	Auger Electron Microscopy (Chap. 6)
AES	Auger Electron Spectroscopy (Chap. 6)
AFM	Atomic Force Microscopy
APS	Appearance Potential Spectroscopy
AR	Angle Resolved (technique prefix)
ASW	Acoustic Surface Wave measurements
ATR	Attenuated Total Reflectance (Chap. 8)
BIS	Bremsstrahlung Isochromat Spectroscopy
CAE	Constant Analyser Energy
CHA	Concentric Hemispherical Analyser (Chaps. 6, 7)
CIS	Characteristic Isochromat Spectroscopy
CL	Cathode Luminescence (Chap. 3)
CMA	Cylindrical Mirror Analyser (Chap. 6)
CPD	Contact Potential Difference (synonymous with work function) (Chap. 1)
CRR	Constant Retard Ratio
CVD	Chemical Vapour Deposition
DAPS	Disappearance Potential Spectroscopy
ΔH_{ads}	Heat of adsorption measurements
$\Delta\emptyset$	Change in work function (Chap. 1)
DIET	Desorption Induced by Electronic Transitions
DRIFT	Diffuse Reflectance Infrared Fourier Transform spectroscopy (Chap. 8)
EDAX	Energy Dispersive Analysis of X-rays (Chap. 3)
EDC	Electron energy Distribution Curve
EDS	Energy Dispersive Spectroscopy (equivalent to EDAX)
EDX	Energy Dispersive X-ray (spectroscopy) (equivalent to EDAX)
EELS	Electron Energy Loss Spectroscopy (Chap. 3)

EID	Electron Impact Desorption (equivalent to ESD)
EIL	Electron Induced Luminescence
EL	ElectroLuminescence
ELEED	Elastic Low Energy Diffraction (Chap. 13)
ELL	Ellipsometry (Chap. 1)
ELS	Energy Loss Spectroscopy (usually low resolution EELS) (Chap. 3)
EM	Electron Microscopy (Chap. 3)
EMP	Electron MicroProbe analysis (equivalent to EDAX) (Chap. 3)
EPMA	Electron Probe MicroAnalysis (see EMP)
ES	Emission Spectroscopy (technique suffix)
ESCA	Electron Spectroscopy for Chemical Analysis (synonymous with XPS) (Chap. 7)
ESD	Electron Stimulated Desorption (may be of ions, neutrals or excited states)
ESDI	Electron Stimulated Desorption of Ions
ESDIAD	Electron Stimulated Desorption Ion Angular Distribution
ESDN	Electron Stimulated Desorption of Neutrals
ESR	Electron Spin Resonance
EXAFS	Extended X-ray Absorption Fine Structure (Chap. 1)
EXFAS	Extended Fine Auger Structures (Chap. 1)
FABMS	Fast Atom Bombardment Mass Spectrometry (Chap. 1)
FABS	Fast Atom Bombardment Spectrometry (see FABMS)
FD	Field Desorption
FD	Flash Desorption (equivalent to TDS/TPD at high ΔT rate)
FDM	Field Desorption Microscopy
FDS	Field Desorption Spectrometry
FDS	Flash Desorption Spectrometry (equivalent to TDS/TPD at high ΔT rate)
FEED	Field Emission Energy Distribution
FEES	Field Electron Energy Spectroscopy
FEM	Field Emission Microscopy (Chap. 1)
FERP	Field Emission Retarding Potential
FIM	Field Ion Microscopy (Chap. 1)
FIM-APS	Field Ion Microscope-Atom Probe Spectroscopy
FIS	Field Ion Spectroscopy (Chap. 1)
FTIR	Fourier Transform InfraRed (Chap. 8)
GDMS	Glow Discharge Mass Spectrometry
GDOS	Glow Discharge Optical Spectroscopy
GIXD	Grazing Incidence X-ray Diffraction
HEED	High Energy Electron Diffraction (Chap. 12)
HEIS	High Energy Ion Scattering (Chap. 9)
HPT	High Precision Translator
HREELS	High Resolution Electron Energy Loss Spectroscopy (equivalent to EELS)

IAES	Ion (excited) Auger Electron Spectroscopy
IBSCA	Ion Beam SpectroChemical Analysis (equivalent to SCANIIR)
ICISS	Impact Collision Ion Scattering Spectroscopy
IETS	Inelastic Electron Tunnelling Spectroscopy
IID	Ion Impact Desorption
IIRS	Ion Impact Radiation Spectroscopy
IIXS	Ion Induced X-ray Spectroscopy
ILEED	Inelastic Low Energy Electron Diffraction (Chap. 13)
ILS	Ionization Loss Spectroscopy
IMFP	Inelastic Mean Free Path (Chap. 1)
IMMA	Ion Microprobe Mass Analysis (Chap. 5)
IMXA	Ion Microprobe X-ray Analysis
INA	Ion Neutral Analysis (equivalent to SMNS)
INS	Ion Neutralization Spectroscopy
IPS	Inverse Photoemission Spectroscopy
IR	InfraRed (technique prefix)
IRAS	Infrared Reflection Absorption Spectroscopy (Chap. 8)
IRRAS	InfraRed Reflection Absorption Spectroscopy (see IRAS)
IRS	Internal Reflectance Spectroscopy (Chap. 8)
IS	Ionization Spectroscopy
ISD	Ion Stimulated Desorption
ISS	Ion Scattering Spectroscopy (Chap. 11)
ITS	Inelastic Tunnelling Spectroscopy
L	Langmuir (10^{-6} Torr sec)
LAMMA	Laser Microprobe Mass Analysis
LEED	Low Energy Electron Diffraction (Chap. 13)
LEELS	Low Energy Electron Loss Spectroscopy (equivalent to EELS)
LEERM	Low Energy Electron Reflection Microscopy
LEIS	Low Energy Ion Scattering (Chap. 11)
LID	Laser Induced Desorption
LIMA	Laser Induced Mass Analysis (equivalent to LAMMA)
LITD	Laser Induced Thermal Desorption
LMP	Laser MicroProbe
LS	Light Scattering
MBE	Molecular Beam Epitaxy
MBRS	Molecular Beam Reactive Scattering
MBSS	Molecular Beam Surface Scattering
MEED	Medium Energy Electron Diffraction
MEIS	Medium Energy Ion Scattering
MOMBE	Metal Organic Molecular Beam Epitaxy
MOSS	Mössbauer Spectroscopy
MS	Magnetic Saturation
NBS	Nuclear Backscattering Spectroscopy (Chap. 9)
NEXAFS	Near Edge X-ray Absorption Fine Structure

NIRS	Neutral Impact Radiation Spectroscopy
NMR	Nuclear Magnetic Resonance
NRA	Nuclear Reaction Analysis (Chap. 9)
PAES	Proton (excited) Auger Electron Spectroscopy
PAS	PhotoAcoustic Spectroscopy
PAX	Photoemission of Adsorbed Xenon
PD	PhotoDesorption
PD	Photoelectron Diffraction
PE	PhotoEmission
PEM	PhotoElectron Microscopy
PES	PhotoElectron Spectroscopy (synonymous with UPS, XPS and ESCA)
PESM	PhotoElectron SpectroMicroscopy
PIXE	Particle (Proton) Induced X-ray Emission
PSD	Photon Stimulated Desorption
PSEE	PhotoStimulated Exoelectron Emission
PSID	Photon Stimulated Ion Desorption
PVD	Physical Vapour Deposition
RAIRS	Reflection Absorption InfraRed Spectroscopy (Chap. 8)
RBS	Rutherford Backscattering Spectroscopy (Chap. 9)
REELS	Reflection Electron Energy Loss Spectroscopy
REM	Reflection Electron Microscopy
RFA	Retarding Field Analyser
RGA	Residual Gas Analyser
RHEED	Reflection High Energy Electron Diffraction (Chap. 12)
SAED	Selected Area Electron Diffraction (Chap. 3)
SALI	Secondary Atom Laser Ionization
SAM	Scanning Auger Microscopy (Chap. 6)
SAXPS	Selected Area X-ray Photoelectron Spectroscopy (Chap. 7)
SC	Surface Capacitance
SDMM	Scanning Desorption Molecular Microscope
SE	SpectroEllipsometry
SEE	Secondary Electron Emission
SEM	Scanning Electron Microscopy (Chap. 3)
SEMPA	Scanning Electron Microscopy with Polarization Analysis
SERS	Surface Enhanced Raman Spectroscopy
SEXAFS	Surface Extended X-ray Absorption Fine Structure (Chap. 1)
SHG	Second Harmonic Generation (optical)
SI	Surface Ionization
SIIMS	Secondary Ion Imaging Mass Spectrometry (Chap. 5)
SIMS	Secondary Ion Mass Spectrometry (Chap. 5)
SLEEP	Scanning Low Energy Electron Probe
SNIFTIRS	Subtractively Normalized Interfacial FTIR Spectroscopy

SNMS	Secondary Neutrals Mass Spectrometry (equivalent to INA)
SP	Spin Polarized (technique prefix)
SP	Surface Potential
SPIES	Surface Penning Ionization Electron Spectroscopy
SPIPE	Spin Polarized Inverse PhotoEmission
SPPD	Spin Polarized Photoelectron Diffraction
SPV	Surface Photovoltage
SRPS	Synchrotron Radiation Photoelectron Spectroscopy
SRS	Surface Reflectance Spectroscopy
SSA	Spherical Sector Analyser
SSIMS	Static Secondary Ion Mass Spectrometry (Chap. 5)
STEM	Scanning Transmission Electron Microscopy (Chap. 3)
STM	Scanning Tunnelling Microscopy (Chap. 10)
STS	Scanning Tunnelling Spectroscopy (Chap. 10)
STS	Surface Tunnelling Spectroscopy
SXAPS	Soft X-ray Appearance Potential Spectroscopy
SXES	Soft X-ray Emission Spectroscopy
TD	Thermal Desorption
TDMS	Thermal Desorption Mass Spectrometry
TDS	Thermal Desorption Spectrometry (equivalent to TPD)
TE	Thermionic Emission
TEAM	Thermal Energy Atomic and Molecular beam scattering
TEAS	Thermal Energy Atom Scattering
TED	Transmission Electron Diffraction
TELS	Transmission Energy Loss Spectroscopy
TEM	Transmission Electron Microscopy (Chap. 3)
TL	ThermoLuminescence
TPD	Temperature Programmed Desorption (equivalent to TDS)
TPRS	Temperature Programmed Reaction Spectrometry
UHV	Ultrahigh Vacuum (Chap. 2)
UPS	Ultraviolet Photoelectron Spectroscopy (Chap. 14)
XAES	X-ray (excited) Auger Electron Spectroscopy (Chap. 7)
XANES	X-ray Absorption Near Edge Structure (equivalent to NEXAFS)
XEM	eXoElectron Microscopy
XES	eXoElectron Spectroscopy
XES	X-ray Emission Spectroscopy
XPD	X-ray Photoelectron Diffraction
XPS	X-ray Photoelectron Spectroscopy (synonymous with ESCA) (Chap. 7)
XRD	X-ray Diffraction
XRF	X-ray Fluorescence
XTEM	cross-sectional Transmission Electron Microscopy
WF	Work Function (Chap. 1)

Surface Science Bibliography

Auciello, O., R. Kelly (eds.): *Ion Bombardment Modification of Surfaces* (Elsevier, Amsterdam 1984)

Bamford, C.H., C.F.H. Tiper, R.G. Compton (eds.): Simple Processes at the Gas-Solid Interface, in *Comprehensive Chemical Kinetics* (Elsevier, Amsterdam 1984)

Barr, T.L., L.E. Davis (eds.): *Applied Surface Analysis* (American Society for Testing and Materials, 1978)

Bauer, E.: LEED and Auger Methods, in *Interaction on Metal Surfaces*, ed. by R. Gomer, Topics Appl. Phys. **4** (Springer, Berlin, Heidelberg 1975) pp. 225–274

Baun, W.C.: Ion Scattering Spectrometry: a Versatile Technique for a Variety of Materials, Surf. Interface Anal. **3**, 243 (1981)

Behm, R.J., W. Höseler: Scanning Tunneling Microscopy – a Review, in *Chemistry and Physics of Solid Surfaces VI*, ed. by R. Vanselow, R. Howe, Springer Ser. Surf. Sci. Vol. 5 (Springer, Berlin, Heidelberg 1986) pp. 361–412

Behrisch, R. (ed.): *Sputtering by Particle Bombardment*, Vols. I and II, Topics Appl. Phys., Vols. 47, 52 (Springer, Berlin, Heidelberg 1981, 1983)

Benninghoven, A.L.: Developments in Secondary Ion Mass Spectrometry and Applications to Surface Studies, Surf. Sci. **53**, 596 (1975)

Benninghoven, A.L., F.G. Rudenauer, H.W. Werner: *Secondary Ion Mass Spectrometry*, Vol. 86 of Chemical Analysis, ed. by P.J. Elving, J.D. Winefordner, I.M. Kothoff (Wiley, New York 1987)

Berkowitz, J.: *Photoabsorption, Photoionization and Photoelectron Spectroscopy* (Academic, New York 1979)

Bianconi, A., L. Inoccia, S. Stipchich (eds.): *EXAFS and Near Edge Structure*, Springer Ser. Chem. Phys., Vol. 27 (Springer, Berlin, Heidelberg 1983)

Blakely, J.M.: *Introduction to the Properties of Crystal Surfaces* (Pergamon, Oxford 1973)

Blakely, J.M. (ed.): *Surface Physics of Materials*, Vols. I and II (Academic, New York 1975)

Blakely, J.M. (ed.): *Surface Physics of Crystal Solids* (Academic, New York 1975)

Boehm, H.P., H. Knözinger: Nature and Estimation of Functional Groups on Solid Surfaces, in *Catalysis – Science & Technology*, Vol. 4 (Springer, Berlin, Heidelberg 1983)

de Boer, J.H.: *The Dynamical Character of Adsorption* (2nd edn.) (Oxford University Press, London 1968)

Briggs, D., M.P. Seah (eds.): *Practical Surface Analysis by Auger and X-ray Photoelectron Spectroscopy* (Wiley, Chichester 1983)

Briggs, D. (ed.): *Handbook of X-ray and Ultraviolet Photoelectron Spectroscopy* (Heyden, London 1977)

Brown, A., J.C. Vickerman: Static SIMS for Applied Surface Analysis, Surf. Interface Anal. **6**, 1 (1984)

Brümmer, D., O. Heydenreich, K.H. Krebs, H.G. Schneider (eds.): *Handbuch Festkörperanalyse mit Elektronen, Ionen und Röntgenstrahlen* (Vieweg, Braunschweig 1980)

Brundle, C.R., A.D. Baker (eds.): *Electron Spectroscopy: Theory, Techniques and Applications*, Vols. 1–4 (Academic, London 1977, 1978, 1979, 1981)

Brundle, C.R.: Elucidation of Surface Structure and Bonding by Photoelectron Spectroscopy, Surf. Sci. **48**, 99 (1975)

Carbonara, R.S., J.R. Cuthill (eds.): *Surface Analysis Techniques for Metallurgical Applications* (American Society for Testing and Materials, 1975)

Cardona, M., L. Ley (eds.): *Photoemission in Solids I (General Principles); Photoemission in Solids II (Case Studies)*, Topics Appl. Phys. **26** and **27** (Springer, Berlin, Heidelberg 1978, 1979)

Carlson, T.A.: *Photoelectron and Auger Spectroscopy* (Plenum, New York 1975)

Chu, K.W., J.W. Mayer, M.A. Nicolet: *Backscattering Spectrometry* (Academic, New York 1978)

Clarke, L.J.: *Surface Crystallography – An Introduction to LEED* (Wiley, New York 1975)

Czanderna, A.W. (ed.): *Methods of Surface Analysis*, Methods and Phenomena Vol. 1 (Elsevier, Amsterdam 1975)

Davis, L.A., N.C. MacDonald, P.W. Palmberg, G.E. Riach, R.E. Weber: *Handbook of Auger Electron Spectroscopy*, 2nd edn. (Perkin-Elmer Corp, Eden Prairie 1979)

Delannay, F. (ed.): *Characterization of Heterogoeneous Catalysts*, Chem. Ind. Ser. **15** (Dekker, New York 1984)

Deviney, M.L., J.L. Grand (eds.): Catalyst Characterization Science, ACS Symp. Ser. **288**, Washington, D.C. (1985)

Dobrzynski, L.: *Handbook of Interfaces and Surfaces*, Vols. 1 and 2 (Garland, New York 1978)

Engel, T., K.H. Rieder: Structural Studies of Surfaces with Atomic and Molecular Beam Diffraction, in *Structural Studies of Surfaces*, G. Höhler (ed), Springer Tracts Mod. Phys. **91** (Springer, Berlin, Heidelberg 1982)

Ertl, G., J. Küppers: *Low Energy Electrons and Surface Chemistry*, Monographs in Modern Chemistry Vol. 4 (Verlag Chemie, Weinheim 1974)

Fadley, C.S.: Angle Resolved Photoelectron Spectroscopy, Progr. Surf. Sci. **16**, 275 (1984)

Fiermans, L., R. Hoogewijs, J. Vennik: Electron Spectroscopy of Transition Metal Oxides, Surf. Sci. **47**, 1 (1975)

Fiermans, L., J. Venniuk, W. Dekeyser (eds.): *Electron and Ion Spectroscopy in Solids* (Plenum, New York 1978)

Gibson, W.M.: Determination by Ion Scattering of Atom Positions at Surfaces and Interfaces, in *Chemistry and Physics of Solid Surfaces V*, ed. by R. Vanselow, R. Howe, Springer Ser. Chem. Phys. Vol. 35 (Springer, Berlin, Heidelberg 1984) pp. 427–454

Gomer, R. (ed.): *Interactions on Metal Surfaces*, Topics Appl. Phys. Vol. 4 (Springer, Berlin, Heidelberg 1975)

Gosh, P.K.: *Introduction to Photoelectron Spectroscopy* (Wiley, New York 1983)

Hannay, N.B.: *Treatise on Solid State Chemistry*, Vol. 6A, Surfaces I, Vol. 6B, Surfaces II (Plenum, New York 1976)

Heiland, W., E. Taglauer: The Backscattering of Low Energy Ions and Surface Structure, Surf. Sci. **68**, 96 (1977)

Heinz, K., K. Müller: LEED-Intensities – Experimental Progress, and New Possibilities of Surface Structure Determination, in *Structural Studies of Surfaces*, G. Höhler (ed), Springer Tracts Mod. Phys. **91** (Springer, Berlin, Heidelberg 1982) pp. 1–54

Hodgson, K.O., B. Hedman, J.E. Penner-Hahn (eds.): *EXAFS and Near Edge Structure III*, Springer Proc. Phys. Vol. 2 (Springer, Berlin, Heidelberg 1984)

Hofmann, S.: Practical Surface Analysis: State of the Art and Recent Developments in AES, XPS, ISS and SIMS, Surf. Interface Anal. **9**, 3 (1986)

Ibach, H. (ed.): *Electron Spectroscopy for Surface Analysis*, Topics Curr. Phys. Vol. 4 (Springer, Berlin, Heidelberg 1977)

Jona, F.: LEED Crystallography, J. Phys. C. **11**, 4271 (1978)

Joyner, R.W., G.A. Somorjai: Recent Trends in the Application of LEED, Surf. Def. Prop. Sol. **2**, 1 (1973)

Kane, P.F., G.B. Larrabee (eds.): *Characterization of Solid Surfaces* (Plenum, New York 1974)

Karchaudhari, S.N., K.L. Cheng: Recent Study of Solid Surfaces by Photoelectron Spectroscopy, Appl. Spectrosc. Rev. **16**, 187 (1980)

Kay, E., P. Bagus (eds.): Topics in Surface Chemistry (Plenum, New York 1978)

Kelley, M.J.: Chemtech **17**, 30 (1987); ibid **17**, 98 (1987); ibid **17**, 107 (1987); ibid **17**, 232 (1987); ibid **17**, 294 (1987); ibid **17**, 490 (1987)

Kemeny, G. (ed.): *Surface Analysis of High Temperature Materials: Chemistry and Topography* (Elsevier, London 1984)

Kimura, K., S. Katsumota, Y. Achiba, T. Yanazaki, S. Iwata: *Handbook of HeI Photoelectron Spectra of Fundamental Organic Molecules* (Japan Sci. Soc. Press, Halsted Press, New York 1981)

King, D.A., N.V. Richardson, S. Holloway: *Vibrations at Surfaces 1985*, Studies in Surface Science and Catalysis (Elsevier, Amsterdam 1986)

King, D.A., D.P. Woodruff: The Chemical Physics of Solid Surfaces and Heterogeneous Catalysis (Elsevier, Amsterdam) Vol. 1 Clean Solid Surfaces (1981); Vol. 2 Adsorption at Solid Surfaces (1983); Vol. 3 Chemisorption Systems (in press); Vol. 4 Fundamental Studies of Heterogeneous Catalysis (1982)

MacIntyre, N.S. (ed.): *Quantitative Surface Analysis of Materials* (American Society for Testing and Materials 1978)

Maradudin, A.A., R.F. Wallis, L. Dobrzynski: *Handbook of Interfaces and Surfaces*, Vol. 3 (Garland, New York 1980)

Marcus P.M., F. Jona (eds.): *Determination of Surface Structure by LEED* (Plenum, New York 1984)

Mayer, J.W., E. Rimini: *Ion Beam Handbook for Material Analysis* (Academic, New York 1977)

McCracken, G.M.: The Behaviour of Surfaces under Ion Bombardment, Rept. Progr. Phys. **38**, 24 (1975)

McGuire, G.E.: *Auger Electron Spectroscopy Reference Manual* (Plenum, New York 1979)

Nizzoli, F., K.H. Rieder, F.F. Willis (eds.): *Dynamical Phenomena at Surfaces, Interfaces and Superlattices*, Springer Ser. Surf. Sci. Vol. 3 (Springer, Berlin, Heidelberg 1985)

Oechsner, H. (ed.): *Thin Film and Depth Profile Analysis*, Topics Curr. Phys. Vol. 37 (Springer, Berlin, Heidelberg 1984)

Pendry, J.B.: *LEED – The Theory and its Application to Determination of Surface Structures* (Academic, London 1974)

Prutton, M.: *Surface Physics* (Clarendon, Oxford 1983)

Rhodin, T.N., G. Ertl (eds.): *The Nature of the Surface Chemical Bond* (North-Holland, Amsterdam 1979)

Roberts, M.W., C.S. McKee: *Chemistry of Metal-Gas Interface* (Clarendon, Oxford 1978)

Roberts, M.W.: Photoelectron Spectroscopy and Surface Chemistry, Adv. Catal. **29**, 55 (1980)

Sickafus, E.N., H.P. Bonzel: *Surface Analysis by Low Energy Electron Diffraction and Auger Electron Spectroscopy* (Academic, New York 1971)

Somorjai, G.A. (ed.): *The Structure and Chemistry of Solid Surfaces* (Wiley, New York 1969)

Somorjai, G.A.: *Principles of Surface Chemistry* (Prentice-Hall, New Jersey 1972)

Somorjai, G.A.: *Chemistry in Two Dimensions: Surfaces* (Cornell University Press, Ithaca 1981)

Spicer, W.E., K.Y. Yu, I. Lindau, P. Pianetta, D.M. Collins: Ultraviolet Photoemission Spectroscopy of Surfaces and Surface Sorption, Surf. Def. Prop. Solids **5**, 103 (1976)

Steele, W.A.: *The Interaction of Gases with Solid Surfaces* (Pergamon, Oxford 1974)

Stöhr, J.: *Surface Crystallography by SEXAFS and NEXAFS*, in R. Vanselow, R. Howe (eds.): *Chemistry and Physics of Solid Surfaces V*, Springer Ser. Chem. Phys. Vol. 35 (Springer, Berlin, Heidelberg 1984)

Teo, B.K., D.C. Joy (eds.): *EXAFS Spectroscopy – Techniques and Applications* (Plenum, New York 1981)

Teo, B.K.: *EXAFS – Basic Principles and Data Analysis* (Springer, Berlin, Heidelberg 1986)

Thomas, J.M., R.M. Lambert (eds.): *Characterization of Catalysts* (Wiley, New York 1980)

Thompson, M., M.D. Baker, A. Christie, T. Tyson: *Auger Electron Spectroscopy* (Wiley, New York 1985)

Tompkins, F.C.: *Chemisorption of Gases on Metals* (Academic, London 1978)

Turner, D.W., C. Baker, A.D. Baker, C.R. Brundle: *Molecular Photoelectron Spectroscopy* (Wiley, New York 1970)

Turner, N.H., R.J. Colton: Surface Analysis: X-ray Photoelectron Spectroscopy and Secondary Ion Mass Spectroscopy, Anal. Chem. **54**, 293R (1982)

Urch, D.S. and M.S.: ESCA-Auger Table, Queen Mary College, Chem. Dept., University of London (1981/2)

van der Veen, J.: Ion Beam Crystallography of Surfaces and Interfaces, Surf. Sci. Rep. **5**(5/6), 199 (1985)

Van Hove, M.A., S.Y. Tong: *Surface Crystallography by LEED – Theory, Computation and Structural Results*, Springer Ser. Chem. Phys. Vol. 2 (Springer, Berlin, Heidelberg 1979)

Van Hove, M.A., S.Y. Tong (eds.): *The Structure of Surfaces*, Springer Ser. on Surf. Sci. Vol. 2 (Springer, Berlin, Heidelberg 1985)

Vanselow, R., S.Y. Tong (eds.): *Chemistry and Physics of Solid Surfaces* (CRC, Cleveland 1977)

Vanselow, R. (ed.): *Chemistry and Physics of Solid Surfaces II* (CRC, Boca Raton 1979)

Vanselow, R., W. England (eds.): *Chemistry and Physics of Solid Surfaces III* (CRC, Boca Raton 1982)

Vanselow, R., R. Howe (eds.): *Chemistry and Physics of Solid Surfaces IV*, Springer Ser. Chem. Phys. Vol. 20 (Springer, Berlin, Heidelberg 1982)

Vanselow, R., R. Howe (eds.): *Chemistry and Physics of Solid Surfaces V*, Springer Ser. Chem. Phys. Vol. 35 (Springer, Berlin, Heidelberg 1984)

Vanselow, R., R. Howe (eds.): *Chemistry and Physics of Solid Surfaces VI, VII, VIII*, Springer Ser. Surf. Sci. Vols. 5, 10, 22 (Springer, Berlin, Heidelberg 1986, 1988, 1990)

Wagner, C.D., W.M. Riggs, L.E. Davis, J.F. Moulder, G.E. Muilenberg (eds.): *Handbook of X-ray Photoelectron Spectroscopy* (Perkin-Elmer Corp., Eden Prairie 1978)

Walls, J.M. (ed.): *Methods of Surface Analysis* (Cambridge University Press, Cambridge 1989)

Wandelt, K.: Photoemission Studies of Adsorbed Oxygen and Oxide Layers, Surf. Sci. Rep. **2**(1), 1 (1982)

Weissmann, R., K. Kümmer: Auger Electron Spectroscopy – a Local Probe for Solid Surfaces, Surf. Sci. Rep. **1**(5), 251 (1981)

Werner, H.W.: The Use of Secondary Ion Mass Spectrometry in Surface Analysis, Surf. Sci. **47**, 301 (1975)

Werner, H.W.: Quantitative Secondary Ion Mass Spectrometry: A Review, Surf. Interface Anal. **2**, 56 (1980)

Werner, H.W., R.P.H. Garten: Comprehensive Study of Methods for Thin-Film and Surface Analysis, Rep. Progr. Phys. **47**, 221 (1984)

Wittmaack, K.: Secondary Ion Mass Spectrometry as a Means of Surface Analysis, Surf. Sci. **89**, 668 (1979)

Williams, R.H., G.P. Srivastava, I.T. McGovern: Photoelectron Spectroscopy of Solids and their Surfaces, Rep. Progr. Phys. **43**, 1357 (1980)

Winograd, N.F., B.J. Garrison: Surface Structure Determination with Ion Beams, Acc. Chem. Res. **13**, 406 (1980)

Woodruff, D.P., T.A. Delchar: *Modern Techniques of Surface Science* (Cambridge University Press, Cambridge 1986)

Subject Index

452